ACKNOWLEDGEMENTS

The valuable assistance of the Organising Committee and Panel of Referees is gratefully acknowledged.

ORGANISING COMMITTEE

Dr. K.R. Dyer (Chairman)	Institute of Oceanographic Sciences
Dr. P.C. Barber	Ceemaid Ltd
M. Crickmore	Hydraulics Research Ltd
Dr. R.A. Falconer	University of Birmingham
Dr. C.A. Fleming	Halcrow Maritime
Dr. R.A. Furness	Cranfield Institute of Technology
Dr. J. Graff	Ceemaid Ltd
Dr. M.J. Green	WRC Engineering
Dr. R.G. Green	Bolton Institute of Higher Education
Dr. R. Herschey	CNS Scientific & Engineering Services
Prof. P. Holmes	Imperial College of Science & Technology
S.B.B. Learoyd	Atkins Research and Development
Dr. R. Rayner	Wimpol Ltd
G.A. Watts	BHRA, The Fluid Engineering Centre
J.H. Pounsford	BHRA, The Fluid Engineering Centre

OVERSEAS CORRESPONDING MEMBERS

Prof. M.B. Abbott	International Institute for Hydraulic & Environmental Engineering, Netherlands
Dr.A. Muller	Swiss Federal Institute of Technology, Switzerland
Dr. A.C.E. Wessels	Delft Hydraulics Laboratory, Netherlands
Prof. R.L. Wiegel	University of California, Berkeley, U.S.A.

International Conference on
Measuring Techniques of Hydraulics Phenomena in Offshore, Coastal and Inland Waters
London, England: 9-11 April, 1986

CONTENTS

International Conference on

Measuring Techniques

of Hydraulics Phenomena in offshore,
Coastal & Inland Waters

London, England:
9-11 April, 1986

**Organised and sponsored by
BHRA, The Fluid Engineering Centre.**

**Co-sponsored by the American Society
of Civil Engineers and the International Association
for Hydraulic Research**

THE FLUID ENGINEERING CENTRE

Editor: J.H. Pounsford

EDITORIAL NOTES

The Organisers are not responsible for statements or opinions made in the papers. The papers have been reproduced by offset printing from the authors' original typescripts to minimise delay.

When citing papers from the volume the following reference should be used:-
Title, Author(s), Paper No., Pages, International Conference on Measuring Techniques of Hydraulics Phenomena in Offshore, Coastal and Inland Waters, London, England. Organised and sponsored by BHRA, The Fluid Engineering Centre, Cranfield, Bedford MK43 0AJ, England (9-11 April, 1986).

Printed and Published by
BHRA, The Fluid Engineering Centre
Cranfield, Bedford MK43 0AJ, England

© BHRA, The Fluid Engineering Centre 1986
ISBN 0 947711 120
ISSN 0269 0888

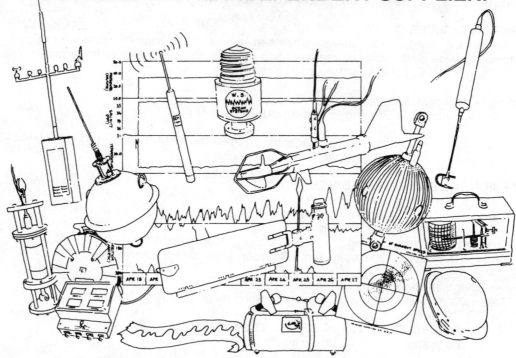

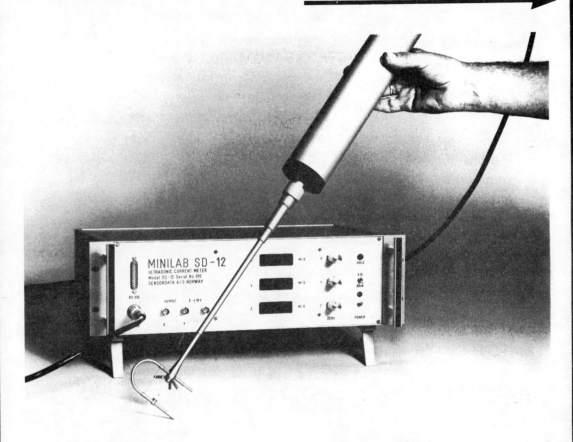

THE USE OF HF RADAR MEASUREMENTS OF SURFACE CURRENTS FOR COASTAL ENGINEERS

D. Prandle and M. J. Howarth

Institute of Oceanographic Sciences, Bidston Observatory,
Birkenhead.

Summary

A new technique for the remote measurement of surface currents has recently been developed. Currents in a region up to 30km from the coast have been measured from the Doppler shift in a backscattered HF radar signal by the OSCR system. The advantages of the system for coastal engineers are discussed and results from several experiments presented in terms of tidal, wind-driven and long term mean components.

Held at Imperial College of Science and Technology, London. Organised and sponsored by BHRA, The Fluid Engineering Centre. Co-sponsored by the American Society of Civil Engineers and the International Association for Hydraulic Research.
©BHRA, The Fluid Engineering Centre, Cranfield, Bedford MK43 0AJ, England 1986.

1. OSCR - OCEAN SURFACE CURRENT RADAR

A prerequisite for many coastal engineering applications - the siting of outfalls and dumps, the construction of sea defences, the movement of pollution - is information about the current field over an area of several kilometres squared near to the shore. A system which provides this information is OSCR - Ocean Surface Current Radar - designed at the Rutherford Appleton Laboratory under the direction of J. H. King (Ref. 1) and based on the CODAR system (Ref. 2). A radar signal with a frequency of 27MHz and wavelength, λ, of 11m is transmitted from the shore and is resonantly backscattered by sea surface waves whose wavelength is $\lambda/2$ - 5.5m. The frequency of the backscattered signal is Doppler shifted by the radial component of the surface current, u, and by the advancing and receding radial surface waves with wavelength $\lambda/2$. This frequency shift is $(\pm 2c+2u)/\lambda$, where c is the phase velocity of the waves. The spectrum of the backscattered signal has two peaks, which in the absence of a background surface current would be symmetrical relative to the transmitted frequency, at ± 0.53Hz. The effect of a surface current is to offset both peaks by Δf, Fig 1; for a radial current component of 1m/s, $\Delta f = 0.18$Hz. The offset is observed and u calculated from $u = \Delta f . \lambda/2$. The measurements of radial velocity components from two OSCR units can be combined to provide a complete two-dimensional pattern of surface current velocities.

The major design difference between the OSCR and CODAR systems is in the receive antenna. In the OSCR system it comprises 16 separate units uniformly spaced over a line 90m long oriented orthogonally to the centre of the radar beam coverage. Signal components from 16 beams each of 6° width can be differentiated and along each beam signals can be separated into 'bin' lengths of, usually, 1.2 or 2.4km. A typical spatial coverage is shown in Fig. 2.

The system operates successfully over a wide range of sea states for distances up to 30km offshore, and larger ranges are possible for restricted conditions. In the May 1985 Red Wharf Bay experiment, Table 1, spectra for a particular beam were recorded over a period of 2 minutes. These data were then analysed in the following 3 minutes before recordings continued for the next beam, thus allowing 12 beams to be recorded every hour at successive 5 minute intervals. Recordings from the second OSCR unit were made in similar fashion with the recording period for unit 2 occurring during the analysis phase for unit 1, thereby eliminating any possible interference between the two units. Successful trials have recently been completed with a similar radar using ranges up to 200km while a shorter range version with much smaller bin lengths suitable for near-shore applications seems technically feasible.

For an engineer the system has many desirable features although it is still being developed and the precise meaning of the measurements is still being studied. The system provides a wide spatial coverage 30 x 30km², with a resolution of 1km, well matched to the spatial variability of currents near the coast. However, the measurement grids are not uniform in space so that errors will vary spatially. Also the grids of the two OSCR units are not co-located, so that the interpolation is necessary to produce a unified current field. The measurements are areal and temporal averages, which, in many cases, is more representative of conditions in the sea than are spot measurements. As a result the observations tend to vary smoothly in space and time, lending them (spurious?) credibility.

The system is shore based and can operate in any weather condition. Hence it is less vulnerable and less likely to suffer data loss due to hardware failures than an in situ moored instrument, which can easily be lost or damaged and whose accuracy may well decrease during a storm. Surface waves with a wavelength of 5.5m appear always to be present sufficient for there to be a backscattered signal, even on 'flat calm' days. OSCR's range is reduced during stormy days because the signal to noise ratio of the return signal deteriorates. OSCR is simple and quick to erect, taking a couple of days to erect and a further one or two days to test. Although the equipment is expensive the cost per measurement is cheap.

The measurement is fundamental - a measurement of frequency shift - and so does not rely on calibration factors. It is inherently stable and is not subject to long term drift. However, both the current estimates and the beam and bin sizes are derived purely from theory and are very difficult to check accurately in practice. The concept of resonant backscattering is simple but some of the physics, both of the radar and of the oceans and, particularly, of their interaction, is poorly understood. The details of the way the 5.5m length waves interact with the rest of the surface wave field and with the currents remains to be understood - in the N. Ireland, March 1984 experiment, see Table 1, the difference between OSCR and in situ measurements appeared to be correlated with whether the wind blew (and hence the waves propagated) in the direction of or against the tidal current during storms (Fig. 3).

Finally the system measures near surface currents, providing direct information about the effect of wind forcing on the sea and the movement of surface waters. This is particularly relevant to the movement of buoyant pollutants and to outfall investigations, where it is the only method of providing this information. The effective depth of the measurement is less than 1m below the sea surface, although there is still debate as to precisely what it is. Where the current profile or a 'depth' mean value is required the OSCR system is complimentary to other techniques - numerical models and, for current profiles, acoustic doppler profiling current meters.

In coastal seas the largest currents are usually tidal (section 3) although storm or wind-driven currents (section 4) are occasionally comparable in strength. For some applications the long term mean current (section 5) will be important.

2. DEPLOYMENTS

Table 1 summarises 9 OSCR deployments starting in March 1984 up to the most recent deployment off Eire in August 1985. Accounts of all deployments in 1984 were described at a meeting held at IOS Bidston in May 1985 by Prandle (Ref. 3). D. Eccles of the Rutherford Appleton Laboratory explained the technical advances incorporated into OSCR which resulted in radial current resolution of the order of 1 cm/s - an order of magnitude better than the CODAR (Ref. 4). The deployment of OSCR from a complete range of coastal sites was noted ranging from sloping meadows, urban concrete promenades to a 100m high limestone cliff shelf; moreover the recent Cumbrian deployment included radar locations several kilometres inland. In subsequent sections, descriptions are restricted to the deployments involving the Institute of Oceanographic Sciences.

3. THE USE OF OSCR FOR MEASURING TIDAL CURRENTS
(Liverpool Bay, April/May 1984; Red Wharf Bay, May 1985)

3.1 Liverpool Bay

With only one OSCR unit available, measurements were made first for 7 days from the Lancashire coast followed, 1 month later, by similar measurements from the Wirral coast. These were subsequently combined to compute current ellipses from the predominant M_2 tidal constituent. These OSCR results were then compared with (i) current ellipses calculated from a 1km grid numerical model and (ii) observed values from 2 conventional current meter moorings located at the centre of the OSCR array. Fig. 4 indicates the close agreement between OSCR results and numerical model values for the major axis of the M_2 current ellipse. Table 2 further indicates the close agreement between values at the centre of the array from (i) OSCR, (ii) the numerical model and (iii) the 2 current meters. A full description of this experiment has been described by Prandle and Ryder (Ref. 5).

3.2 Red Wharf Bay

Despite the close agreement found between OSCR results and comparable data in the Liverpool Bay experiment, the relatively small spatial gradients associated with the M_2 constituent were not sufficient to provide evidence of the focussing precision of OSCR (focussing is used here to denote the accuracy to which an OSCR measurement from one bin value reflects only currents pertaining to that delineated area). The shortness of the record length (as little as 24 hours along some beams) prevented accurate determination of higher harmonic tidal constituents such as M_4 and M_6. By conducting a second experiment, using two OSCR units, with each beam recording hourly for 1 month, an accurate tidal analysis could be made with precise resolution of constituents such as M_4 and M_6. Thence, the more pronounced spatial structure anticipated for such constituents should indicate the degree of focussing obtained with OSCR.

Fig. 5 shows the distribution of surface current ellipses for the M_2 constituent measured by OSCR. The sweep of this predominent tidal constituent through this region is particularly smooth and regular despite the proximity of coastal headlands. etc. This regularity perhaps explains why numerical models with coarse grid representations can accurately simulate the propagation of this constituent. By contrast Fig. 6 shows the equivalent distribution for the M_4 constituent. The distribution now varies markedly from one position to the next, indicating significantly different results from adjacent bins and hence suggesting strong 'focussing'. This irregular distribution for M_4 accords with the anticipated complexity of local generation and subsequent propagation. Studies are progressing aimed at relating the measured patterns for M_2 with those for M_4 and M_6 etc. Studies are also progressing into the distribution of instantaneous residuals, Fig. 7 shows the net residual distribution over the month with current directions aligned with the net propagation of tidal energy towards the River Conway (in the south east corner of the diagram).

This study formed part of a collaborative experiment which, in addition to OSCR, involved an air-borne multispectral scanner, a profiling acoustic current meter, conventional current meter moorings and CTD surveys. Scientists from IOS and RAL were joined by groups from the Universities of Bangor and Cambridge, inter comparison of results is continuing and relevant interpretations will be discussed.

4. THE USE OF OSCR FOR MEASURING SURFACE WIND-DRIVEN CURRENTS

Estimates of tidal currents calculated from OSCR measurements form a synoptic picture over an area of $30 \times 30 km^2$ which can be interpreted in terms of existing measurements and theories. For wind-driven currents, however, the OSCR measurements of near surface current provide new information which should aid investigation of the processes involved. The effective depth of the measurement is within 1m of the sea surface, where it is very difficult to make accurate in situ measurements, especially during storms. Since the technique is new, the significance and meaning of the measurements is still being investigated. However, comparisons with in situ measurements are complicated; differences between the observations may be real i.e. have an oceanographic origin, may have arisen from failures in either the OSCR or the in situ measurements or may be due to differences in sampling - OSCR's areal and depth average compared with an in situ, discrete, measurement. Moreover the effective depth of the OSCR measurement is uncertain - electro-magnetic theory predicts it to be a weighted and integrated measurement over a depth of a small fraction of the radio wavelength (λ). The fraction depends on the assumed current profile and values for the vertical averaging scale of 0.08λ (Ref. 6) and 0.022λ (Ref. 7) have been calculated. Both estimates suggest an effective depth of less than 1m, for OSCR's wavelength of 11m, but the differences are large enough to affect significantly the estimation of current profiles.

Small errors or biases that might not be important in a tidal analysis of a data set lasting 14 or 29 days assume greater significance in the study of wind-driven currents. Three experiments during OSCR deployments have compared estimates of wind-driven currents made from OSCR measurements with those made from in situ recording current meters. In the first a single OSCR unit was deployed at St. John's Point in Northern Ireland in March 1984 and in addition tracked drogue measurements were made, details are given in Ref. 8. In the second, by SMBA, two OSCR units were deployed on the Mull of Kintyre in October 1984 overlooking the Sound of Jura (Ref. 9) and in the third two OSCR units were deployed on the south coast of Eire, to the west of Cork, in August 1985. In these experiments the recording current meters were moored in a variety of ways; measurements within the top 1m were made by special instruments mounted integrally into a surface following buoys whilst measurements in the depth range 1-5m below the surface were made by commercial vector averaging current meters moored either beneath a surface following buoy or beneath a buoy referenced to the wave trough level. The behaviour of the various moorings under the action of surface waves will differ, suggesting that the observations might also be different.

However, the comparisons show that the mean difference between the OSCR and in situ measurements was small (probably less than 1cm/s) and that the standard deviation of the difference was of the order of 3 cm/s. The residual current estimated from the OSCR measurements was correlated with the wind, with a slope for the linear regression between 1 and 2%. Also the difference between the OSCR and in situ measurements at 2 or 3m was correlated with the wind, implying the presence of a sheared wind-driven current and that the effective depth for the OSCR measurements was certainly less than 2m. This also implies that the standard deviation of 3 cm/s referred to above represents an upper bound on the difference.

Clearly OSCR measures near surface current and calculation of the wind-driven current from its measurements will not provide an indication of what the currents at depth without further assumptions.

5. LONG TERM MEAN CURRENTS

The length scale of variations of long term mean currents is small and related to the local topography, so that the spatial coverage provided by OSCR measurements is useful (Fig. 7 and 8). Note, however, that present deployments of OSCR, maximum duration a month for the Red Wharf Bay deployment, have been too short for the calculation of a reliable estimate of the mean. Also OSCR measures near surface currents - the mean current at depth will probably be different from that at the surface.

6. CONCLUSIONS

In this paper we have discussed the uses of OSCR measurements in actions on tidal, wind-driven and long term mean currents. In addition, particularly for real time applications, although OSCR measures the Eulerian current at each of the bins, Lagrangian currents can be estimated by moving a tracer with the current from bin to bin.

The technique is in its infancy and it is still being validated in a variety of operating conditions. The best methods of operating it and analysing the results are being developed. Two of the experiments discussed here (Red Wharf Bay May 1985 and the south coast of Eire, August 1985) are recent and the results are still being studied.

ACKNOWLEDGEMENTS

Some of the experiments discussed in this paper were funded by the U.K. Dept. of Energy.

REFERENCES

1. King, J.W., Bennett, F.D.G., Blake, R., Eccles, D., Gibson, A.J.,
    Howes, G.M. and Slater, K.: "OSCR (Ocean Surface Current Radar)
    observations of currents off the coasts of Northern Ireland, England,
    Wales and Scotland". In: Proc. Current Measurements Offshore (17 May 1984)
    Society for Underwater Technology, 1984, 38pp.

2. Barrick, D.E., Evans, M.W. and Weber, B.L.: "Ocean surface currents
    mapped by radar". Science, 198, 1977, pp. 138-144.

3. Prandle, D.: "Measuring currents at the sea surface by H. F. radar
    (OSCR)". Underwater Technology, 11(2), 1985, pp. 25-27.

4. Leise, J.A.: "The analysis and digital signal processing of NOAA's
    surface current mapping system". IEEE Journal of Oceanic Engineering,
    OE-9(2), 1984, pp. 106-113.

5. Prandle, D. and Ryder, D.K.: "Measurement of surface currents in
    Liverpool Bay by high-frequency radar". Nature 315, No. 6015,
    9 May 1985, pp. 128-131.

6. Stewart, R.H. and Joy, J.W.: "HF radar measurements of surface currents".
    Deep-Sea Research, 21(12), 1974, pp. 1039-1049.

7. Ha, E.-C.: "Remote sensing of ocean current and current shear by H.F.
    backscatter radar". Ph.D. dissertation, Stanford University.

8. Collar, P.G., Eccles, D., Howarth, M.J. and Millard, N.: "An
    intercomparison of HF radar observations of surface currents with
    moored current meter data and displacement rates of acoustically
    tracked drogued floats". In: Proc. Ocean Data Conference - Evaluation,
    Comparison and Calibration of Ocean Instruments, 4 & 5 June 1985,
    Society of Underwater Technology, 24 pp.

9. Griffiths, C.R., Booth, D.A., Eccles, D. and Bennett, F.D.G.:
    "Comparison of near-surface current measured by acoustic and
    electromagnetic current meters with HF radar current measurements".
    Unpublished manuscript. International Council for the Exploration
    of the Seas. ICES C.M. 1985/C:25, pp 18.

TABLE 1 SUMMARY OF OSCR DEPLOYMENTS

March 1984   Island of Mull, Scotland. Comparison
                  of OSCR 1 with meters deployed by
                  the Scottish Marine Biology Association.

March/April   St. Johns Point, Northern Ireland.
1984   Comparison of OSCR 1 with meters
                  deployed by IOS, Wormley. Operations
                  staff supplied by IOS, Bidston.

April/May   OSCR 1 operated at Crosby then Hoylake
                  (Liverpool Bay) to simulate vector
                  observations. Operations staff supplied
                  by IOS Bidston.

August 1984   Angle, Wales. Comparison of OSCR2 with
                  meters deployed by the Scottish Marine
                  Biology Association.

September/   Swansea Bay, Wales. First vector trial
October   of OSCR1 and OSCR2. Support of the
1984   RAE SIR-B experiment on the challenger
                  Space Shuttle. Meters deployed by
                  Swansea University.

October/   Peninsula of Kintyre, Scotland. Second
November   vector trial of OSCR1 and OSCR2.
1984   Meters deployed by the Scottish Marine
                  Association.

May   Red Whard Bay, Anglesey. 30 days
1985   continuous monitoring. Collaborative
                  experiment, IOS, RAL, Universities of
                  Bangor & Cambridge.

July/Aug   Cumbrian Coast. Total of 4 sites. Data
1985   to be used for siting offshore pipelines.
                  Marex, RAL, IOS, North West Water Authority.

August   Southern Coast of Eire. 14 days data
1985   obtained for comparison with offshore
                  measurements using moored current meters,
                  buoys etc. Marex, IOS.

TABLE 2 : M_2 tidal ellipse properties at 53°28.3'N,
3°15.4'W. Direction measured a-c due east.

	Amplitude (major axis)	Direction	Phase	Eccentricity +ve anticlockwise
(i) Numerical model	58	350°	243°	-0.02
(ii) Current meters (1)	54	351°	237°	-0.02
(2)	50	356°	231°	-0.04
(iii) H.F. Radar (surface)	61	342°	232°	+0.03
(iv) H.F. Radar corrected to mid-depth value	55	344°	229°	+0.08

$cm\ s^{-1}$

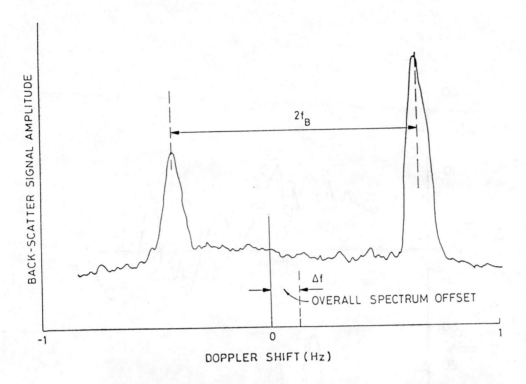

Figure 1   The spectrum of the backscattered radar signal. The two peaks are spaced apart by $2f_b = \frac{4c}{\lambda}$ are offset by $\Delta f = \frac{2u}{\lambda}$

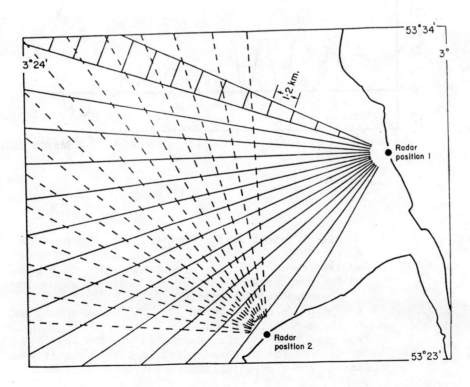

Figure 2   Typical spatial coverage given by the OSCR system. The deployment was in Liverpool Bay March/April 1984. (Figure reproduced from Underwater Technology 11(2)).

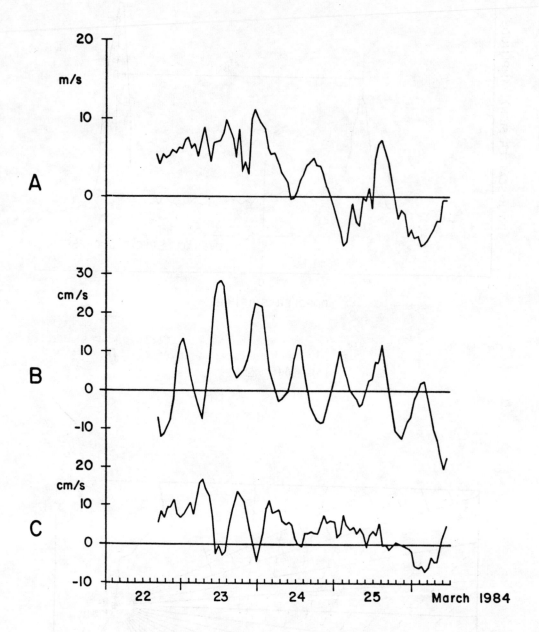

Figure 3   Single OSCR unit at St. John's Point, Northern Ireland,
                  March 1984
                  a) The component of the wind towards the OSCR site.
                  b) The component of the current measured at 3m depth
                     towards the OSCR site.
                  c) The difference between the OSCR measurements and
                     those plotted in b).
                  When the wind and the maximum tidal flow are in opposite
                  directions the difference (c) is large, whilst when they
                  are in the same direction the difference is small.

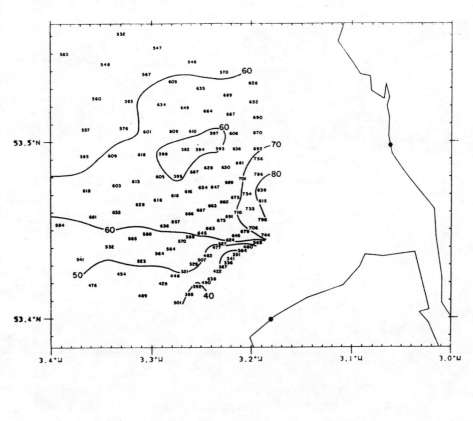

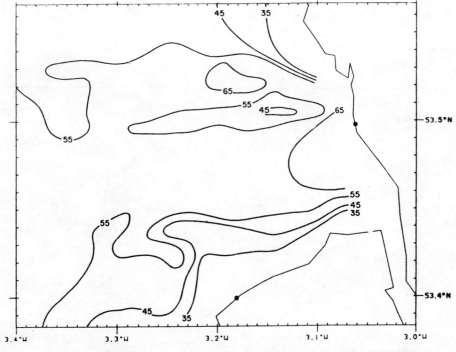

Figure.4 M_2 major axis (top) OSCR measurements (mm/s) (bottom) numerical mode (cm/s)

(Figure reproduced from Nature 315 (6015))

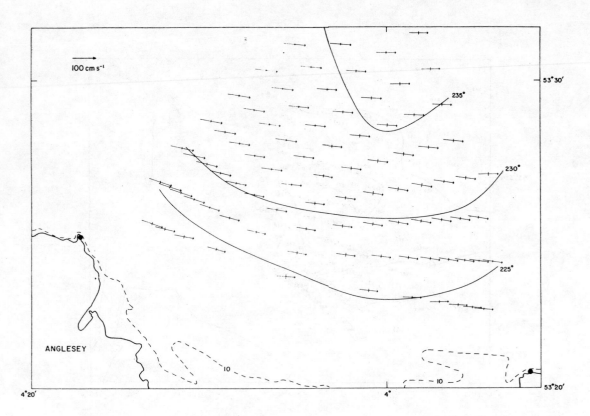

Figure 5   M$_2$ current ellipses. Results from 1 month's continuous hourly measurements at each position during May 1985.

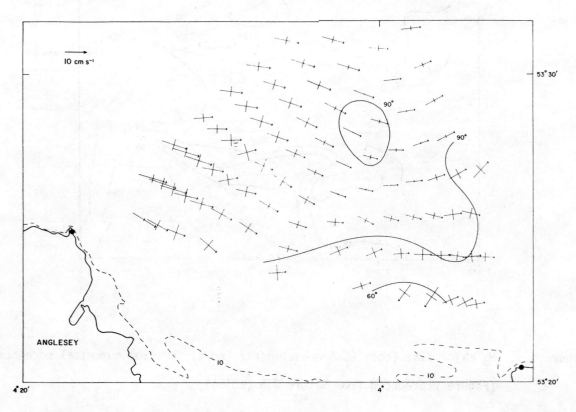

Figure 6   M$_4$ current ellipses. (see also Fig. 5)

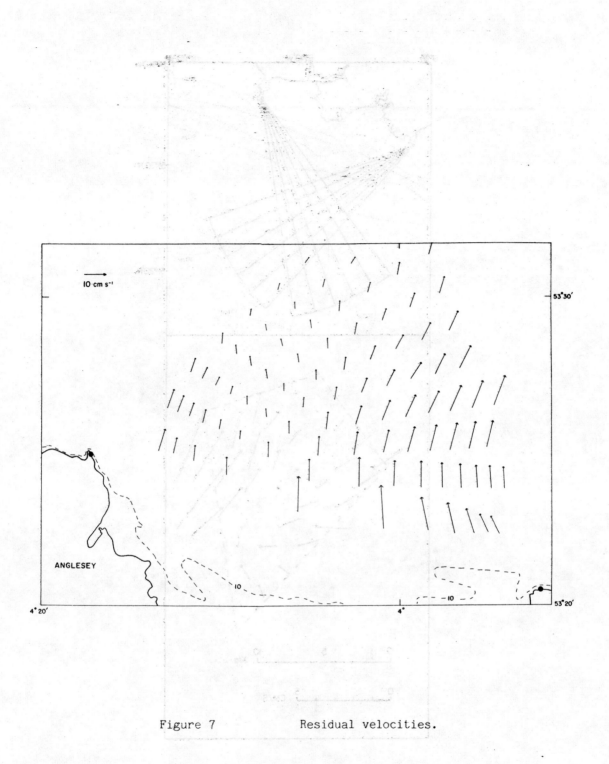

Figure 7   Residual velocities.

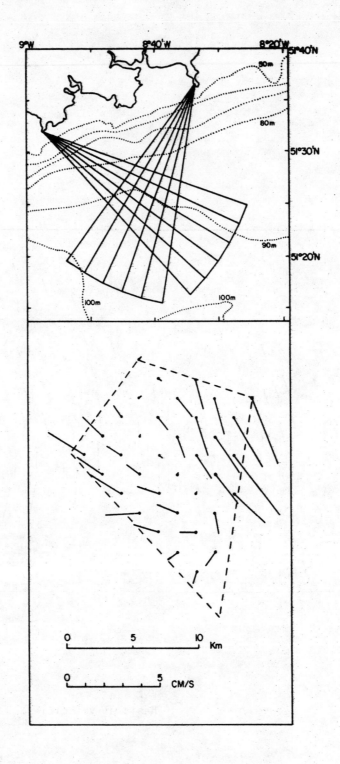

Figure 8   Twin OSCR units at Old Head of Kinsale and Galley Head,
                to the west of Cork, Eire, August 1985.
                a) Map showing OSCR configurations.
                b) Mean currents over a period of 14 days plotted on a
                   2km grid for the region of overlap of the two OSCR
                   beams.

OCEAN SURFACE WAVE AND CURRENT MEASUREMENT WITH HF GROUND-WAVE RADAR

L.R. Wyatt,   J. Venn,   G.D. Burrows

Department of Electronic and Electrical Engineering,
University of Birmingham,
P.O. Box 363, Birmingham B15 2TT,
England.

and   M.D. Moorhead

Neptune Radar Ltd.,
Gardiners Farm, Sandhurst, Gloucester, G12 9NW,
England.

Summary

An HF radar has advantages over conventional oceanographic instrumentation.
It is shore-based and can provide simultaneous measurement over a large area with good
spatial resolution.   Thus it represents a substantial enhancement over single wave-
buoy or current meter deployment.   An HF radar system that is optimised for ocean
surface wave measurement, developed by the Department of Electronic and Electrical
Engineering at the University of Birmingham, is operated from the University's field
station at Angle, Pembroke, S.W. Wales.   This radar surveys an area 200km in radius
by 90 degrees in azimuth with 10 degree azimuthal and 7.5km range resolution.   It has
provided measurements of surface current, significant waveheight, wave period, the
directional spectrum of swell components and an indication of wind speed and direction.
In this paper we discuss results from intercomparisons held in 1983, NURWEC1, 1984,
SIR-B, and 1985, an I.O.S. current measuring exercise.   There will be a demonstration
(NURWEC2) of a dual-radar system in Spring 1986 that will be supported by the Dutch
Rijkswaterstaat who are planning to deploy three wave-buoys and two current meters to
provide ground-truth for the radar measurements.   We indicate improvements in accuracy
and performance that should result from the new hardware and software developments
and those that should result from the use of a two radar system.

Held at Imperial College of Science and Technology, London. Organised and sponsored by BHRA, The
Fluid Engineering Centre. Co-sponsored by the American Society of Civil Engineers and the
International Association for Hydraulic Research.
©BHRA, The Fluid Engineering Centre, Cranfield, Bedford MK43 0AJ, England 1986.

NOMENCLATURE

f_b	= first order Bragg frequency
Hs	= significant waveheight
$\underline{k}$	= ocean vector wavenumber
k_o	= radio wavenumber
$S(\underline{k})$	= directional spectrum of wave energy
T1	= a representative wave period derived from the first moment of the frequency spectrum
Tz	= wave period using the second moment

1. INTRODUCTION

Barrick (1971,1972) has shown that the Doppler spectrum of the backscattered radar echo from the ocean surface can be related to the ocean-wave directional spectrum through a non-linear 2-dimensional integral equation. The Doppler spectrum clearly divides into first and second-order contributions (see figure 1). The first order Bragg peaks are echoes from ocean wave components that have exactly half the radar wavenumber and are travelling directly towards (positive shift) or away from the radar. The second order spectrum is a result of non-linearities in the ocean wave field as well as double scattering processes. The first order peaks are used for current and wind direction measurements and are needed to calibrate the second order returns which have the potential to be used to derive the full directional spectrum of wave energy.

Wyatt (1984,1985a) has developed techniques that use the integral equation to provide measurements of significant waveheight, Hs, first and second moment wave period, T1 and Tz, and the longwave directional spectrum, $S(\underline{k})$. Hs, T1 and Tz are found by taking moments of sections of the second-order sidebands of the Doppler spectrum, surrounding the larger first order peak and using a regression analysis. $S(\underline{k})$ is found using a model-fitting technique to solve the integral equation. All four sidebands of the spectrum are required for this analysis and this imposes a strict signal-to-noise requirement on any data. At least 10dB is necessary in the lowest sideband which can mean up to 60dB overall dynamic range in certain conditions (this requirement is particularly sensitive to wind direction). For Hs,T1 calculations, 10dB signal-to-noise in the upper sidebands is required which means an overall dynamic range of 40dB, and sometimes only 30dB, is sufficient. The model-fitting technique has further limitations. Simulation work shows that it provides good estimates of amplitude only when the swell is propagating at an angle of less than 90 degree to the radar beam and good estimates of direction only when they are perpendicular. In addition it restricts measurement to periods greater than 9 secs at 9MHz and 6 secs at 20MHz.

An HF radar system that is optimised for ocean surface wave measurement has been developed by the Department of Electronic and Electrical Engineering at the University of Birmingham with funding from S.E.R.C. and the Wolfson Foundation. This radar, which is operated from the University's field station at Angle, Pembroke, S.W. Wales (see figure 2), is capable of surveying an area 200km in radius by 90 degrees in azimuth with 10 degree azimuthal and sub km range resolution. In practice the range resolution is limited to 7.5km by the University's radio-frequency bandwidth allocation. Doppler spectra with the required signal-to-noise have been obtained with this system (see Wyatt et al 1985). In this paper we will discuss results from an intercomparison experiment held in 1983 (NURWEC1) as well as more recent short experiments (SIR-B and an I.O.S. current measuring exercise).

An improved radar is under construction at Birmingham and a commercial prototype has been developed by Neptune Radar Ltd. The new Birmingham radar will be deployed at Pembroke and the commercial radar will be installed in N.Devon to complete the two radar system which, it has been shown theoretically, is necessary for accurate measurement of the directional spectrum of wave energy. There will be a demonstration (NURWEC2) of this dual-radar system in Spring 1986 which will be supported by the Dutch Rijkswaterstaat. This experiment is discussed in section 3.

2. INTERCOMPARISON EXPERIMENTS

2.1 NURWEC

The Netherlands Rijkswaterstaat (RWS) have taken a great interest in the Birmingham University HF radar work. They have a particular requirement for wave monitoring in the North Sea for flood prevention purposes. The first Netherlands, U.K., Radar, Wave-buoy, Experimental Comparison (NURWEC) began in October 1983. RWS deployed a Datawell Wavec buoy 30km from the radar site. As a result of unusually calm conditions throughout the first three weeks of November the experiment was continued until December 10th which allowed monitoring of two storms. The results of the intercomparison are described in detail in Wyatt, 1985b. Radar measurements of Hs,

T1, shortwave direction, and S($\underline{k}$) were made with varying degrees of success. T1 was measured with a standard deviation of only 0.6 secs over the range 3 to 10 seconds. Estimates of Hs were found to have a standard deviation of 15% if they were less than 2m, and 20% if greater although in this case there was a positive bias of 41%. The correlation between Wavec and radar Hs estimates is good during high sea-states and the advantage of radar over a single wave-buoy measurement is illustrated in figure 3 which shows the spatial and temporal development of Hs during the two storm periods.

Measurement of the directional spectrum of the longwave components was severely limited by interference. However, the radar results that were obtained did display the same properties as the simulations. Conditions were favourable for good amplitude estimates during the development of the December storm and indeed, where comparison was possible and not limited by transmission problems with the buoy, good agreement was found. Frequency spectra as a function of range and time are plotted in figure 4. The spectra have been extended beyond the maximum frequency that can be measured by the longwave technique, to 0.2Hz using a simple, high frequency model-fitting technique.

The procedure involved in the calculation of Hs is very similar to that used for T1 for which good estimates were obtained. Indeed the Hs calculation makes use of a smaller part of the second order Doppler spectrum and, while this may increase the random errors, the method should be less sensitive to ship or interference signals which can appear anywhere in the spectrum. There are two differences between the Hs and T1 calculations which could be causing problems. The power in the first order part of the spectrum is used in the estimation of Hs but not in T1. This part of the spectrum has not been studied in great detail but since it isolates a single component out of the entire two-dimensional wave spectrum it will be more sensitive to local departures, in space and time, from the model that has been assumed in the algorithm development. The first order peaks are also used in the model-fitting technique to derive the long-wave directional spectrum. It is worth noting that the bias in amplitude estimates for the longwave components during NURWEC1 was found to be smaller than that for Hs. This suggests that another source of error may lie in that part of the second-order spectrum used in the Hs calculation but not in the model-fitting. One result that supports this hypothesis is a measurement of Hs obtained by applying the combined long and shortwave modelfitting techniques. This leads to much more accurate Hs estimates during the final NURWEC1 storm. Unfortunately the calculation depends on the availability of longwave data which was very limited. Because of the success of these few calculations and in order to provide an algorithm that could be applied to a larger proportion of the NURWEC1 data, the regression analysis was re-peated but restricted now to the same region of the Doppler spectrum used in the long-wave measurements. This has the required effect of reducing the bias at high sea-states at the expense of increased scatter for Hs less than 2m presumably because of increased statistical variability. The improvement achieved during the December storm with this reduced range algorithm, is shown in figure 5. The search for a more robust algorithm continues. The fact that the T1 calculation involves a ratio of moments of the second-order part of the spectrum may make it less sensitive to the effects implied here.

2.2   SIR-B

Contours of significant waveheight have been sketched in figure 6 to illus-trate once again the potential of HF radar for measuring spatial and temporal varia-bility. The data was obtained during a 60 hour exercise in October 1984 to provide ground support for the interpretation of SIR-B (Shuttle Imaging Radar) imagery obtained during a flight of the space shuttle, Challenger. The data for each of the maps in figure 6 was collected over a period of about 2 hours during which time stationarity in the wave field has been assumed. During this time wind speed increased from about 6 to 10m/sec blowing from SSW. There was no ground-truth as such for this experiment and because of operational difficulties during the shuttle flight, the SIR-B data is unlikely to be useful in any quantitative comparison. There is some indication by comparison with the I.O.S. Isles of Scilly wavebuoy and Met. Off wave model products, that the radar measurements are, as in NURWEC, overestimating waveheight.

2.3   Current Measurement

A single radar can give no azimuthal current information since only the radial component is measured as the frequency shift of the 1st order peak of the Doppler spectrum.   A low signal to noise ratio does not degrade this measurement and current information can be obtained at further ranges than wave information and from shorter data sequences.

Radar data was obtained during the IOS Celtic Sea current measurement program in September 1985 as an instrumental comparison exercise.   Ten minute long data sequences and a Hanning window were used for the 30 hour time series of figure 7. It is a plot of surface current at 30km distance from the radar on the 270 degree beam and is therefore a measure of the Easterly component of the current vector.   The comparative data is taken from Admiralty chart 1123 at a position only half this distance from the coast.   Accordingly, the tidal stream amplitudes are greater and have been phase-shifted by plus half-hour for a better fit to the radar data.

Clearly the major part of this surface current variation is due to the tidal stream.   Spatial variation with range up to 200km is also consistent with the tidal stream atlas, showing a positive change of phase at the further ranges.   Data from 60 km range shows the expected lower velocity amplitudes but these signals are, in some cases, offset from the zero current line.   It is presumed that this is a result of a wind drift current.

3.   DEMONSTRATION

Birmingham University is collaborating with Neptune Radar Ltd. in an exercise with RWS, to be known as NURWEC2 planned for Spring 1986.   The main aim of the experiment is to demonstrate that HF radar has the potential to provide round the clock near real time wave data over a wide area extending up to 200km offshore for some parameters and at least 100km for all parameters.

NURWEC1 probably raised as many questions as it answered.   This was not surprising since it took place at a time when the group had had little experience of this scale of operation, and had no experience of applying the wave parameter extraction techniques to real radar data.   It was an invaluable exercise for gaining experience in both these areas and for identifying where further research, software development and improvements in radar design were needed.   NURWEC2 will be supported once again by RWS who plan to deploy three wavebuoys as well as two current meters to provide ground truth beyond the range where the availability of good radar data has been proven (i.e. about 30km from the radar) in order to test the range capability of the radar and the accuracies achievable with a dual-radar system.

Apart from the overall aim of NURWEC2 described above there are some additional aims:

1.   To demonstrate that improvements to the radar design and operating procedures provide usable radar data in an acceptable range of operating conditions.

2.   To show that inaccuracies of the analysis technique observed in certain sea conditions can be significantly reduced by a dual radar installation.

3.   To show that the fundamental left/right ambiguity presently resolved using external data can also be resolved independently by a dual radar installation.

Perhaps the most disappointing result from NURWEC1 was the 40% overestimation of significant waveheight in high seas.   This could be due to a number of factors, many of which have been investigated since the experiment.   The most obvious problem was probably due to intermodulation distortion within the radar system.   This may have arisen as a result of inadequate monitoring of receive signal level, a faulty computer interface or degradation within the aerial system.   A new receiver has been developed to have overall improved performance as well as less susceptability to intermodulation.   This receiver is currently undergoing tests and will be in operation during NURWEC2.

One factor that delayed the analysis of NURWEC1 data was the necessity to identify all data contaiminated by ship and interference signals.   This was achieved by inspecting plotted Doppler spectra, identifying such signals by eye and appropriately marking the data files.   A simple algorithm has been developed that performs this task automatically for ship signals as long as there is sufficient range information. It produces a cubic spline interpolant for the sea-state parameters at the range cells that are identified. This operates as a post processing algorithm. More work is required to develop techniques for identifying and removing ship echoes from the Doppler spectrum before sea-state parameter extraction proceeds. Interference signals cannot be rejected in this way because they appear at all ranges. These are likely to require hardware developments and are under investigation.

An additional operational constraint identified after NURWEC1 was the requirement to operate above a critical wind speed that is dependent on radio frequency. This criterion arises as a result of the requirement that the wave components causing the first order Bragg peaks should be in the Phillip's equilibrium regime. The effect of this constraint is to limit the available operating frequencies in low sea-states. NURWEC1 was dominated by such conditions and many measurements did not satisfy the criterion. This is not a restriction that is important for most operationally important sea-states and will become less of a problem as extraction techniques with more general application are developed.

The extraction of longwave parameters during NURWEC1 was found to be severely limited by interference and by the limited range of oceanographic conditions for which the method was known to work. A new technique has been developed (Wyatt, 1985c) which to a large extent overcomes both problems. Data from two radar systems is combined to produce what is essentially an eight sideband model-fitting technique. The simulation study shows that good estimates of longwave amplitude, direction and spread can be obtained if the radar look directions are separated by at least 30 degrees no matter how the radar systems are configured relative to the ocean wave field. In addition, and perhaps more importantly, the analysis also produces good estimates if a four sideband technique is used with the best two sidebands from each radar thus reducing the signal-to-noise requirements to those used in Hs estimation. NURWEC2 will provide the first data set for testing this method. Figure 2 shows the minimum expected coverage of longwave measurements using the two (from each) or four sideband technique. The use of two radars also gives vectored current and allows a resolution of the ambiguities in both long-wave and wind direction estimates.

There is no dount that NURWEC2 will identify further areas where software development is required. Techniques for measuring the high frequency part of the ocean wave spectrum are being developed and this work will continue. NURWEC2 will provide a further data set for testing such techniques.

The aims of the experiment which have been outlined above are shared by both Birmingham University and Neptune Radar. Each group has additional specific objectives which relate to the eventual commercialisation of the hardware and software.

For the radar system to find commercial use it must be designed to operate with nominal or no support staff. The major hardware development implemented by Neptune radar has been to automate the radar control function. This automation has allowed a limited degree of automatic frequency management and in this way the radar modulation parameters can be adaptively set to obtain the best compromise with the co-channel interference environment.

Aims for the experiment that are specific to Neptune Radar are:

- to demonstrate that the company has the ability to produce a radar with adequate performance for sea wave measurement,
- to advance the development of the unattended radar design through further operating experience.

The further aims of the Birmingham University group are:

- to demonstrate that the software package for sea-state parameter extraction is capable of being run on site to produce near real time evaluation of the directional ocean wave data,
- to identify those areas needing further research.

There are differences between the two radar systems. The Neptune Radar system includes passive receive array elements in place of the C&S active loops used by Birmingham University. Although larger, 6m instead of 2m in height, these elements offer considerable operational advantages. The reduction in vertical lobe gain yields a reduction in ionospheric echoes and in skywave interference.

The radar data processing will be carried out in two parts. An on-line data collection system (at both sites) runs when the radar is illuminating the ocean surface. This carries out spectral analysis of the sea-echoes within each radar resolution cell. The output consists of an average Doppler spectrum for each radar cell. This software is well understood and Neptune Radar will be making their output data files available to the University for further processing at the East Blockhouse site. The second stage processing consists of the Birmingham University sea-state parameter extraction package. It is envisaged that Hs, T1, S(k), wind parameters and the radial current component will be calculated on site immediately following each data collection with the Birmingham University radar. A period of about three hours will be needed for data collection on five beams and the production of maps of various parameters. This amounts to a near real time operation. The Neptune Radar data second processing will be delayed by the transport time from the N.Devon site. The processing of the dual-radar data will be carried out either on site when there is a suitable gap in the data collection or back at Birmingham University on the mainframe. Processing and communications software development is required before this becomes a near real time operation. For the prototype system Neptune Radar is employing a computer system which is compatible with the Birmingham University system. This allows easy transfer of software and data. Once the computer resources required for the ocean surface analysis software have been determined there is considerable scope for employing a simpler computer system in the production radar.

4.   CONCLUDING REMARKS

HF radar has demonstrated two important advantages over conventional oceanographic instrumentation. It is shore-based which allows for ease of access for maintenance and control and it can provide simultaneous measurement over a large area with good spatial resolution. Thus it represents a substantial enhancement over single wavebuoy or current meter deployment. Although the accuracy of wave measurement with HF radar is, at the present time, less than that achievable with a wavebuoy, the ability to measure spatial variability with one instrument provides a more representative data set. In general this will allow a consumer of wave data to make a more accurate assessment of the impact of a particular wavefield.

The indications from the intercomparison experiments and from computer simulations are that recent progress made on both hardware and software are sufficient to aim directly at the production of a working system for the near real-time extraction of sea-state parameters. The Spring 1986 demonstration should indicate the potential for oceanographic investigation.

It is now time for us to consider the ramifications of a proliferation of such systems. Discussion is needed on the oceanographic needs for frequency allocations within the crowded HF band. Representations will have to be made to regulatory bodies and the case for oceanographic usage must be justified by scientific and commercial users of wave and current data.

ACKNOWLEDGMENTS

We are extremely grateful to the Dutch Rijkswaterstaat and to the Institute of Oceanographic Sciences for all their support and encouragement. We acknowledge the financial support of S.E.R.C. and the Wolfson Foundation.

REFERENCES

Barrick, D.E.: "Theory of HF/VHF propagation across the rough sea, Parts I and II". Radio Science 6, 517-533, 1971.

Barrick, D.E.: "Remote sensing of sea state by radar, in Remote Sensing of the Troposphere". ed. V.E. Derr, U.S. Government Printing Office, Washington, D.C., 12-1 to 12-46, 1972.

Wyatt, L.R.: "The measurement of Significant Waveheight and Mean Period from HF Radar Doppler Spectra". Departmental Memo. 508, Department of Electronic & Electrical Engineering, University of Birmingham, U.K. 1984.

Wyatt, L.R.: "The measurement of the Ocean Wave Directional Spectrum from HF Radar

Doppler Spectra".   Departmental Memo. 509, Department of Electronic & Electrical Engineering, University of Birmingham, U.K., 1985a.

Wyatt, L.R.: "Netherlands/UK. Radar, Wave-buoy Experimental Comparison (NURWEC) Oct-Dec. 1983".   Report on the radar results, Department of Electronic and Electrical Engineering, University of Birmingham, March 1985b.

Wyatt, L.R.: "Ocean Wave Parameter Measurement Using a Dual-Radar System: a simulation study".   Departmental Memo. 514, Department of Electronic & Electrical Engineering, University of Birmingham, U.K., September 1985c.

Wyatt, L.R., Burrows, G.D. and M.D. Moorhead: "An assessment of a FMICW ground-wave radar for ocean wave studies".   Int. J. Remote Sensing, 6, 275-282, 1985.

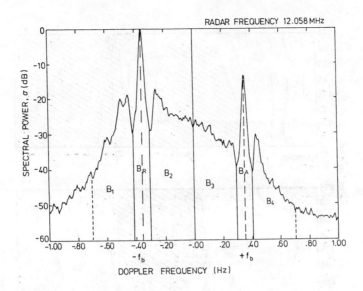

Figure 1: A radar doppler spectrum showing first (B_R, B_A) and second $(B_i, i=1,4)$ order regions used in wave parameter determination. $+f_b$ - frequency of the ocean waves of twice the radio wavenumber, k_o, moving along the radar look direction. $f_b = \sqrt{2gk_o}$ for deep ocean gravity waves.

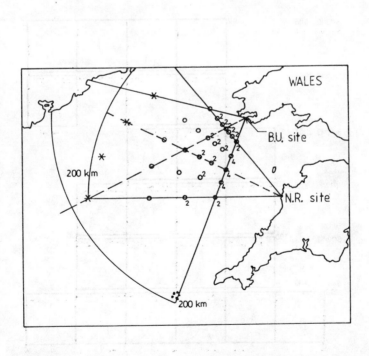

Figure 2: Map of the Celtic Sea showing the coverage area of the Birmingham University (BU) radar and the combined (with Neptune Radar Ltd., NR) coverage for the demonstration planned for Spring 1986. Seven beams will be in operation from East Blockhouse and five from the North Devon site providing 29 intersection points within 200km of either radar. The intersection points between the radar beams where the angle is greater than 30° are shown with a circle and with a * where the angle is less than 30°. Where these are at ranges which would have provided measurements during NURWEC1 (i.e. a maximum range on both beams of 100km at 7MHz), the number of sidebands that could have been used (from each radar) is shown. All the remaining intersection points with angles less than 30° are within 150km of both radars.

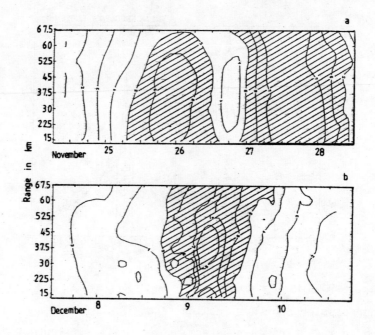

Figure 3:   Contours of significant waveheight as a function of range and time during
(a) November storm (b) December storm, waveheights greater than 3m are hatched.

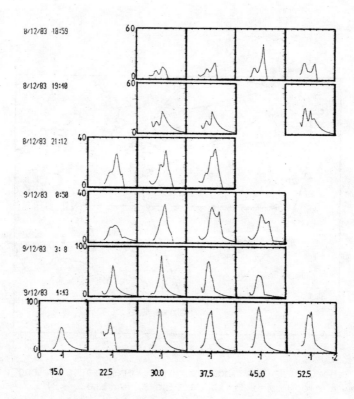

Figure 4:   The variation with range and time of the radar-measured frequency spectrum
during the development of the December storm.

The vertical scale is the amplitude in m^2/Hz and the horizontal scale is frequency in
Hz.

The date and time of each observation is indicated on the left-hand side of the plots
and the range, in km, below the plots.

24

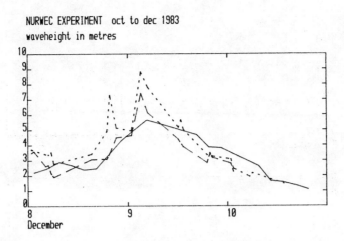

NURWEC EXPERIMENT oct to dec 1983

Figure 5: Significant waveheight measured during NURWEC1. WAVEC data is indicated with a solid line; radar data using the original regression algorithm, with a short-dashed line; radar data using the reduced range regression, with a long-dashed line.

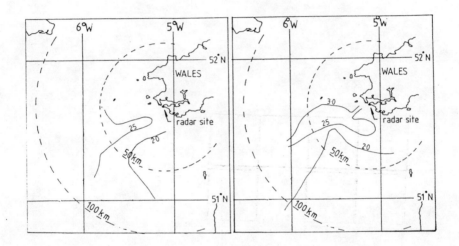

Figure 6: Waveheight contours (in metres) obtained during the SIR-B experiment.

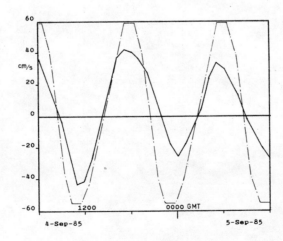

Figure 7: Easterly component of surface current obtained from
a) radar data from lat 51°49.9N long 5°33.2W on 4/5th September 1985
b) Admiralty tidal stream data from lat 51°36.5N long 5°17.0W (dashed line)

25

PAPER A3

Measuring Techniques
of Hydraulics Phenomena in offshore, Coastal & Inland Waters
London, England: 9-11 April, 1986

THE DISTORTION OF CURRENT METER MEASUREMENTS DUE TO
UNUSUAL HYDRAULIC PHENOMENA

J. Graff

CEEMAID Limited

Summary

The purpose of this paper is to illustrate the importance of a systematic approach
to the analysis and interpretation of current meter measurements in order to identify
hydraulic phenomena which may be overlooked in the course of a normal analysis
procedure. The lesson is relevant to any other area concerning the measurement and
interpretation of flows in the natural environment. The paper illustrates a recent
case where abnormal distortions in current meter records from the North Sea have
been reinterpreted to provide evidence of extreme flows arising from unusual
hydraulic phenomena caused by the interaction between the tidal current and a drift
current. It highlights the importance of post processing procedures and the need
for a critical objective approach in translating the resulting information.

The author was formerly Chief Marine Scientist at WIMPOL Ltd.

Held at Imperial College of Science and Technology, London. Organised and sponsored by BHRA, The
Fluid Engineering Centre. Co-sponsored by the American Society of Civil Engineers and the
International Association for Hydraulic Research.

NOMENCLATURE

h_i = amplitude of harmonic tidal constituent

σ_i = constituent speed (degrees/solar hr)

d_i = phase lag of constituent

$H(t)$ = sea level heights

A_o = average sea level

$H_T(t)$ = tidal component of sea levels

$H_R(t)$ = surges and non tidal residuals

$v(t), u(t)$ = orthogonal components of current velocity

v_o, u_o = average drift current

$v_T(t), u_T(t)$ = tidal component of current

$v_R(t), u_R(t)$ = surges and non tidal residuals

1. INTRODUCTION

The accurate measurement of currents using a buoyed mooring system is notoriously difficult to achieve especially in the near surface zone or in shallow waters. The measurements are influenced by the movement of the mooring, the relative position of the meter in the water column, its form of attachment to the mooring line, and the type of meter used. Under certain conditions the above factors can introduce a marked distortion of the recorded data leading to difficulties in resolving the true nature of the underlying flow and generally resulting in the data being discarded as poor quality. On the other hand, similar and often more dramatic distortions in the recorded data can arise as a consequence of unusual hydraulic phenomena that occur in nature and which are indeed characterised by a distorted form of the current meter records. The results presented here show that events have been detected in the North Sea which are associated with the generation of a particular type of extreme flow which can be localised and can recur in a cyclic fashion over a period of several days. In view of such results the paper explores the methodology required to improve the quality of analysis and interpretation of current meter measurements.

2. CURRENTS

The structure of ocean currents is extremely complex and their measurement poses considerable difficulties. The primary motion of ocean currents in the horizontal plane can be defined (ref. 14) as an oscillatory tidal flow which decays from a near surface maxima to the seabed.

Surface currents alone are more strictly defined as;

$$V = V_S + V_G + V_D$$

Where V_S is due to Stokes drift arising from wave action, V_G are currents arising from sea level gradients and propagating swells, and V_D is the current produced by direct action of the wind. The transfer of wind energy downwards through the water column can make a significant contribution to the vertical flow structure (ref. 2,11) especially in shallow water regions (less than 100m) where frictional effects and bed topography become important. Temperature fronts and stratification which can arise equally in coastal waters or the open seas also induce flows (ref. 10) which contribute further to the disturbance of the vertical current structure.

Current measurements are defined as Lagrangian or Eularian, which simply refers to the reference frame (moving or static) against which the measurements are made. Bearing in mind that tidal currents can be adequately described in an Eularian frame, whereas surface currents strictly require a Lagrangian frame, we readily begin to see the pitfalls in analysis and interpretation of current meter (Eularian) measurements. Even currents recorded by identical instruments at closeby locations, and at identical depths, will reveal differences (ref. 1,7,9) which may be as much a function of instrument performance as a function of extraneous factors such as winds and topography, which induce temporal and spatial changes in the current environment itself.

In the case discussed here we restrict ourselves to the more obvious composition of currents and the ways in which these can be resolved using established analysis techniques.

3. ANALYSIS

After the current meter data are translated and calibrated they are resolved into east and north (u, v) component time series of speed values. Cursory examination of these time series plots, together with parallel plots of the current vector magnitude and direction, provides a reasonably good preliminary basis of quality control. The usual practice is then to analyse the separate u and v time series to resolve the set of harmonic tidal constituents which define the tidal variations of the current. These constituents may be used to predict the (u, v) component time series of tidal current speed which are combined to produce the total tidal current vector. The residual difference between the measured and predicted current represents the influence of non tidal flows and mooring motion. The analysis procedure (ref. 4,5) follows the same principle as used for tide gauge sea levels which are represented as a time series of sea surface elevations (wave effects being filtered out) consisting of primary tidal effects and residual non tidal perturbations due mainly to meteorological effects.

Primary tidal energy is contained in the frequency bandwidth 1cpd - 6cpd with most of the energy centred around the semi diurnal (2cpd) and diurnal (1cpd) frequencies, the former representing the dominant lunar M_2 and solar S_2 tide generating harmonics responsible for the twice daily flood and ebb observed in European waters. Low amplitude tidal harmonics are also identifiable in the low frequency range 1cpd to 3 x 10^{-3} cpd. These are difficult to resolve explicitly since they are closely associated with the response to seasonal and interseasonal changes in meteorological effects. At the high frequency end of the tidal spectrum the most significant low amplitude harmonics arise from non linear interactions in shallow water, and in essence can be viewed as higher order corrections to the envelope of the primary tidal motion. A period of 30 days covering a full spring and neap cycle is normally adequate to resolve a primary set of 29 tidal harmonics suitable to synthesize and predict the primary tidal signal. Any long period harmonics below 1cpd resolved in the analysis are normally excluded because the data are insufficient to provide any confidence in the "repeatability" of these terms which are sensitive to local meteorological anomalies and other non tidal events. Similarly for some of the higher order harmonics in shallow water regimes. The influence of the long period ($<$ 1cpd) harmonic terms is instead absorbed into the mean residual value computed by the analysis for the full data length.

From tidal theory, sea levels $H(t)$ can be expressed as;

$$H(t) = A_o + H_T(t) + H_R(t) \tag{1}$$

where the tidal component $H_T(t)$ is defined as;

$$H_T(t) = \sum_{i}^{N} h_i \cos(\sigma_i t + \alpha_i) \tag{2}$$

The selected set of N tidal harmonics defined as the parameter pair (h, g) [g is defined from σ_i and α_i] are resolved in a Fourier fashion by the least squares method used to minimise $H_R(t)$ in equation (1). In the case of currents, which are vector quantities, their description can be reduced into a pair of orthogonal time series (u, v) of the same form as equation (1). It is usual to define these components in the North and East directions as;

North : $$v(t) = v_o + v_T(t) + v_R(t) \tag{3}$$

East : $$u(t) = u_o + u_T(t) + u_R(t) \tag{4}$$

The series then represents the component vectors of the total current vector. The tidal harmonics defined in the component time series are strictly speaking pseudo current harmonics since the theoretical description of a tidal current harmonic (eg M_2) takes the form of a pair of counter rotating harmonic vectors. The vector pair is derived from the associated orthogonal (u, v) harmonics and traces the form of an ellipse which conveniently describes the physical geometry of the true tidal current harmonic. The semi-major axis gives the true amplitude of the tidal current harmonic and its orientation gives the direction of maximum flow (Figure 3).

The above stages give a simple outline of basic steps normally used in decomposing a current meter time series into the component time series normally used for subsequent interpretation and post-processing. These stages will be reviewed later in the light of the case study illustrated in the next section.

4. MEASUREMENTS

During the period March - April 1984 a current measurement programme was undertaken on behalf of SHELL EXPRO UK in the vicinity of the central North Sea Block 21 to

provide information on environmental design criteria related to offshore engineering
activities in the Gannet region. Currents were recorded at a number of stations in
water depths ranging from 80-100m. The primary instrument used was the proven
Aanderaa RCM4. In addition the Aanderaa RCM4S (modified rotor head to limit pumping
effects due to surface wave action) was deployed together with the INTEROCEAN S4
vector averaging electromagnetic current meters. The latter is a novel "golfball
design" instrument which presented many potential advantages, and it is notable that
SHELL accepted to support the first UK trials of these instruments in the North Sea.
The measurement locations A and B are indicated in Figure 1 which also shows the
primary seabed contours in the surrounding region.

The measurements discussed here were recorded at location A in 95 metres of water by
two RCM4's and an S4 moored in the configuration shown in Figure 2. The mooring and
instruments were recovered after 30 days and a replacement set was redeployed for a
further 30 days at a location approximately 500m from the original position.
Simultaneous measurements from similar instruments deployed at equivalent depths were
recorded at a position B located about 30km away. The current meters were set to
sample speed and direction at 10 minute intervals. The RCM4 computes current
velocity in the form of speed as a rotor count over the 10 minute interval, and
direction as the spot 10 minute value. The S4 computes current velocity in the form
of east and north component speeds, the respective values being the integrated
vector average of 0.5 second data sampled over a 1 minute period at 10 minute
intervals. In the case of steady flows the instruments respond in a similar manner
whereas under highly variable flow conditions, one must be careful to distinguish
(ref. 1,9) between the character of the "equivalent" measurements.

5. INTERPRETATION

Figures 4a, b illustrate the time series resolved in the form of equations (3) and
(4) from analysis of the near surface currents recorded by the S4 current meter at
location A. The tidal flow is dominant in a North-South direction with minimum tidal
energy identified in the East-West direction. Figure 4a indicates that the analysis
results are satisfactory and that a strong well behaved tidal current has been
resolved against a low level of residual effects. The diagram itself indicates
little evidence of any significant mean residual flow. Figure 4b, which is a further
section of the same data set presents a different picture whose interpretation is not
immediately obvious.

Here the observed current shows features characteristic of diurnal flow which are in
fact reproduced in the synthesized tidal motion based on the tidal harmonics resolved.
Furthermore there is no significant difference in the residual flow characteristics
compared with the earlier period shown in Figure 4a although the observed current
direction has changed from the semi-diurnal reversal in North-South direction to an
approximately constant South-West direction. The reference tidal trace for the
Standard Port of Aberdeen shows that this event coincides with a period of low Neap
tidal range.

Since the event was reproduced in the data recorded by the other current meters
(Figures 6,7) and since cursory examination of the resolved tidal harmonics suggested
no unusual features (the principal harmonics agreeing favourably with published data
(ref. 3,13) for the region) the event was assumed to be real. However it was of
interest to confirm whether similar events had been observed elsewhere in the North
Sea, in particular the interesting feature of a pronounced diurnal current arising
during a period of weak Neap flow. No evidence of similar events was identified
except in the Irish Sea region which was discounted since it represents a wholly
more complex and different current regime (ref. 15) than the central North Sea.

In an attempt to obtain a clearer insight into the unusual event the data were
reanalysed in a more rigorous manner. The essential difference from the earlier
analysis being to prefilter the recorded u and v velocity components using a low
pass tidal filter having a cut off at 1cpd. The objective being to provide a much
"cleaner" time series for harmonic tidal analysis and to obtain a clearer description
of the non tidal flow (low pass data) which occurred during the period. The filter
(ref. 12) used was the 39 hour centrally averaging Doodson X_0 tidal filter designed

to resolve variations of mean sea level from tide gauge records. Using this prefilter, the equation (3) is reexpressed as :

$$U = U^{HP} + U^{LP} \tag{5}$$

$$U^{HP} = U_T^{HP} + U_R^{HP} \tag{6}$$

similarly for the component u in equation (4) where HP denotes high-pass data (frequency components $\geqslant$ 1cpd) and LP denotes low-pass data (frequency components $<$ 1cpd). Under normal circumstances the prefiltering is not applied since the harmonic tidal analysis method performs such that :

$$A_0 = \frac{1}{T} \int H^{LP} dt \quad \text{and} \quad H_T = H_{T\,LP}^{HP}$$

In the case considered here the vector $V(t)$ describes the resolved form of the non tidal current, and Figures 5a, b show clearly its development into a unidirectional constant flow over the five day period 24-29 March covering the unusual event.

During this period the non tidal current has a mean speed of 25cm/sec in direction approximately 220° (from true North). Figure 5b also shows that revised tidal components U_T^{HP} and u_T^{HP} are now of the correct semi-diurnal form we would expect.

During the earlier period when the non tidal current is weak and showed no defined character, the results of the separate analyses shown in Figures 4a and 5a are virtually identical as expected.

The diurnal form of the observed current during the five day period is now explained directly as the result of interaction between the residual current and the tidal neap current. The tidal current in the form shown in Figure 3 attains maximum amplitude in the direction 10° and 180° + 10°, at approximately 6 hourly intervals in accordance with the dominant semi-diurnal nature of the tide. During the neap period the mean maximum tidal current amplitude (M_2-S_2) is approximately 18cm/sec. It is now obvious that the flow interaction will be such that the maximum semi-diurnal tidal flow in direction 10° will be totally masked by the residual current every 12 hours and reinforced in the reverse direction, every other 12 hours. The resultant flow will thus take on the characteristic of a purely diurnal flow, synchronised (by shear chance), with the weak diurnal harmonic component of the tidal current (see also Figure 3). As a consequence a high level of non tidal energy becomes systematically trapped within the diurnal tidal energy band over a period of several hours every tidal cycle. The response of the harmonic analysis procedure is to recognise this signal as being of tidal origin and to recreate it in an optimal way by subtle readjustment of the phases resolved for the set of tidal harmonic terms specified. Because the distortion was focused on a period of particularly weak neap flow the misalignment of phases resolved is marginal and not easily detected by examining the component tidal harmonics alone.

Subsequently the data recorded by all other current meters were reanalysed in the same fashion. The results show that the disturbance was recorded at current meter stations approximately 30km apart (Figure 6). At the station A where the maximum effect was recorded the disturbance can be traced throughout the full water column (Figure 7) with little evidence of any significant variation in current speed except near the seabed. Cursory examination of daily weather charts for the period did not indicate the presence of any localised directional wind field which might be associated with the prolonged residual flow observed. Equally, by examining the limited temperature and pressure data recorded by some of the current meters there was no evidence to suggest the presence of any stratification which might contribute to "slab" like flows (ref. 6,16) throughout depth.

Beyond a simple qualitative description of the disturbed flow it becomes necessary to give closer consideration to factors such as wind effects, bed profile, bed friction and eddy viscosity (ref. 2) in order to gain a closer insight into the vertical current profile observed. The non tidal current itself is probably related to more distant events occurring in the North Sea.

6. CONCLUSION

The example presented in this paper is typical of many examples where measurements contain a small segment of abnormal disturbance which is often discounted as instrumental error or simply left unexplained. As shown here it requires a particularly objective approach to explain the cause of some distorted results which subsequently serve to identify a phenomena of primary interest. In this case the event is due to significant tide-current interactions which are of importance to offshore subsea engineering operations. The unusual nature of the disturbance is such that the measurements alone are insufficient to explain the evolution of the event. This is probably related to far field disturbances of meteorological origin which would require the use of a hydrodynamic North Sea model to investigate it more fully. With hindsight, there probably exists a large volume of North Sea current measurements which contains invaluable signatures of similar events (identified also in ref. 6,16) which have gone unnoticed to date. In view of the importance of such events it would seem appropriate to reitterate the need (ref. 8) for a more standardised approach to the processing of current measurements.

The example here also serves to illustrate a number of important aspects concerned with the analysis of current meter time series data. The sampling interval of the original data was 10 minutes which allows one to identify variations of the order of 30 mins per cycle. These might describe real transient disturbances due to unusual forms of surge-tide interaction; however they may also be due, wholly or partially, to the dynamic response of the mooring and meter under less unusual flow influences. An interpretation of such velocity transients would also be dependent on whether they were recorded by a solid state vector averaging meter or a conventional rotor-vane meter. For the purpose of harmonic tidal analysis the recorded data are resampled to provide the hourly interval data which are a sufficient requirement for the analysis. This means that in order to recover any description of the high frequency non tidal events which may be contained in the original 10 min records it is necessary to translate the predicted hourly tidal values into the required 10 min interval (by polynominal interpolation) to provide the correct basis to compute the non tidal residuals. Hourly residuals alone will not contain a meaningful signature of intermediate high frequency events.

In the case of currents, residuals are particularly sensitive to the quality of tidal harmonics resolved independently in the orthogonal directions. In the sample here the East component contains a very low level of tidal energy and consequently only the few principal harmonics will be adequately resolved from the background noise. The resulting residual current vectors reconstituted from the orthogonal pairs may therefore contain unrealistic values of the true residual vectors, especially in cases of high frequency disturbances.

Although prefiltering of the current meter records is a helpful aid to discriminate between tidal and non tidal effects it is still important to bear in mind the expected characteristics of the environment being monitored. For example in the open ocean the tidal current is symetrical with respect to tidal phase, whereas in coastal waters an asymetry often occurs due to topographic effects. In the latter case the use of a standard filter might in fact contribute to problems rather than help to resolve them.

7. ACKNOWLEDGEMENTS

Permission to reproduce data presented in the Figure diagrams was provided by SHELL Exploration UK. The current meter measurements were made by WIMPOL Ltd.

8. REFERENCES

1. Beardsley R.C. et al. CMICE: A near-surface current meter intercomparison experiment. Deep Sea Research. Vol 28A, No 12, 1981, 1577-1603.

2. Davies A.M.: Application of sigma coordinate sea model to the calculation of wind induced currents. Continental Shelf Research. Vol 4, No 4, 1985, 389-423.

3. Davies A.M. & Furnes G.K.: Observed and computed M tidal currents in the North Sea. Journal of Physical Oceanography. Vol 10, 1980, 237-257.

4. Foreman M.G.G.: Manual for tidal currents analysis and prediction. IOS Patricia Bay, Sidney B.C., Pacific Marine Science Report 78-6. 1978, 70pp.

5. Godin G.G.: The analysis of tides. University of Toronto Press. 1972, 264pp.

6. Gould W.J.: The current regime on the continental slope north and west of the United Kingdom. Proceedings of Conference on Current Measurements Offshore, May 1984 (London), Society for Underwater Technology.

7. Gould W.J. & Sambuco E.: The effect of mooring type on measured values of ocean currents. Deep Sea Research. Vol 22, 1975, 55-62.

8. Graff J.: A standardised method for the analysis of current meter measurements for engineering applications. Proceedings of Conference on Current Measurements Offshore, May 1984 (London), Society for Underwater Technology.

9. Halpern D. et al.: Intercomparison tests of moored current measurements in the upper ocean. Journal Geophysical Research, Vol 86, No C1, 1981, 419-428.

10. Heathershaw A.D.: Some observations of internal wave fluctuations at the Shelf-edge and their implication for sediment transport. Continental Shelf Research. Vol 4, No 4, 1984, 485-493.

11. Howarth M.J. & Huthnance J.M.: Tidal and residual currents around a Norfolk sandbank. Estuarine, Coastal and Shelf Science. 19, 1984, 105-117.

12. Lennon G.W.: A note on the routine reduction of tidal records to give mean sea level. Extrait des Cahiers Oceanographiques, XVII, 6. 1985.

13. Pingree R.D. & Griffiths D.K.: S tidal simulation on the north-west European Shelf. J. mar.biol.Ass. UK. 61, 1981, 609-616.

14. Prandle D.: The vertical structure of tidal currents and other oscillatory flows. Continental Shelf Research. Vol 1, No 2, 1982, 191-207.

15. Robinson I.S.: The tidal dynamics of the Irish and Celtic Seas. Geophys. J.R.astr. Soc. 56, 1979, 159-197.

16. Weller R.A. & Halpern D.: The velocity structure of the upper ocean in the presence of surface forcing and mesoscale eddies. Phil.Trans.R.Soc.Land.A. 308, 1983, 327-340.

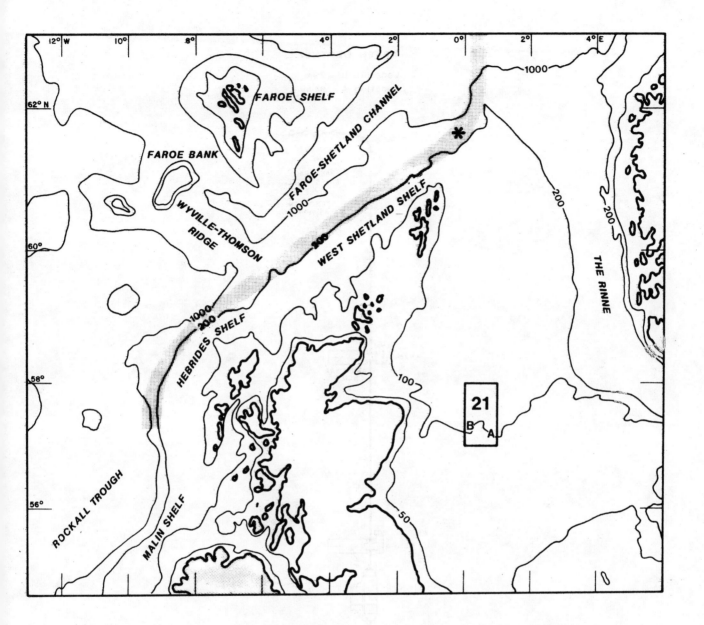

FIGURE 1   Current measurement locations in North Sea Block 21
during March–May 1984. A represents location with three
meter mooring shown in Figure 2 and B represents location
where simultaneous data were recorded at mid-depth.
The shaded portion shows the region of 1982/83 CONSLEX
current measurement experiment where (*) large current
anomalies extending over 3 days were recorded at 100m level
in 500m depth around April.

*   The distorion of current meter measurements due to unusual hydraulic
phenomena.

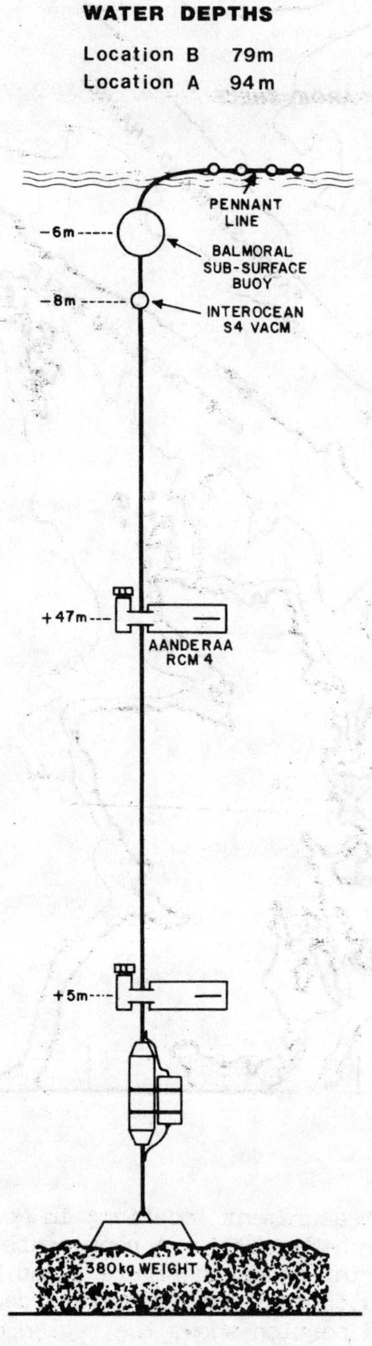

WATER DEPTHS

Location B 79m
Location A 94m

FIGURE 2   Current meter mooring used at location A.

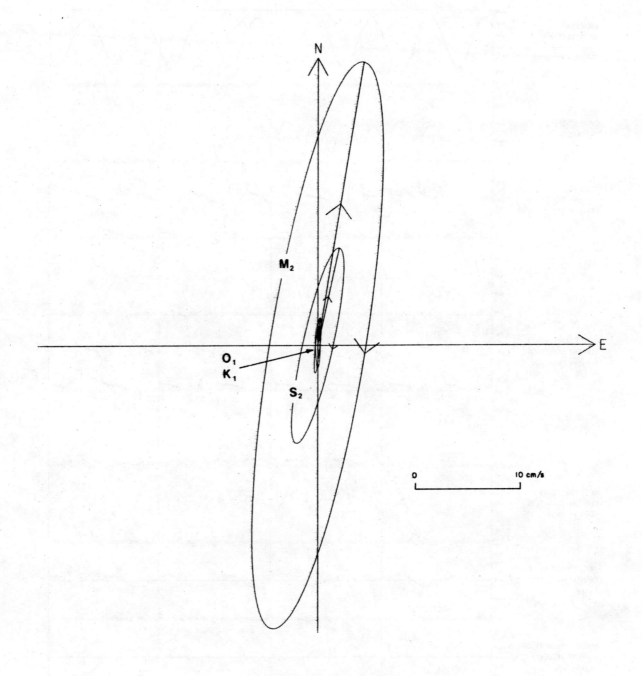

FIGURE 3   Principal diurnal (O_1, K_1) and semi-diurnal (M_2, S_2) harmonic tidal constituents represented in current ellipse form for mid-depth location A.

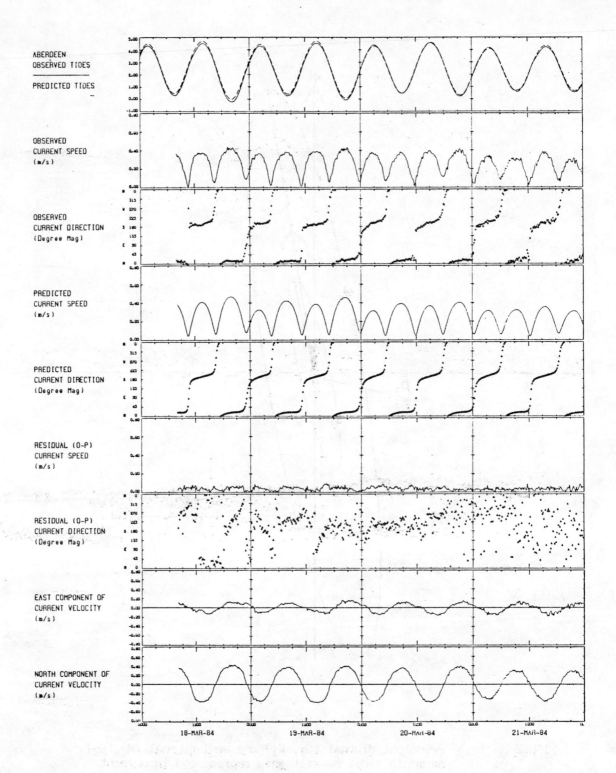

FIGURE 4a   Standard analysis of upper meter records at location A
             showing observed and decomposed current during early period
             when no flow anomalies are evident. Note the typical
             semi-diurnal character of tidal current component.

38

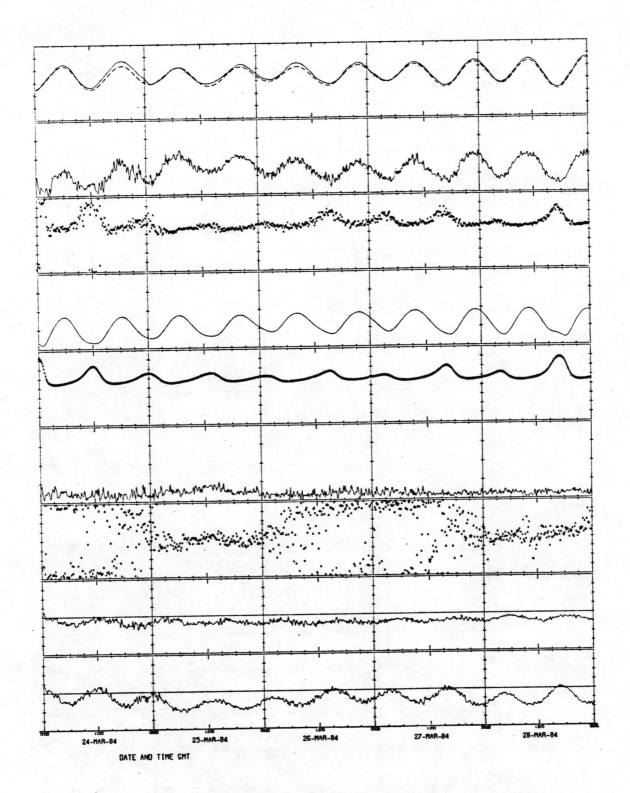

24-MAR-84   25-MAR-84   26-MAR-84   27-MAR-84   28-MAR-84

DATE AND TIME GMT

FIGURE 4b   Later section of same analysis results shown in Figure 4a
but covering the period during which a diurnal flow anomaly
was recorded at neaps. Note the diurnal character of the
tidal current component.

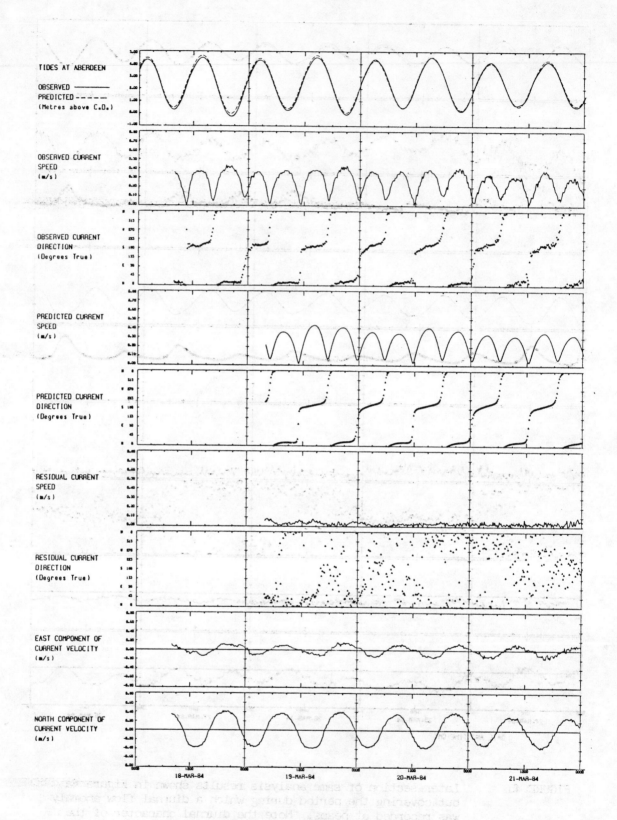

FIGURE 5a   As per Figure 4a except data have been prefiltered to remove
                non tidal variations before harmonic analysis. Note close
                similarity with 4a results.

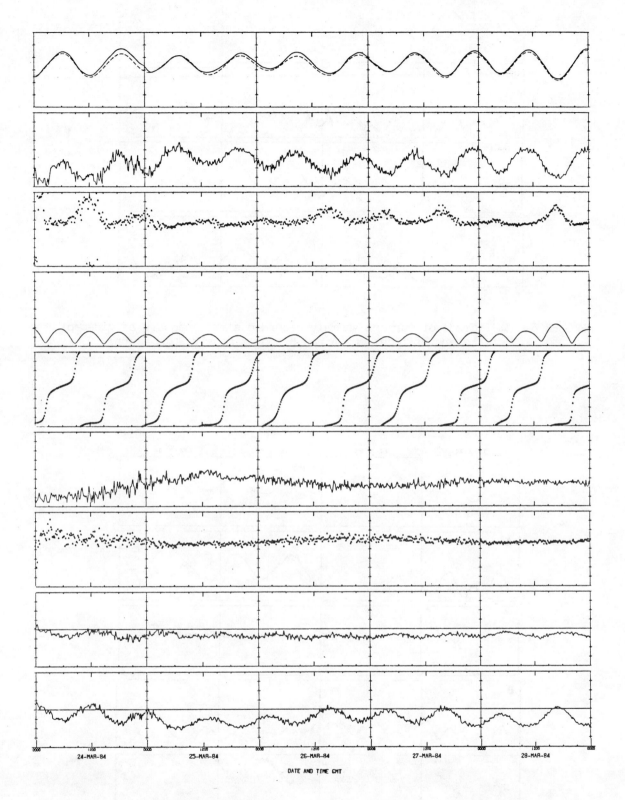

24-MAR-84   25-MAR-84   26-MAR-84   27-MAR-84   28-MAR-84

DATE AND TIME GMT

FIGURE 5b   As per Figure 4b except data have been prefiltered to
                remove non tidal variations before harmonic analysis. Note
                that the semi-diurnal character of the tidal current
                component has now been resolved and also the unidirectional
                nature of the non tidal flow has been identified.

41

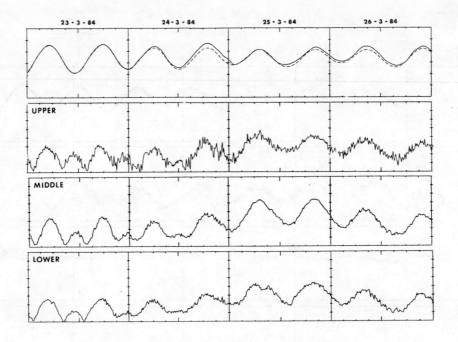

FIGURE 6   Trace of current anomaly recorded simultaneously at mid-depth meters at locations A and B approximately 30km apart.

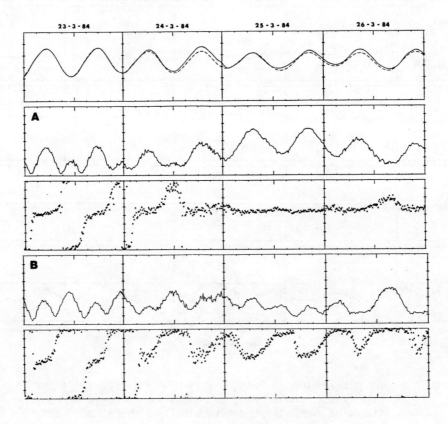

FIGURE 7   Trace of current anomaly recorded simulateneously at all meters deployed through 95m depth at location A. Note absence of any marked attenuation between near-surface and near-bed currents.

International Conference on

Measuring Techniques

of Hydraulics Phenomena in offshore, Coastal & Inland Waters

London, England: 9-11 April, 1986

INTEGRATED FIELD MEASUREMENTS AND NUMERICAL MODEL

SIMULATIONS OF TIDAL EDDIES

R.A. Falconer, Lecturer, Department of Civil Engineering, University of Birmingham, Birmingham.

E. Wolanski, Principal Research Scientist, Australian Institute of Marine Science, Townsville, Queensland, Australia.

L. Mardapitta-Hadjipandeli, Research Student, Department of Civil Engineering, University of Birmingham, Birmingham.

Summary

A detailed field study has been undertaken by the Australian Institute of Marine Science, Queensland, to investigate the tide induced eddies occurring in the wake of Rattray Island and to try and improve the existing knowledge of eddies shed by such headlands. Rattray Island, located just east of Bowen, North Queensland, is approximately 1.5km long, 300m wide and lies in well mixed water having a typical depth of 25m to 30m. In the field study twenty six current meters were deployed at various sites in the south eastern lee of the island, with time series sea level recordings being taken around the island. Visual observations of the tidal eddy were also made using Landsat imagery and aerial photography, and the surface temperature field was measured at the time of aerial observations. The tidal range was generally about 3m and a strong south eastward current was observed and measured during the rising tide. This flood tide current gave rise to a strong clockwise rotating eddy in the south eastern lee of the island, which was observed to be stable and about twice the size of the island for most of the rising tide.

In conjunction with the field measurement study, a two-dimensional depth averaged numerical model was refined and applied to the site with the aim of accurately modelling the main features of the measured eddy. In the model, particular emphasis was focussed on the modelling of the lateral mixing in the free shear layer of the island's wake, together with the treatment of the advective accelerations. The comparisons between the measured and the predicted velocity fields have proved to be encouraging, with the eddy dimensions and circulation strengths being similar for all phases of the tides

Held at Imperial College of Science and Technology, London. Organised and sponsored by BHRA, The Fluid Engineering Centre. Co-sponsored by the American Society of Civil Engineers and the International Association for Hydraulic Research.

International Conference on

Measuring Techniques

of Hydraulic Phenomena in offshore, coastal & inland waters

London, England, 9-11 April 1986

NOMENCLATURE

C	=	Chezy roughness coefficient
d	=	particle size
f	=	Coriolis parameter
g	=	acceleration due to gravity
H	=	total depth of flow
n	=	Manning roughness coefficient
R	=	experimental constant
t	=	time
U,V	=	depth average velocity components in the x,y co-ordinate directions
U_o	=	free stream velocity
U_*	=	shear velocity
x,y	=	co-ordinate axes in the horizontal plane
x',y'	=	longitudinal and lateral co-ordinate axes in the mixing zone
β	=	correction factor for non-uniformity of vertical velocity profile
γ	=	ratio of y':x'
η	=	water surface elevation above mean sea level
$\nu_t\vert_b$	=	bed generated eddy viscosity
$\nu_t\vert_f$	=	free shear layer eddy viscosity
ρ	=	fluid density
$\left.\begin{array}{l}\sigma_{xx},\tau_{xy}\\\tau_{yx},\sigma_{yy}\end{array}\right\}$	=	Reynolds stress components in the x,y co-ordinate directions

1. INTRODUCTION

In recent years increasing interest has been expressed in understanding the flow mechanisms of tide induced circulation in the lee of headlands, promontories and breakwaters. Such eddies are driven and maintained by the generation and transfer of vorticity by the mixing zone - immediately downstream of the promontory tip - to the re-circulation zone in the downstream wake. With increasing use being made of numerical models to predict flow and transport rates in estuarine, coastal and offshore waters, it is important that these tidal eddies are included and accurately reproduced in such numerical models. These predictive models are typically of the two-dimensional depth averaged type and are used for such environmental and civil engineering problems as:-
(i) modelling radioactive waste, sewage or heat discharges into coastal waters,
(ii) modelling sediment transport fluxes, including erosion and deposition rates, and
(iii) determining ideal locations of fisheries and fish farms, for example, along coastal reaches. In all of these examples the existence of large scale tidal eddies within the modelling domain can be extremely important. The eddies formed can trap water, pollutants and suspended particulate matter, with the result that any contaminants etc. may be inadequatly dispersed by the tidal stream.

In view of the lack of knowledge concerning the hydrodynamic phenomena associated with tidal eddies shed by such headlands, an integrated study was undertaken to measure and model the re-circulating tidal flow field in the wake of Rattray Island, North Queensland, Australia, see Fig.1. Apart from the free shear mixing zone which can be seen in Fig.1, the tidal eddy can also be visibly identified by the upwelling that occurs near the centre of the eddy. Rattray Island is approximately 1.5km long, 300m wide and lies in well mixed water, having a typical depth of between 25m and 30m (see Fig.2). Its longer axis lies at about 60° into the direction of the semi-diurnal tidal current, with the strong tides passing the island giving rise to separation at the most northerly tip of the island and circulation in he downstream wake.

The detailed field measurements and observations of the downstream dominant eddy were undertaken by the Australian Institute of Marine Science, see Wolanski et al (Ref.1). Current meter recordings and temperatures were taken at twenty six sites in the south eastern lee of the island (see Fig.3), with water level recordings being taken at several locations around the island. Visual observations of the tidal eddy were also made using Landsat imagery and aerial photography, with the site being particularly suited to this approach because the waters were turbid. No measurements were taken in the north western lee, although some visual observations were recorded.

The numerical model predictions of the depth averaged velocity and vorticity fields, and the various shear stress distributions around Rattray Island were undertaken at the University of Birmingham, with the preliminary study being outlined by Falconer et al (Ref.2). The hydrodynamic model was of the two-dimensional depth integrated type, with the surface slope and the depth mean velocity fields being evaluated from the conservation equations of mass and momentum. Included in the momentum equations were the effects of the earth's rotation, bed and surface shear forces and a simple turbulence model representing both bed and free shear layer generated turbulence. The vorticity fields were indirectly evaluated from the velocity field predictions, although vorticity was not strictly conserved in the model.

2. FIELD MEASURING TECHNIQUES

The main series of field measurements and observations of the tidal eddy in the south eastern wake of Rattray Island were taken during periods of calm weather conditions, in June, November and December 1982. Hence, meteorological conditions were not significant at the time of measurement.

2.1 Current measurements.

The point velocities were measured every 10 minutes at sites 1 to 24 using Aanderaa current meters. The meters were suspended 5m above the bed, using a subsurface buoy located 2m above the meter. In addition, continuously recording Marinco current meters were located at sites 25 and 26. These meters were suspended 5m below the surface from two-point-anchored rubber rafts and were set to average the current data at 1 minute intervals. Sail drogues (2 x 2m) were also deployed on various occasions, with the current being measured at 5m depth and the drogue being tracked by radar.

A coastal ocean surface radar was also used to map the radial component of the surface current field. The radar was operated for a full sweep of $30°$ during two tidal phases, one during ebb and one during the flood tide on the 4th December 1982. Further details of the current measuring procedure are given by Wolanski et al (Ref.1), together with an analysis and detailed description of the corresponding results.

2.2   Sea Level Recordings.

The water elevations were measured around the island using three Aandraa model WLR5 tide gauges. These gauges were tied to concrete blocks, set in 10m of water, and were located at the island tips and approximately mid-way between both tips along the southern coastline.

2.3   Aerial Photographs and Remote Sensing.

Aerial photographs were obtained at an altitude of 1 to 1.6km for the rising tide (i.e. south eastern current) on the 4th December 1982. Similarly, aerial photographs were obtained for a falling tide (i.e. north western current) on the 21st and 26th June 1982. Only ultra-violet filters were used for the aerial photographs since strong turbidity currents were present for much of the rising and falling tides.

One Landsat image print was available. This was taken during a rising tide, and the image was enhanced using the water depth algorithm of Jupp et al(Ref.3). Although Landsat cannot normally distinguish between bathymetry and turbidity currents, the image obtained for this site included both effects but with the turbidity patterns dominating.

2.4   Temperature and Salinity Measurements.

At the time of aerial observations the surface temperature and salinity were measured from a ship using a thermosalinograph and a CTD-probe. The track of the ship was perpendicular to the uni-directional current, and a total of five transects were undertaken in the wake region, at intervals of approximately 1km. The resulting salinity gradients were found to be too small to be measured reliably and hence the corresponding data was discounted.

3.   FIELD MEASUREMENT RESULTS

3.1   Tidal Data.

Rattray Island is open to the shelf waters to the north and by only a narrow channel to the south. Hence, the currents are primarily controlled by the tides. The peak measured spring tidal range was about 3.5m, with the free stream currents flowing north west and south east for the rising and falling tides respectively. The well documented effects of non-linearities and bed friction on tidal currents, e.g. Proudman (Ref.4), caused a large time lag of up to 2 hours between slack water and current reversals.

3.2   Measured Currents.

From the detailed stick plots obtained for the 26 sites, synoptic maps of the measured currents were produced at hourly intervals for the duration of the period of measurement. As an example of the hourly maps, the results are shown in Fig.4 for the largest tidal range during the period of measurement. Further details of the hourly maps are given in Wolananski et al (Ref.1).

The resulting stick plots at site 2 confirmed that the free stream currents closely followed the tidal elevations, though lagging by 2 hours, and that these oscillatory currents were generally uni-directional. For the maximum and minimum tidal ranges, the corresponding peak velocities were approximately $0.6ms^{-1}$ respectively.

From the detailed synoptic maps, and the aerial photographs, it was observed that the currents were deflected around the island, with circulation cells - or eddies - being created on the flood tide by the moments of the effective stresses associated with flow separation past the island's tips. Typically, it could be seen from the synoptic maps that the current direction in the lee of the island started to rotate clockwise after one hour into the flood tide cycle. The resulting dominant

46

eddy progressively grew in size until it reached its maximum size at close to high tide (see Fig.4). The width of the eddy was roughly the same as that of the island (parallel to the longest axis of Rattray Island) and the maximum length of the eddy was about twice that of the island width. The current structure in the eddy was found to be stable, with there being no signs of Karman vortices.

The vertical variation of the velocities were measured near site 24 using a Marinco model Q-15 current meter, with no marked vertical gradients being measured. Furthermore, comparisons were made of the hourly current speeds and directions measured using the drogues and the current meters, which were located approximately 5m below the surface and 5m above the bed respectively. The resulting comparisons showed no systematic differences, which implied that a clockwise rotation prevailed throughout the water column of the main eddy.

Drogue recordings were also undertaken at sites 2 and 13, with the specific aim of estimating the local horizontal diffusion coefficient and the corresponding mixing scale. At the turn of the tide the root-mean-square of the relative separation between the various drogues was about 100m to 150m, which resulted in an estimated diffusion coefficient of order $0.5m^2s^{-1}$, see Wolanski et al (Ref.1). This diffusion coefficient indicated a corresponding mixing scale of only about 500m and a peak Reynolds number of about 1000, with the velocity and length scales of the Reynolds number being assumed to be the free stream velocity and the maximum island width respectively.

3.3 Aerial Photographs and Remote Sensing Records

The main aerial photographic survey, undertaken on the 4th December 1982, was commenced at 11:00 hours, i.e. at about 1 hour into the south eastern current cycle of a rising tide. The flow was seen to divide into two around the island, with a mass of turbid water observed in the island's wake. The pronounced turbidity fronts in the wake indicated a dominant eddy which was about equal in size to the island's width. An hour later the length of the wake was estimated to be twice that of the island, with the width of the wake having increased by approximately 300m. A single clockwise-rotating eddy extended across most of the wake, and was visible as spiral bands of alternately clear and turbid waters, all converging towards the centre of the eddy. An hour later, close to high tide at 13:00 hours, the eddy size appeared much the same, with the observations being similar to the results of the synoptic map shown in Fig.4. Pictorial comparisons of the aerial surveys are given by Wolanski et al (Ref.1).

In June 1982 aerial photographs were taken for a north western current, i.e. during a falling tide. Observations showed that two counter-rotating eddies were present in the northern wake. The more pronounced eddy was located over the scour hole to the north east of the island, with the other eddy being located north west of the island and in the lee of the south west tip. It was also observed that during the subsequent rising tide the north eastern eddy was advected to the east of the island, which implied that at least some of the wake waters may oscillate from one side of the island to the other. This phenomenon implied that some of the water around the island was probably trapped locally for long term periods.

The Landsat imagery of Rattray Island was taken approximately 3 hours into the rising tide on the 26th May 1975. The resulting image, illustrated and described by Wolanski et al (Ref.1), confirmed the existence of a wake in the lee of the island. This was highlighted in particular by observing a zone of intermediate water - relative to the highly turbid near shore waters and the less turbid off-shore waters - the waters appear to have divided into two intruding lenses as the intermediate zone flowed past the island. These intruding lenses did not then merge downstream of the island, but left a mass of more turbid water in the island's wake.

3.4 Temperature Field Measurements

The surface temperatures were measured between about 12:00 and 13:00 hours on the 4th December, at the time of aerial observations and current and water level measurements. The resulting surface temperatures are shown in Fig.5, together with the corresponding measured currents. The results show the presence of a relatively cold region of water just east of the northernmost island tip, surrounded by warmer intruding shelf waters. Furthermore, the temperature in the island wake can be seen to be slightly lower than the surrounding temperatures, which could be due to enhanced mixing in the shear zone and three-dimensional effects due to the existence of the eddy.

The vertical temperature and salinity distributions were also measured, with the latter being negligible. The temperature distributions are illustrated in Fig.6 for the flooding tide, and confirm the existence of a weak shallow thermocline. However, the relatively small temperature variations through the depth indicate that the water column appears to be reasonably well mixed. Hence, the temperature gradients were regarded as being large enough for temperature to be used as a tracer, but small enough for baroclinic effects to be ignored.

3.5   Summary of Results

The results confirm the existence of a dominant eddy in the south eastern wake of Rattray Island. The characteristics of this wake are similar to those observed around many other islands on the Great Barrier Reef continental shelf, see Wolanski and Thompson (Ref.5). The eddy rapidly develops in response to the tidal current, with a stable horizontal circulation for most of the flooding tide. The eddy was found to have a maximum downstream extent of twice the island's width, which suggests an effective Reynolds number of an equivalent two-dimensional flow around a bluff body of about 10, see Batchelor (Ref.6). Such a Reynolds number suggests an effective horizontal momentum exchange coefficient of the order of $90 m^2 s^{-1}$. However, this value is much greater than that estimated from the field data.

4.   MATHEMATICAL MODEL DETAILS

4.1   Governing Hydrodynamic Equations

The equations of mass and momentum conservation, as used in the mathematical model to predict numerically the wake features in the lee of Rattray Island, were of the two-dimensional depth integrated type. The corresponding three-dimensional equations of mass and momentum, in both the x and y horizontal co-ordinate directions, were integrated over the depth to give (see Falconer (Ref.7)):-

$$\frac{\partial \eta}{\partial t} + \frac{\partial UH}{\partial x} + \frac{\partial VH}{\partial y} = 0 \qquad \ldots \ldots (1)$$

$$\frac{\partial UH}{\partial t} + \beta \left[\frac{\partial U^2 H}{\partial x} + \frac{\partial UVH}{\partial y} \right] - fVH + gH\frac{\partial \eta}{\partial x} + \frac{gU(U^2+V^2)^{\frac{1}{2}}}{C^2} - \frac{1}{\rho}\left[\frac{\partial H \sigma_{xx}}{\partial x} + \frac{\partial H \tau_{xy}}{\partial y} \right] = 0 \ldots \ldots (2)$$

$$\frac{\partial VH}{\partial t} + \beta \left[\frac{\partial UVH}{\partial x} + \frac{\partial V^2 H}{\partial y} \right] + fUH + gH\frac{\partial \eta}{\partial y} + \frac{gV(U^2+V^2)^{\frac{1}{2}}}{C^2} - \frac{1}{\rho}\left[\frac{\partial H \tau_{yx}}{\partial x} + \frac{\partial H \sigma_{yy}}{\partial y} \right] = 0 \ldots \ldots (3)$$

where η = water surface elevation above mean sea level, t = time, U,V = depth average velocity components in the x,y co-ordinate directions, H = total depth of flow, β = correction factor for non-uniformity of the vertical velocity profile (= 1.016 for assumed seventh power law velocity distribution), f = Coriolis parameter, g = acceleration due to gravity, C = Chezy roughness coefficient (= $\frac{H^{1/6}}{n}$ where n is the Manning roughness coefficient), ρ = fluid density, σ_{xx}, τ_{xy}, τ_{yx}, σ_{yy} = Reynolds stress components in the x,y co-ordinate directions respectively.

Before eqs. (1) - (3) could be solved numerically it was necessary to "close" eqs. (2) and (3) either by neglecting the Reynolds stresses or by representing them in terms of the unknown variables η, U and V. For this purpose the direct stresses (σ_{xx} and σ_{yy}) were firstly neglected, since these terms are generally small in comparison with the lateral shear stresses (τ_{xy} and τ_{yx}), see Kuipers and Vreugdenhil (Ref.8). For the lateral shear stresses, consisting of both bed generated and free shear layer turbulence, use was made of semi-theoretical and empirically based relationships to represent large scale eddy motion in jets and wakes, see Townsend (Ref.9). These relationships are based on experimental and field measured evidence and allow a more realistic representation of the relatively high velocity gradients occurring in the island's wake, as compared with the coarser representation governed by the numerical model grid spacing.

The universal velocity profile assumed in the mixing zone for the representation of the lateral shear stress τ_{xy} was therefore similar to that given by Townsend (Ref.9), and is of the form:-

$$U = \frac{U_o}{2} \left\{ 1 + \text{erf} \left[\left(\frac{R}{2}\right)^{\frac{1}{2}} \frac{y'}{x'} \right] \right\} \qquad \ldots \ldots (4)$$

where U_o = free stream velocity, R = experimental constant $\cong$ 288 (see Ref.9), and x'y' = longitudinal and lateral co-ordinate axes in the mixing zone beyond the island's tip, see Falconer et al (Ref.2). Using the Boussinesq representation for the lateral shear stress and differentiating eq.(4) with respect to y' yields (see Falconer et al (Ref.2)):-

$$\tau_{xy} = \rho(\nu_t|_f + \nu_t|_b) \, U_o \left(\frac{R}{2\pi}\right)^{\frac{1}{2}} \frac{1}{x'} \, e^{-0.5R\gamma^2} \qquad \ldots \ldots (5)$$

where $\nu_t|_f$ = free shear layer eddy viscosity, $\nu_t|_b$ = bed generated eddy viscosity and $\gamma = \frac{y'}{x'}$. For the free shear layer eddy viscosity, Townsend (Ref.9) suggests taking a mean value approximated to that at the centreline (i.e. y' = 0) giving:-

$$\nu_t|_f = \frac{U_o x'}{2R} \qquad \ldots \ldots (6)$$

Similarly, for the bed generated eddy viscosity, the relationship used in the model is based on an assumed logarithmic velocity profile in the vertical plane and field data, see Fischer (Ref.10), giving:-

$$\nu_t|_b = 0.16 U_* H \qquad \ldots \ldots (7)$$

where U_* = shear velocity $(=(\frac{\tau_o}{\rho})^{\frac{1}{2}}$ where τ_o is the bed shear stress).

The form of the lateral shear stress used in this study differs from that used previously by Falconer et al (Ref.2), in that the lateral velocity gradient has also been calculated using the universal velocity profile given by Eq.(4). In the previous study (Ref.2), the velocity gradient, arising from the Boussinesq representation of the lateral shear stresses, was evaluated numerically using the finite difference scheme and the velocities predicted at the grid points. Hence, the representation of the lateral shear stresses outlined herein is, in principle, more accurate within the mixing zone than that used previously by the authors (Ref.2).

In comparing the production of turbulence due to the free shear layer in the mixing zone with that generated by the bed, Lean and Weare (Ref.11) have shown that free shear layer turbulence dominates within the mixing zone in the region:-

$$0 < x' < \frac{2C^2H}{\pi g} \qquad \ldots \ldots (8)$$

with bed generated turbulence dominating outside of these limits. Hence, in closing Eqs.(2) and (3) the lateral shear stress component was assumed to be given by Eqs.(5), (6) and (7) within the region specified by Eq.(8), and by Eqs.(5) and (7) only outside these limits, i.e. $\nu_t|_f = 0$. It is interesting to note at this stage that using typical values, from the field data, in Eqs.(6) and (8) gives mean values for the eddy viscosity in the wake of approximately $2m^2s^{-1}$. This value for the eddy viscosity is inline with Wolanski's (Ref.1) estimated diffusion coefficient of about $0.5m^2s^{-1}$, as outlined in Section 3.2.

In an analysis of the vorticity transport equation corresponding to the momentum eqs. (2) and (3), Falconer et al (Ref.2) have shown that the generation of vorticity in the island's wake is predominantly governed by the advection terms, the lateral shear stresses and bathymetric effects due to the use of a quadratic friction law, see Pingree and Maddock (Ref.12). An order of magnitude analysis of the various terms of the vorticity transport equation - for typical data pertaining to Rattray Island - also indicated that the lateral shear stress component τ_{xy} is a dominant term in the mixing zone. This result is in line with Flokstra's (Ref.13) findings and confirms the need to include and model realistically the lateral shear component in this study.

4.2   Numerical Scheme and Boundary Conditions

The governing differential equations (1)-(3), together with eqs.(5),(6) and (7) for the lateral shear stress components, were solved using an alternating direction implicit finite difference scheme. All terms were fully centred in both time and space, with the non-linear advective accelerations being centred by iteration. Full details of the difference scheme are given by Falconer (Ref.7 and 14), including details of the upwind differencing approach used to represent the cross-product advective accelerations. This upwinding had the advantage in that:- (i) the corresponding momentum flux was evaluated nearer to its position of origin, see Williams and Holmes (Ref.15), and (ii) grid scale oscillations were eliminated via numerical diffusion, see Edwards and Preston (Ref.16). The actual shape of the island was included in the model using an irregular second order grid representation in the immediate vicinity of the island, see Smith (Ref.17).

In the model simulations a mesh of 60 x 43 grid squares was used, with a grid spacing of 200m, which is a refinement on the preliminary study where a mesh of 40 x 29 grid squares, and a spacing of 300m, covered the same plan-form area. The bathymetric data was obtained from Admiralty Charts and most of the simulations were undertaken for an assumed Manning's roughness value n of 0.025. The boundary conditions for the model were obtained from the field measured water elevations and currents recorded over a period of four tides in December 1982. At the northernmost boundary, water elevations were defined using the data recorded around the island's perimeter - together with an appropriate phase lag. The normal velocity component was assumed to be zero along this boundary and hence the water elevation was defined at the most north westerly grid square, with the other boundary values being obtained from the corresponding geostrophic water surface slope. Along the western and eastern boundaries the normal velocity component was assumed to be zero and a free slip boundary was assumed, i.e. the boundary was originally orientated along the direction of the oscillatory free stream current. Finally at the southernmost boundary, velocities were defined from field measurements recorded at site 23 (see Fig.3).

The model simulations were always started from rest, with the initial mean water elevation across the domain being set to the elevation closest to the tidal phase of current reversal. In all of the numerical simulations the same water elevation and depth integrated velocities were applied at the open boundaries, with the model being run for up to 50 hours, i.e. approximately four tides.

4.3   Mathematical Model Results

The mathematical model has been used in this integrated study with two objectives in mind. The main objective has been to model as accurately as possible the eddy characteristics and the corresponding current measurements as observed and recorded in the wake of Rattray Island.   However, the model has also been used as a means of obtaining a better understanding of the hydrodynamics of such tidal eddies.

In the first of these objectives a number of model simulations, each having different features, have been undertaken and compared with the field data. These differences include:- various grid spacings and time steps, different bed roughness characteristics, various lateral shear stress representations and different finite difference representations of the island's geometry. The resulting numerically predicted currents at the 25 measuring sites are shown in Fig.7, for a grid spacing of 200m, a time step of 90 seconds, a bed roughness coefficient (n) of 0.025, and with the lateral shear stresses and the island's geometry being represented as outlined previously. On the whole this particular model simulation gave the closest agreement with the field measured results, and the velocity field predictions in Fig.7 can be compared directly with the corresponding field measured results at 9hr, 11hr and 13hr in Fig.4. When the resulting velocity fields are compared closely, it can be seen that the tidal eddy features predicted in the mathematical model show an encouraging degree of similarity with the field measured results. The current speeds and directions within the eddy agree favourably on the whole, with the eddy first being detected by Wolanski et al (Ref.1). Similar comparisons have been made at other tidal phases and for different tides, with much the same degree of agreement being obtained as for Figs. 4 and 7. In all cases, the predictions obtained in this study were in closer agreement with the field measurements than for the previous study by Falconer et al (Fig.3). Tests and comparisons showed that the improved accuracy was attributed to both the finer grid and the use of the universal velocity distribution of Eq.(4) for the velocity gradient component of the lateral shear stress.

The complete tidal flow field is illustrated just before high tide in Fig.8, with the corresponding vorticity field being shown in Fig.9. From a series of similar graphical representations, at various tidal phases, the predicted eddy can be seen to be stable throughout the flood tide and grows progressively to a maximum width of about 1½ times the island's width and a maximum length of twice the island's width. These geometric features are in line with those described earlier for the field measured observations, with the predicted circulation in Fig.8 showing a close similarity to the measured surface currents shown in Fig.5, at a similar tidal phase.

In using the mathematical model as a means of obtaining a better understanding of the hydrodynamics of the dominant eddies in the island's wakes, a series of simulations were undertaken and compared for different flow and boundary conditions. Initially, the lateral shear stress terms were excluded from the momentum equations, with the resulting circulation strength and dimensions in the southern lee being reduced by as much as 20%. However, when the advective accelerations were also excluded no re-circulation was predicted, as shown in Fig.10. This implied that in the numerical model the generation of vorticity in the island's wake was predominantly due to the advective accelerations. This conclusion is in contrast to Flokstra's theoretical analysis of the generation of two-dimensional re-circulating flow (Ref.13), where an extended analysis of the vorticity transport equation showed that inclusion of the lateral shear stresses and a no-slip boundary condition were the governing requirements for modelling such re-circulating flows. This disparity between these numerical tests and Flokstra's theoretical studies – concerning the importance of the advective accelerations in modelling re-circulating flows – can be partly accounted for by numerical dispersion, see Lean and Weare (Ref.11), and is presently the focus of much research interest.

In a specific analysis of the vorticity transport equation, relating to length and velocity scale parameters associated with Rattray Island, Falconer et al (Ref.18) have shown that bathymetric effects are important. Hence, to investigate the influence of the bathymetry on the numerical model predictions, simulations were undertaken for a horizontal bed and for uniform bed slopes of 2.5×10^{-3} in the x and y co-ordinate directions respectively. The resulting predictions for a horizontal bed showed that a much more symmetric eddy structure was observed in the northern and southern island wakes for the ebb and flood tides respectively, as compared with the previous comparisons outlined in Ref.(2). This confirmed that the dominant single eddy structure in the southern wake (e.g. Fig.8) was predominantly governed by bathymetric features, see Fig.11. When the uniform bed slopes were considered, the velocity fields were not found to differ substantially from those predicted for the horizontal bed, although the corresponding vorticity fields showed that the vorticity was increased and decreased for the lateral and longitudinal slopes respectively. These results are in agreement with Pingree and Maddock's analysis (Ref.13) and confirm the scope for vorticity production and dissipation through water depth variations, arising as a consequence of using a quadratic friction law. Furthermore, for the lateral bed slope in particular, the stagnation point occurring in the island's lee was observed to shift clockwise around the island and up the slope. This effect, which was also in agreement with Pingree and Maddock's findings, was due to rotational effects, although the clockwise rotation was not so pronounced in this study as in that undertaken by Pingree and Maddock. However, this was not entirely surprising since the Rossby number in the study outlined herein was markedly higher than that in Pingree and Maddock's study, with typical values of about 7.0 and 0.1 respectively.

The main observations from the remaining tests concerned the influence of the bed friction roughness coefficient and the effects of the earth's rotation. Comparisons were undertaken for various Manning's roughness coefficients, including n values of 0.015, 0.025 and 0.035 for various bed slopes, with the results indicating marked variations in the magnitudes and distributions of the bed shear stress. However, although the bed shear stress distribution varied considerably, the corresponding variation in the vorticity structure was much less marked. This observation confirmed that the bed slope component, introduced via the quadratic friction law, influenced the vorticity structure much more than the choice of resistance coefficient.

When the effects of the earth's rotation were considered in conjunction with bottom friction effects, it was found that the earth's rotation was only noticeable at slack tide. However, this result was also not surprising since the Ekman number had a maximum value of only about 0.8 during the flood tide. Further details of some of these comparisons are given by Falconer et al (Ref.18).

5. CONCLUSIONS

The paper describes an integrated study to measure and predict numerically the tidal eddy features in the lee of Rattray Island, Australia. Field measurements were taken of the currents, water levels, temperatures and salinities, with observations of the dominant tidal eddy being made using aerial photography and Landsat imagery. The main results from the field measurements were that the large eddy in the lee of the island develops and decays rapidly, soon after slack water at low and high tide respectively. The maximum length of the eddy was found to be about twice the island's width and the width of the eddy was similar to that of the island. The eddy was stable throughout the flood tide and the currents within the eddy had similar magnitudes to those in the free stream.

The mathematical model used to predict the field measured eddy was of the depth integrated type and included an empirically based zero-equation turbulence model to represent more accurately the free shear layer turbulence occurring in the mixing zone of the island's wake. The agreement between the predicted and the field measured currents at 25 sites was generally favourable and the mathematical model was also used to confirm the importance of the advective accelerations and the lateral shear stresses in accurately modelling the tidal eddies in the wake of Rattray Island. The bathymetry, and to a lesser extent the bed roughness, were also found to influence the eddy geometry and its vortex strength, with rotational effects only being dominant at slack tide.

Finally, in addition to undertaking the field measurements, Wolanski et al (Ref.1) have also carried out a simple vorticity analysis of the vertical re-circulation. The results of this study have shown that an analogy with laboratory Reynolds scaling simulations is not valid because of the effects of bottom friction and self-driven Ekman suction.

6. ACKNOWLEDGEMENTS

The extensive field data obtained for this study was collected by the Australian Institute of Marine Science, Townsville, Australia. The mathematical model developments were undertaken at the Department of Civil Engineering, of the University of Birmingham, with the study being partially funded by a Science and Engineering Research Council research studentship.

The hydrography and field data reproduced in Figs. 2 to 6 is reproduced with the kind permission of the Journal of Geophysical Research.

7.   REFERENCES

1.   Wolanski, E., Imberger, J. and Heron, M.L., "Island wakes in shallow coastal waters", Journal of Geophysical Research, Vol.89, No.C6, November 1984, pp.10553-10569.

2.   Falconer, R.A., Wolanski, E. and Mardapitta-Hadjipandeli, L., "Numerical simulation of secondary circulation in the lee of headlands", Proceedings of the Nineteenth Coastal Engineering Conference, ASCE, Houston, U.S.A., Vol.III, September 1984, pp.2414-2433.

3.   Jupp, D.L.B., Mayo, K.K., Kendall, S. and Hegger, S., "The use of Landsat data to assess bathymetry and topographic structure in the great barrier reef region", Technical Memorandum, CSIRO Division, Water Land Resources, Canberra, 1983, pp.1-60.

4.   Proudman, J. "The effect of friction on a progressive wave of tide and surge in an estuary", Proceedings of the Royal Society, Series A, Volume 232, No.8, pp.407-418.

5.   Wolanski, E, and Thompson, R.E., "Wind-driven circulation on the northern great barrier reef continental shelf in summer", Estuarine Coastal and Shelf Science, Vol.18, 1984, pp.271-289.

6.   Batchelor, G.K., "An introduction to fluid dynamics", Cambridge University Press, Cambridge, 1967, pp.1-515.

7.   Falconer, R.A., "Mathematical modelling of jet-forced circulation in reservoirs and harbours", Ph.D. Thesis, University of London, 1976, pp.1-237 (unpublished)

8.   Kuipers, J. and Vreugdenhil, C.B., "Calculations of two-dimensional horizontal flow", Report No.S163, Part 1, Delft Hydraulics Laboratory, Delft, October 1973.

9.   Townsend, A.A., "The structure of turbulent shear flow", Cambridge University Press, Cambridge, 1956.

10.   Fischer, H.B. "Longitudinal dispersion and turbulent mixing in open channel flow", Annual Review of Fluid Mechanics, Vol.5, 1973, pp.59-78.

11.   Lean, G.H. and Weare, T.J., "Modelling two-dimensional circulating flow", Journal of the Hydraulics Division, ASCE, Vol.105, No.HY1, January 1979, pp.17-26.

12.   Pingree, R.D. and Maddock, L., "The tidal physics of headland flows and offshore tidal bank formation", Marine Geology, Vol.32, 1979, pp.369-389.

13.   Flokstra, C., "The closure problem for depth averaged two-dimensional flow", Paper A106, 17th IAHR Congress, Baden-Baden, Germany, 1977, pp.247-256.

14.   Falconer, R.A., "Temperature distributions in a tidal flow field", Journal of environmental engineering, ASCE, Vol.110, No.6, December 1984, pp.1099-1116.

15.   Williams, J.W. and Holmes, D.W., "Marker-and-cell technique, a computer programme for transient stratified flows with free surfaces", Report No.134, Hydraulics Research Station, Wallingford, 1974, pp.1-54.

16.   Edwards, N.A. and Preston, R.W., "Grid scale oscillation in 'flow' - the CERL shallow water solver", Report No.TPRD/L/2779/N84, Central Electricity Research Laboratory, CEGB, Leatherhead, 1985, pp.1-22.

17.   Smith, G.D., "Numerical solution of partial differential equations", Oxford University Press, Cambridge, 1956.

18.   Falconer, R.A., Wolanski, E. and Mardapitta-Hadjipandeli, L., "Modelling tidal circulation in an island's wake", Journal of Waterway, Port, Coastal and Ocean Engineering, ASCE (in press), 1985 .

Fig.1   Illustration of the Eddy in the Lee of Rattray Island during the Flood Tide

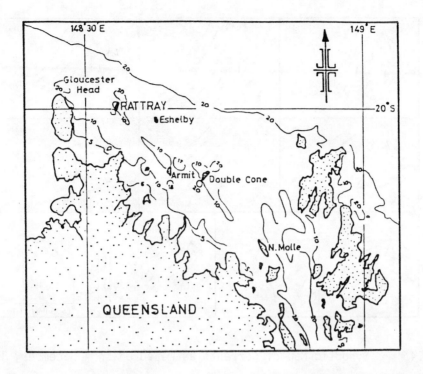

Fig. 2   Location of Rattray Island and bathmetry (in fathoms)

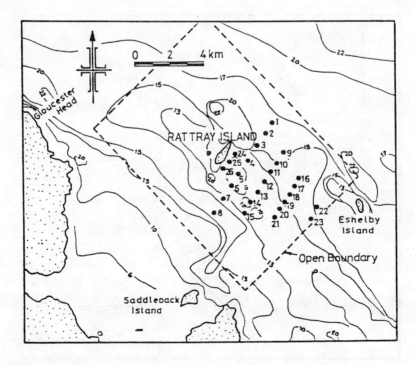

Fig.3   Area around Rattray Island showing location of current meters and model
boundary

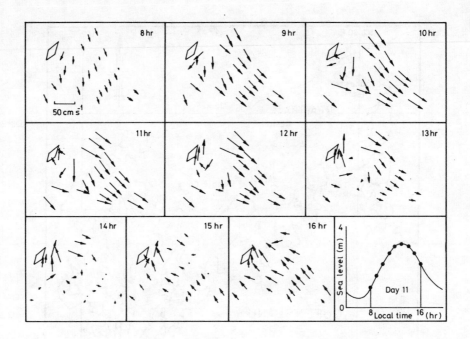

Fig.4   Synoptic Map of Currents as measured on 4 December, 1982

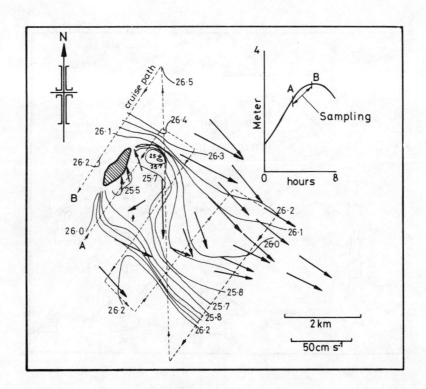

Fig.5   Distribution of surface temperature (^{O}C) and currents on December 4, 1982

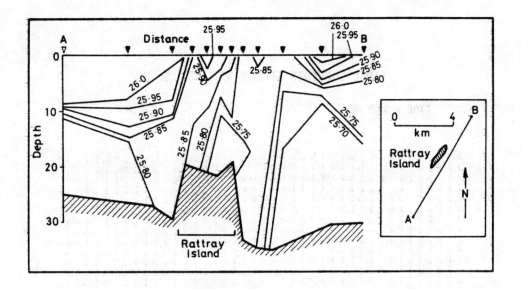

Fig.6   Cross-sectional distribution of temperature (^{O}C) across the wake on
         December 3, 1982

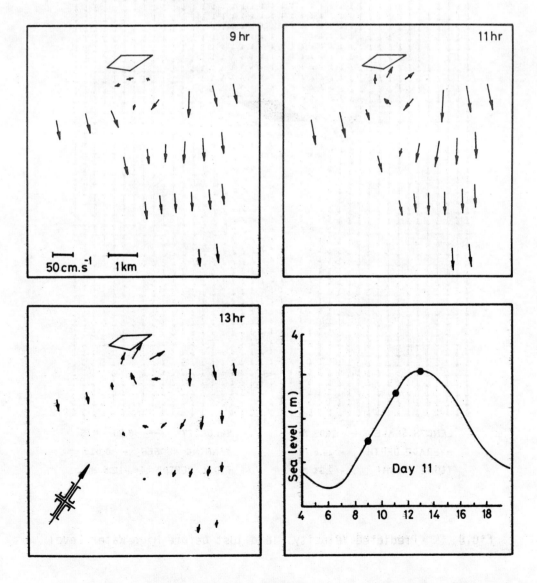

Fig.7   Numerically Predicted Velocities at Measuring Sites

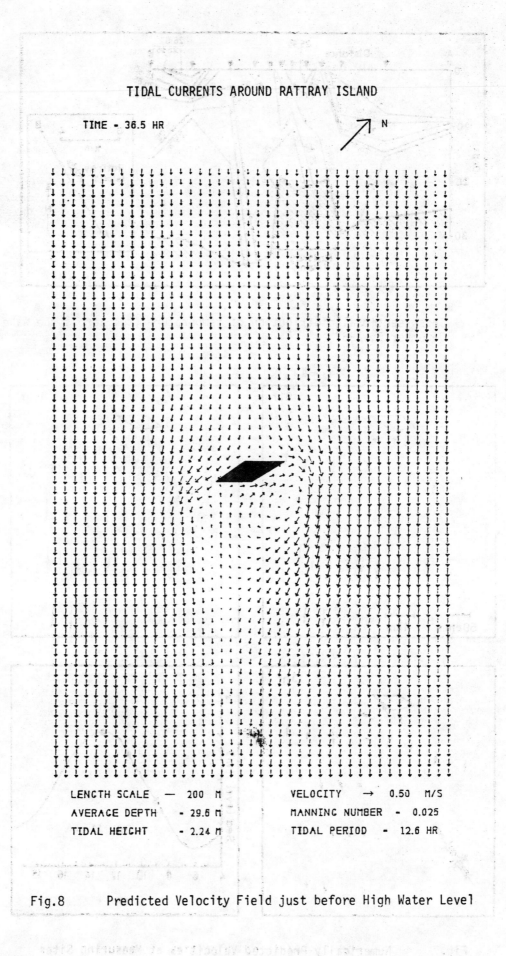

Fig.8   Predicted Velocity Field just before High Water Level

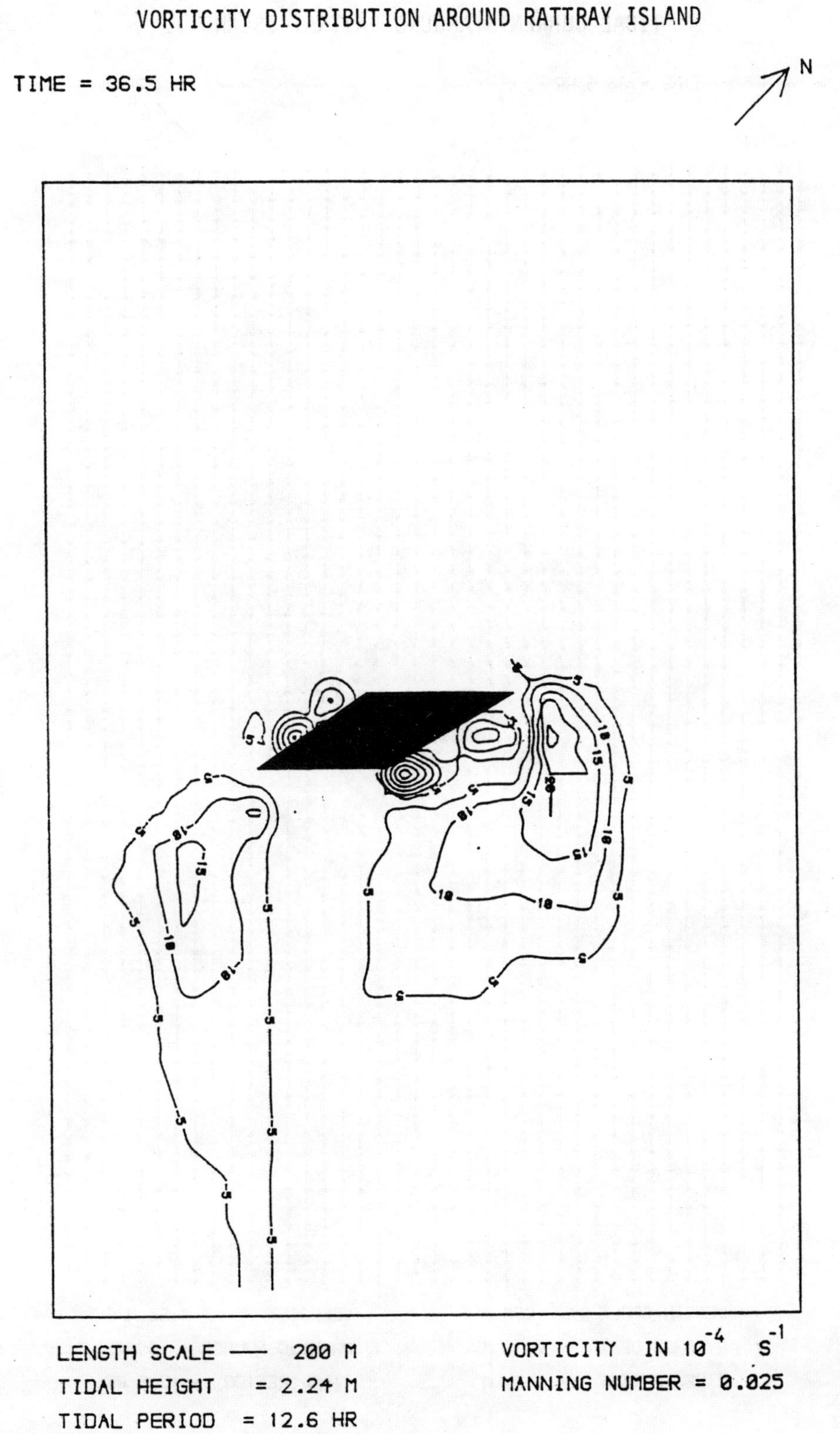

VORTICITY DISTRIBUTION AROUND RATTRAY ISLAND

TIME = 36.5 HR

N

LENGTH SCALE ──── 200 M   VORTICITY IN 10^{-4} S^{-1}
TIDAL HEIGHT   = 2.24 M   MANNING NUMBER = 0.025
TIDAL PERIOD = 12.6 HR

Fig.9   Predicted Vorticity Field just before High Water Level

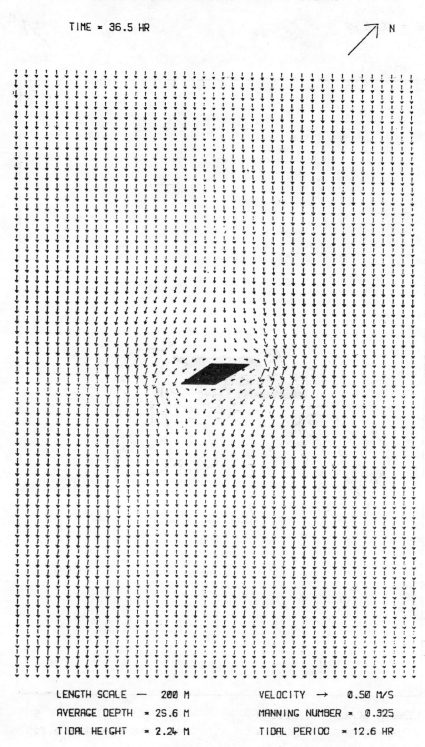

TIDAL CURRENTS AROUND RATTRAY ISLAND

TIME = 36.5 HR

LENGTH SCALE — 200 M   VELOCITY → 0.50 M/S

AVERAGE DEPTH = 25.6 M   MANNING NUMBER = 0.325

TIDAL HEIGHT = 2.24 M   TIDAL PERIOD = 12.6 HR

Fig.10   Numerically Predicated Velocity Field just before High Water Level Excluding Advective Accelerations and the Lateral Shear Stresses

Measuring Techniques

of Hydraulics Phenomena in offshore, Coastal & Inland Waters

London, England: 9-11 April, 1986

A COMPREHENSIVE WAVE DATA ANALYSIS PACKAGE

E.P.D. Mansard, Dr-Ing. and E.R. Funke, M.Sc., DIC
Hydraulics Laboratory
National Research Council of Canada
Bldg. M-32, Montreal Road
OTTAWA, Ontario, K1A 0R6
Canada

SUMMARY

The paper is intended to contribute towards the establishment of a consensus on the definition of wave parameters and the computational methods applied to their analysis.

Short descriptions with explanatory figures and examples are given for the following operations:

- the preprocessing of wave data for the removal of trends and of two common forms of noise, and also for the dynamic compensation of transducer induced distortions. Examples for several cases are given.

- the zero-crossing analysis by both zero-up and zero-down crossing methods of known time domain wave parameters. For zero-crossing parameters where existing or potential ambiguities prevail, definitions and computation methods are given.

- the spectral analysis which may be carried out by two popular methods. The suitability of their application is described briefly.

- the analysis of both the measured long wave and the theoretical long wave component. The latter may be used for the reconstitution of the low frequency spectrum of Waverider buoy data.

- the SIWEH and groupiness spectrum analysis. A comparison with the expected groupiness spectrum is being made for one example case.

An example plot with flow chart for a comprehensive analysis of wave data is included. References for the application of some non-conventional wave parameters are given. The definitions conform substantially to the recently completed IAHR List of Sea State Parameters.

Held at Imperial College of Science and Technology, London. Organised and sponsored by BHRA, The Fluid Engineering Centre. Co-sponsored by the American Society of Civil Engineers and the International Association for Hydraulic Research.

1. INTRODUCTION

The analysis of wave data has evolved from the subjective evaluation of visual observations, to the wave by wave inspection of strip-chart traces, to the computer processing of digitally recorded wave data, and finally to the on-board processing of wave data on wave recording buoys and their transmission to repeater satellites. The extensive development over the past ten years of fixed and floating offshore structures for the exploration of petroleum resources and the simultaneous evolution of low cost and fast computers and digital data acquisition systems stimulated the development of improved and more efficient methods for the analysis of wave data. Some researchers feel that it is premature to process wave data on board of wave recording instruments prior to their transmission or storage because this leads to a loss of information which may contribute to the understanding of higher order processes.

Concurrent with the efforts to improve the description of the natural sea state are the attempts to refine the reproduction of natural sea states in laboratory wave basins and flumes. Wave data analysis is therefore not only a matter of describing the natural sea state but also of confirming that the laboratory simulations have successfully reproduced the conditions which were to be modelled. This explains the active role taken by coastal and marine dynamic laboratories in the analysis of wave data.

Because it is generally much easier to obtain accurate point observations of the water surface elevations in the laboratory and because it is also possible there to check analysis results by alternate methods, the opportunity exists to verify and refine computational algorithms. The wave data analysis package which is described here is the product of a hydraulics laboratory and has been used extensively over the past five years in support of wave dynamic research. It has also been applied to various full scale data recordings which originated in the Canadian coastal zone.

Last but not least, research into non-linear wave dynamics provided the opportunity to advance the state-of-the-art both in the synthetic generation as well as the analysis of non-linear wave data. This has led to the definition of several new wave parameters many of which have not as yet found wide acceptance.

In order to promote a wider interchange of ideas and to encourage standardization of terminology and methodology in random wave generation and analysis, the International Association for Hydraulic Research (IAHR) formed in 1981 a working group on wave generation and analysis under the Maritime Hydraulics Section. This working group brought together participants from some of the major hydraulics laboratories from nine different countries and provided thereby a forum for comparative laboratory studies and for the formulation of a standard list of wave parameters. This list of IAHR-recommended symbols and definitions is now completed and will be published shortly by the IAHR (Ref. 1). The notations and definitions used by the NRCC wave data analysis programs conform as much as possible with the IAHR-recommended list.

This paper, which gives highlights from a more detailed report (Ref. 2), is intended to contribute towards the achievement of an international concensus on algorithmic methodology in wave analysis.

The computer programs which are part of the NRCC Advanced Wave Data Analysis Package were not always coded with the object of maximum execution efficiency. A testing laboratory is obliged to accommodate on short notice a wide range of client preferences in data analysis. This demands a maximum amount of flexibility. Furthermore, researchers in oceanography as well as in hydrodynamics are interested in exploring all aspects of the characteristics of waves. Therefore, a considerable portion of the computational activity is directed towards the evaluation of unusual and unconventional wave parameters. The programs are therefore not, in their present form, a replacement for the run-of-the-mill wave data processors but are rather seen as an analysis tool in a research environment.

All programs run currently on Hewlett-Packard HP 1000 computers under an NRCC developed operating system called GEDAP. Conversion of the GEDAP system to Digital Equipment VAX computers is under way and considerable portions of the software have already been transferred. Ref. 2 gives a summary description of the GEDAP system and describes the modular nature of analysis programs, the conversational interaction between user and program and lastly the manner by which different programs may be linked to construct an analysis procedure.

2. PREPROCESSING OF WAVE DATA

Most data collected at sea suffer from shortcomings of one form or another. The preprocessors are designed to eliminate or minimize these defects and thereby enhance the quality of the data.

2.1 Mean and trend removal.

Most data have an offset and some also contain a trend of the mean level. Both are usually caused by instrumentation problems. However, the trend may also be the recording of a tidal variation by a staff gauge or pressure cell on which is superposed the wave activity of interest.

Trends may be of first or second order. The former is a straight line from the beginning to the end of the record which describes the drift of the mean value. The latter, which is generally much smaller, describes a parabolic drift of the mean value. The removal of the mean and occasionally the removal of the drifts can be important to the outcome of a zero-crossing analysis. It is often advisable to undertake this preprocessing for spectral analysis as well, because this extremely low frequency energy can cause significant leakage into the adjacent frequency bands during analysis.

2.2 Removal of periodic noise.

At times, data may be corrupted with periodic noise. The source of such noise can be a wobble on a capstan of an FM tape recorder, or the interaction between the sampling frequency, some electronic hum or a modulation carrier frequency.

The periodic noise represents a pronounced peak in the spectrum which should appear as a line. However, because of the finite resolution of spectral analysis techniques, this peak appears generally in the shape of the spectral analysis filter. Its energy can be significant and, as a result may completely destroy the usefulness of otherwise perfectly good data.

The periodic noise may be compared to a mean value offset which would appear in the spectrum as a line at zero frequency. Whereas it is relatively easy to compute and then to remove a mean value offset, the same cannot be said for the computation of the periodic noise. The program SCREN carries out this task by optimally fitting in a least squares sense the amplitude, phase and frequency of the dominant sinusoidal function and by removing this pollutant from the record through subtraction in the time domain. It is important to appreciate that such an operation cannot be achieved by band-pass filtering since no filter is ideal and always leads to undesirable distortions. The method used by the program SCREN gives usually completely satisfactory results. Fig. 1 gives "before and after" examples of time histories and spectra of a current record which was corrupted by a modulation noise.

2.3 The removal of spikes.

Spikes or "glitches" are a frequent curse which typically result from wireless radio transmission of data. The energy of glitches is generally distributed over a wide range of frequencies and is, in this sense, equivalent to white noise. Linear filtering operations can diminish the spike amplitudes but at the expense of serious distortions of the wave data itself. The program GLTFX was designed to remove such glitches without altering the data during the intervals which are glitchfree.

The procedure adopted by GLTFX involves the conversion of a copy of the corrupted wave record into its first derivative. The sharp spikes will then appear as even more pronounced double spikes which are obvious statistical outliers among data with an otherwise normal distribution. A probability analysis of the derivative data leads to the computation of a threshold for the detection of derivative spikes that must be assumed to be caused by glitches. This information is used to remove spikes from the data by linear interpolation. Fig. 2 gives an example of such an operation.

2.4 Compensation for transducer transfer characteristics.

Waverider buoys and pressure cells do not produce the same output for all frequencies of the water surface elevation. Whenever the attenuation and phase shift characteristics of these transducers are dependent only on the frequency, they may be considered linear transfer functions and the data may be suitably compensated.

Compensation of data involves the use of the Fourier transform, the division of the complex transform by the complex transfer function and the subsequent inverse Fourier transform of the complex quotient. This is a fairly straight forward operation. A difficulty results from the difference between data measured in the laboratory and data recorded in nature. In the former case, the record length is typically the basic recycling length of the synthesized wave data. It is therefore correct to operate on the data with the Fourier transform without concern about the implications of the potential spill-over or wrap-around effects. However, when data is obtained from measurements in nature, as is always the case with Waverider buoy data, then data does not recycle. Any one data record represents only a short sample of a process which exists for extended periods of time. In other words, the data that preceded the record and the data that followed are unknown. Compensation techniques based on the Fourier transform have the effect of wrapping around the end pieces. In other words the beginning of the record may affect the end and vice versa. Therefore, it is strictly speaking necessary to discard the beginning and end portions of the compensated wave data record. The length of these throw-away sections depend on the compensation filter. A future version of the analysis package will include a program which will compensate data without an appreciable loss of valuable data.

Pressure cells can be used in the laboratory for the recording of wave data. Fig. 3 gives an example of such a pressure record which was compensated by a Fourier technique to give both the equivalent wave record and the corresponding spectral density.

3. ZERO CROSSING ANALYSIS

The program AZCRA performs zero-crossing analysis on a wave record. After the identification of the zero crossing events, which are defined by linear interpolation between adjacent sample intervals, the program extracts the lists of consecutive

- wave heights
- wave periods
- wave crests
- wave crest periods
- wave troughs
- wave trough periods
- wave steepnesses
- horizontal asymmetries
- crest front steepnesses
- crest rear steepnesses and
- vertical asymmetries.

Whenever applicable, these lists are computed for both the zero-up and the zero-down crossing method and the lists are then used to compute most of the presently known statistical wave parameters.

The concept of zero-crossing is prone to some confusion. There is potentially the question if a zero-up crossing wave is defined by two successive zero-up crossings or if it is the wave which includes the transition from a trough to a following peak. Both could reasonably be called zero-up crossing waves and this confusion has prevailed for many years. It is worth noting here that the IAHR list of wave parameters defines the up-crossing wave as being defined by two successive zero-up crossings and, conversely, the down-crossing wave by two successive zero-down crossings. Fig. 4 has been taken from the IAHR list of wave parameters in order to clarify this concept.

It is not necessary to describe the various statistical wave parameters here, since most of them are well known and their definitions are given in other publications (Refs. 1 and 2). However, there are five concepts to be clarified which can lead to differences in computational results of the respective wave parameters.

3.1 The wave height.

When the water surface elevation is measured by digital data recording systems, the resultant time discrete representations of the continuous function may not include the precise amplitudes and locations of the function's maxima. The severity of this omission depends on the sampling rate; the longer the sample interval, the larger the error. The sampling effect is noticed particularly in the calculation of wave asymmetries which rely on an accurate placement of the wave's crest. In order to overcome this problem a technique of second order interpolation is practiced.

All crest heights are computed by fitting a parabola to the three adjacent highest samples of a wave crest. The highest point of the parabola gives the maximum value of the crest and its location. A similar operation determines the magnitude and location of the trough. Therefore, the wave height, which is the sum of the crest and the absolute value of the trough height, is dependent on this parabolic interpolation technique. As a consequence it is quite possible to compute a maximum wave height for a wave record which exceeds the difference between the tabulated largest sample and the nearest minimum sample of a wave record.

3.2 The significant wave.

This parameter is defined as the average of the highest one-third of all waves in a record. This implies algorithmically a sorting of all waves in the order of their magnitude. From this set of ordered heights, the upper third is averaged.

However, sorting is a time consuming operation and therefore the significant wave height is computed from the probability density function of wave heights. This procedure is open to some variations, all of which can give different results. The AZCRA program subdivides the interval from the highest to the lowest wave height into 50 equal intervals in which the various wave height occurrences are counted. In order to avoid the quantization error resulting from the finite cell width of the histogram, the lower bound of the upper 33% area of this histogram is determined by linear interpolation between the cells which are adjacent to the one containing the boundary.

Once this boundary has been established, then the average is computed by the evaluation of the first moment of the upper one-third area of the histogram. The same algorithm is used in the calculation of the average of the highest 1/10th wave heights which is considerably more sensitive to the resolution of the histogram analysis.

3.3 Steepness estimations.

Since wave records are usually functions of time, the wave length is not known and therefore the wave steepness and the crest front or rear steepnesses are not directly available from the time data. Instead, periods must be converted to wave lengths.

The present version of the program applies linear dispersion theory to the zero-up or zero-down crossing wave period as the case may be. From this, either the zero-up or the zero-down crossing wave length is obtained.

To estimate crest front or crest rear steepnesses, the horizontal distances between the location of a crest peak to the preceding zero-up crossing on the one hand and to the following zero-down crossing on the other must first be calculated. These two distances shall be referred to as the crest front length and the crest rear length respectively. It seems reasonable to assume that the ratio of the crest front length to the wave length is the same as the ratio of the crest front period and the wave period. A similar argument applies to the crest rear length. From this principle the crest front and rear lengths can be calculated.

Although the crest itself is independent of the zero-up or zero-down crossing convention, for the purpose of calculating the crest steepnesses a choice must be made between the zero-up or the zero-down crossing periods which are associated with a particular crest. The NRCC analysis program performs this calculation for both the zero-up and the zero-down crossing. The average crest front steepness is then obtained from the average of the zero-up and the zero-down crossing crest front steepness. A similar average is computed for the crest rear steepness.

3.4 The vertical asymmetry.

Fig. 5 defines the various wave asymmetry parameters which were introduced by Kjeldsen and Myrhaug (Refs. 3 and 4). There is now increasing evidence that the asymmetrical profile of waves can explain the unexpected structural damage and capsizes which were not predicted from laboratory investigations (Ref. 5).

The vertical asymmetry is defined as the ratio of the crest front steepness to the crest rear steepness. The progra AZCRA computes this parameter for every wave crest, again using both the zero-up or the zero-down crossing wave periods. In the calculation of the average vertical asymmetry the program departs significantly from the definition used by Kjeldsen and Myrhaug (Ref. 4). According to the Kjeldsen and Myrhaug definition the average vertical asymmetry is the average of all vertical asymmetries. This ratio has been shown to have a strong bias (Ref. 6) so that even for a perfectly symmetrical Gaussian process this ratio is appreciably larger than one. Instead, the program computes the average vertical asymmetry by obtaining first the individual crest front and crest rear steepnesses averaged over the zero-up and the zero-down crossing wave periods. The average vertical asymmetry is then the ratio of the average crest front to the average crest rear steepness. This has an expected value of 1 for Gaussian processes.

It is clear that the results obtained by this program are not directly comparable to those obtained by Kjeldsen and Myrhaug. They are, however, a more meaningful measure of this parameter, particularly if attempts are being made to synthetically produce in the laboratory asymmetric waves by wind shear or by phase control. Attention is drawn to the fact that the mooring of Waverider buoys will erroneously give the impression that wave data are steeper in the rear than in the front (Ref. 6).

3.5 The period threshold for zero-crossing detection.

It is not uncommon to find ripples on a wave train which create multiple crossings of the mean level. For example, a sample at one instance of time may be negative. The next sample which occurs one sample interval later may be slightly positive and the next sample is again negative. The zero-crossing analysis would then recognize a wave crest of small amplitude and with a very short crest period which may be substantially less than one sample period.

Ripples of this size can seriously spoil the distribution of wave properties. In particular, if the associated wave period is converted to wave length as part of a calculation of wave steepness or crest front and crest rear steepness, then the results usually become quite extreme.

To overcome this problem, one might resort to digital low pass filtering for the purpose of smoothing the profile of the wave. This method works to some extent, depending somewhat on the quality of the filter and on the cut-off frequency selected. However, as a consequence of the filter, the maximum and the significant wave heights are reduced.

An alternate and very much superior method rejects any wave crest or wave trough period which is smaller than a user selected period threshold and simply treats the ripple crossing as if it did not exist. A default threshold of 1.5 times the sample interval has generally proved adequate for the control of extreme steepness calculations.

4. SPECTRAL ANALYSIS

As mentioned in Chapter 2.4, a distinction should be made between the analysis of laboratory generated wave data which recycle over their record length and the analysis of data recorded in nature which are samples of a long continuous process. This has a bearing on the manner in which spectrum analysis may be applied to the data. Spectral density analysis programs such as FSPDW and FSPDP are based on the application of the Fourier transform over the entire record length and apply to staff gauge or pressure cell data respectively. On the other hand, programs such as VSPDW, VSPDP and VSPDB use the modified Welch method and are applied to nature recorded data from staff gauges, pressure cells and Waverider buoys respectively. Ref. 7 gives examples of how the two different methods can affect the outcome of a spectral analysis. The programs FSPDG and VSPDG are used for general purpose spectral analysis. A more detailed description of these programs is given in Ref. 2.

All spectral density programs in the NRCC package which apply to wave data analysis compute most of the known spectral wave parameters. These are:

- the peak frequency
- the peak frequency by the Delft method
- the Goda peakedness factor
- the zeroth, first, second and fourth moments
- the broadness factor
- the wave power.

Most of these variables are defined by Ref. 2 and the wave power is computed by the spectral integration formula, compensated for water depth (Ref. 8).

4.1 The 80% confidence band.

Based on a Chi-square distribution, the 80% confidence band is computed for each of the spectral estimates. This information is organized in such a manner that line spectra of confidence limits may be plotted as is shown in Fig. 6.

5. LONG WAVE ANALYSIS

There are typically two requirements. First it is of interest to determine the long wave activity which could be measured and second, one may wish to compare this to the theoretical long wave activity which may be expected to prevail as a result of the measured wave grouping.

Waverider buoy data do not contain meaningful long wave components. The instrument is not sensitive enough in this very low frequency range. Therefore, there is no point in attempting a low pass filtering operation on the data. However, pressure cell or staff gauge data can be used for this purpose. Because the amount of energy contained in this low frequency band is small compared to the first order wave energy, it is important to utilize a low pass filter with adequate attenuation characteristics outside the pass band. Single precision Fourier transform filters are not sufficient for this task.

The filter program FILTW which was developed for this purpose uses a 201 point Kaiser filter and operates on the data in the time domain. The time domain filter is used to compute, what is referred to as a running average and must therefore be moved into the record in the beginning and then out of the record at the end. Therefore, the first and the last 100 points of a filtered record must be considered suspicious and should be ignored.

For comparison, the theoretical long wave components can be calculated from radiation stress theory which states that the group bound set down under a wave is proportional to the square of the wave height where the proportionality constant is a function of the depth of water and the wave period. The program BLWAV carries out this calculation which involves the Fourier transformation of the wave data, the cross-multiplication of complex frequency pairs with common difference frequencies and their subsequent summation. This final sum constitutes a component of the long wave spectrum. Its inverse Fourier transform is the theoretical long wave which is associated with the original wave data record.

6. SIWEH ANALYSIS AND THE GROUPINESS SPECTRUM

The SIWEH is defined by Ref. 9. It is a smoothed elevation squared envelope function of the wave data record. It may be regarded as the primary excitation for structures which respond mainly to the square of the water surface elevation (Ref. 10). It has also proved of importance in the study of the stability of rubble-mound breakwaters (Ref. 11).

The SIWEH may be calculated in three ways, -by low pass filtering of the square of the wave record (Ref. 9), -by summing combinations of pairs of complex Fourier coefficients (Ref. 9) or -by combining the square of the wave record with the square of its Hilbert transform (Ref. 12). The program SIWEH performs either the first or the third of these options. It must however be pointed out that the two methods do not yield completely identical results. The difference is mainly in the high frequency content of the two differently computed SIWEHs.

Spectral density analysis of the SIWEH leads to what has been referred to as the Groupiness Spectrum (Ref. 13). This function has an expected value which is proportional to the auto-correlation of the expected variance spectral density. In nature, this expected variance spectral density is not known very reliably and it is therefore difficult to estimate the expected groupiness spectrum. However, the opposite applies to laboratory generated wave records. Fig. 7 gives an example of the expected and the measured groupiness spectral densities derived from laboratory generated wave data. The line spectrum indicates upper and lower limits of the 80% confidence band derived from a chi-square distribution.

The Groupiness Factor is defined by Ref. 9. It may be considered either as the standard deviation of the SIWEH normalized by the record variance or as the square root of the zeroth moment of the groupiness spectrum normalized by the zeroth moment of the first order variance spectral density. From this it will be apparent that the groupiness factor depends on the arbitrary cut-off frequency used for the computation of the groupiness spectrum. This explains also the difference between the two methods of the SIWEH calculation stated above. However, the groupiness factor is a useful measure of the wave grouping activity in the very low frequency range and has been shown to affect the second order response of large floating structures (Ref. 10). The Groupiness Factor was found to range from 0.45 to 1.15 for over 1000 wave trains of 16 minute length recorded at three stations off the east coast of Newfoundland during the period from January 16 to February 16, 1982.

7. EXAMPLE OF A COMPREHENSIVE ANALYSIS

The various wave data analysis programs are linked together by the operating system which facilitates the transfer of intermediate results by disk files. Fig. 8 gives an example flow-chart for the comprehensive analysis illustrated in Fig. 9.

It may be noticed that the figure includes tabulations of most of the parameters calculated by the program package. The naming convention of the parameters follows the IAHR List of Sea State Parameters and is summarized in Table I.

8. CONCLUSION

Several unusual wave parameters and techniques have been described for the purpose of promoting a wider concensus on wave analysis methodology. The wave analysis package described in the paper has been in use for several years. It is expected that future versions will include the following features which are currently under development: Kimura run length analysis, non-linear dispersion relations for the calculation of wave lengths, the optimal fitting of spectral models and the analysis of second harmonic bound wave components.

9. REFERENCES

1. "I.A.H.R. List of Sea State Parameters". To be published jointly by the IAHR Section on Maritime Hydraulics and PIANC, 1986.

2. Mansard, E.P.D. and Funke, E.R.R.:"The NRCC Advanced Wave Data Analysis Package". National Research Council of Canada, Hydraulics Laboratory Technical Report TR-HY-003, 1986 (to be published).

3. Kjeldsen, S.P. and Myrhaug, D.:"Kinematics and Dynamics of Breaking Waves". NHL Report No. STF 60, A78100, Norwegian Hydrodynamics Laboratory, Trondheim, 1978.

4. Kjeldsen, S.P. and Myrhaug, D.:"Formation of Wave Groups and Distributions of Parameters for Wave Asymmetry". NHL Report No. STF 60, A79044, River and Harbour Laboratory, Trondheim, 1979.

5. Kjeldsen, S.P.:"Dangerous Wave Groups". Norwegian Maritime Research, No.2, 1984, vol.12 pp. 4-16.

6. Funke, E.R.R. and Mansard, E.P.D.:"The Control of Wave Asymmetries in Random Waves". In: Proc. 18th Int. Conf. on Coast. Engg., Cape Town, South Africa, 1982.

7. Mansard, E.P.D. and Funke, E.R.R.:"Statistical Variability of Wave Parameters" (Variabilité statistique des paramétres de vagues). In: Proc. International Symposium on Maritime Structures in the Mediterranean Sea, National Technical University of Athens, Athens, Greece, Sep. 1984 (in French).

8. Mogridge, G.R., Funke, E.R., Baird, W.F. and Mansard, E.P.D.:"Analysis and Description of Wave Energy Resources". In: Proc. of the 2nd Int. Symp. on Wave Energy Utilization, The Norwegian Institute of Technology, Trondheim, Norway, 1982.

9. Funke, E.R.R. and Mansard, E.P.D.:"On the Synthesis of Realistic Sea States". National Research Council of Canada, Hydraulics Laboratory Technical Report, LTR-HY-66, Aug. 1979.

10. Mansard, E.P.D. and Pratte, B.D.:"Moored Ship Response in Irregular Waves". In: Proc. 18th Int. Conf. on Coast. Engg., Cape Town, South Africa, 1982.

11. Johnson, R.R., Mansard, E.P.D. and Ploeg, J.:"Effects of Wave Grouping on Break-water Stability". In: Proc. 16th Int. Conf. on Coastal Engg., Hamburg, Germany, 1978.

12. Bitner-Gregersen, E.M. and Gran, S.:"Local Properties of Sea Waves Derived from a Wave Record". Applied Ocean Research, 1983, No.4, vol5, pp 210-214.

13. Pinkster, J.A.:"Numerical Modelling of Directional Seas". Proc. Symposium on Description and Modelling of Directional Seas, Technical University of Denmark, June, 1984.

TABLE I

LIST OF SYMBOLS IN FIG. 9

XXXU =	Stands for a parameter computed by the zero-up crossing method
XXXD =	Stands for a parameter computed by the zero-down crossing method

H13	= the significant wave height
HMAX	= the maximum wave height
HMAX/H13	= the ratio of the maximum to the significant wave height
TAV	= the average wave period
TH13	= the average period of the significant waves
THMAX	= the period of the maximum wave
S13	= the significant value (i.e. average of the highest one-third of the steepnesses)
SH13	= the average steepness of the significant waves
SMAX	= the maximum steepness
SHMAX	= the steepness of the maximum wave
SH13MAX	= the maximum steepness of the significant waves
SCFAV	= the average crest front steepness
SCF13	= the significant value of the crest front steepnesses
SCFH13	= the average crest front steepness of the significant waves
SCFMAX	= the maximum wave steepness
SCFHMAX	= the steepness of the maximum wave
SCRAV	= the average crest rear steepness
SCR13	= the significant value of the crest rear steepnesses
SCRMAX	= the maximum crest rear steepness
MYAVH	= the average horizontal asymmetry
MY13H	= the significant value of the horizontal asymmetries
HM0	= the significant wave height computed from the zeroth moment of the variance spectral density
TP	= the peak period, $= 1/f_p$
TPD	= the peak period computed by the Delft method, $= 1/f_{pD}$
GF	= the groupiness factor
QP	= the Goda peakedness factor
HLW13	= the significant long wave height
HLWMAX	= the maximum long wave height
TLWAV	= the average period of the long wave
THLWMAX	= the period of the maximum long wave

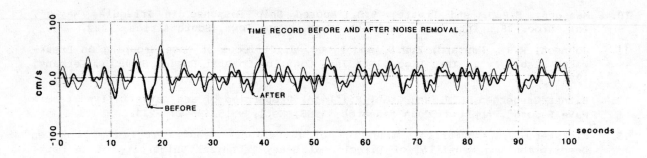

TIME RECORD BEFORE AND AFTER NOISE REMOVAL

BEFORE

AFTER

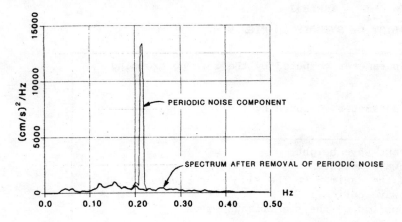

PERIODIC NOISE COMPONENT

SPECTRUM AFTER REMOVAL OF PERIODIC NOISE

REMOVAL OF A
PERIODIC NOISE COMPONENT
FROM A CURRENT RECORD

FIG. 1

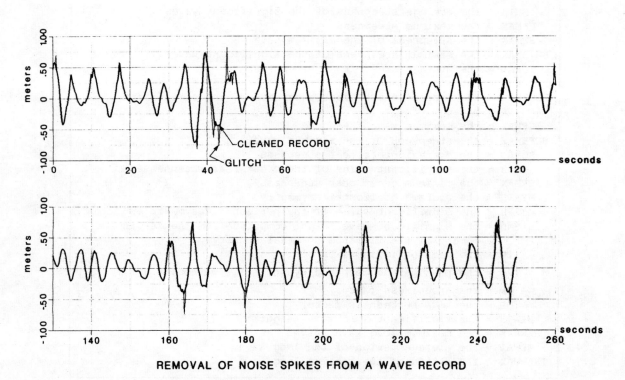

CLEANED RECORD

GLITCH

REMOVAL OF NOISE SPIKES FROM A WAVE RECORD

FIG. 2

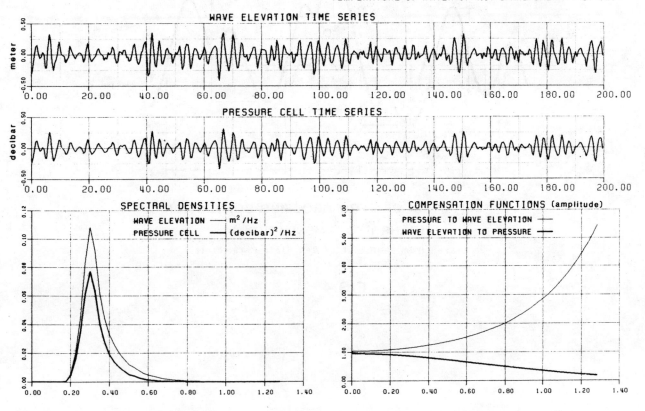

DYNAMIC COMPENSATION OF SIMULATED PRESSURE CELL RECORDED WAVE DATA

FIG.3

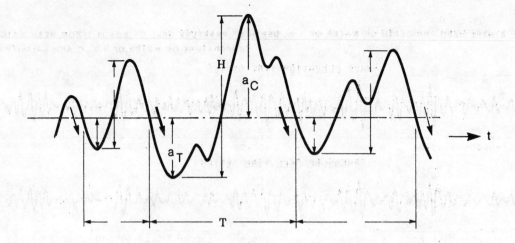

DEFINITION OF ZERO CROSSING CONVENTION
(IAHR RECOMMENDATION)
(Taken from IAHR List of Sea State Parameters)

FIG.4

DIRECTION OF WAVE PROPAGATION

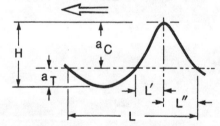

WAVE STEEPNESS:	$S_z = H/L$
CREST FRONT STEEPNESS:	$S_C = a_C/L'$
CREST REAR STEEPNESS:	$S_C = a_C/L''$
VERTICAL ASYMMETRY:	$\mu_V = L''/L'$
HORIZONTAL ASYMMETRY:	$\mu_H = a_C/H$

DEFINITION OF WAVE ASYMMETRY PARAMETERS
(Taken from IAHR List of Sea State Parameters)

FIG.5

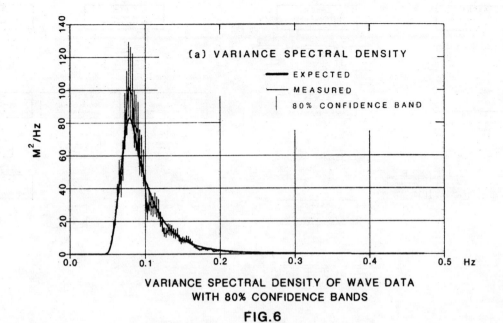

VARIANCE SPECTRAL DENSITY OF WAVE DATA
WITH 80% CONFIDENCE BANDS

FIG.6

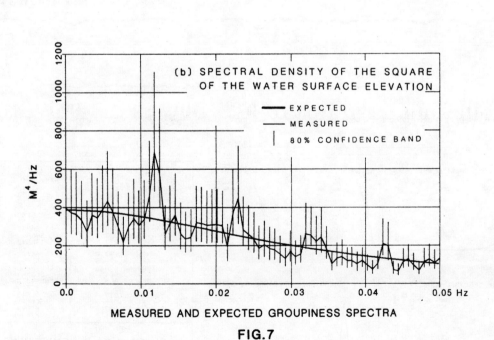

MEASURED AND EXPECTED GROUPINESS SPECTRA

FIG.7

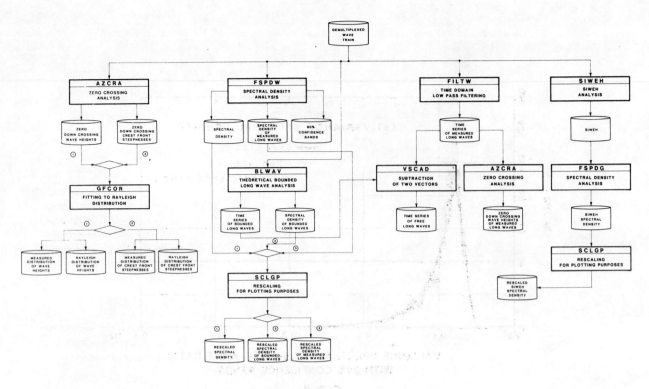

EXAMPLE FLOW CHART OF A COMPREHENSIVE

WAVE DATA ANALYSIS

FIG.8

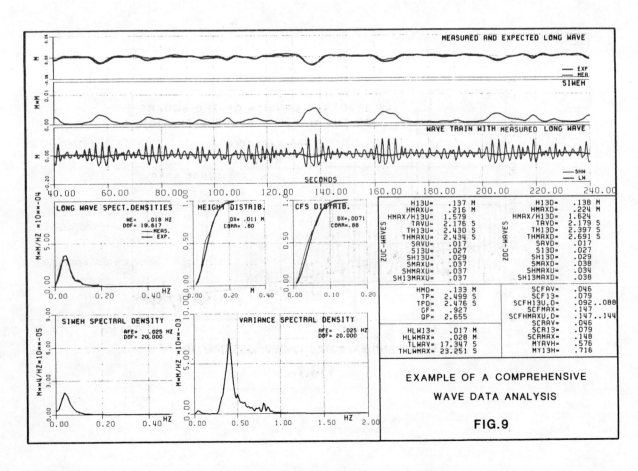

EXAMPLE OF A COMPREHENSIVE

WAVE DATA ANALYSIS

FIG.9

Measuring Techniques

of Hydraulics Phenomena in offshore, Coastal & Inland Waters

London, England: 9-11 April, 1986

A SECOND GENERATION WAVE RECORDING NETWORK

A.J. Allan A.W. Bolton A.T. Webb

N.S.W. Public Works Department
Manly Hydraulics Laboratory, Manly Vale, Sydney, N.S.W. 2093

Summary

The evolution of the wave recording network used by the N.S.W. Public Works Department is traced, over a period of ten years, and it is shown how operational experience and use of the network has prompted additions and redesign to arrive at the system currently used. The importance of software and a high productivity program development environment is detailed, and an example of the scope of the wave data access software now in use is given.

Held at Imperial College of Science and Technology, London. Organised and sponsored by BHRA, The Fluid Engineering Centre. Co-sponsored by the American Society of Civil Engineers and the International Association for Hydraulic Research.

1.   INTRODUCTION

Wave climate data is needed by the N.S.W. Public Works Department in connection with the design of structures and the investigation of coastal processes. In order to collect such data a network of sites had to be installed and operated over a long period. In order to prepare for this a pilot wave recording site was set up in 1974 at Port Kembla by the Department's Hydraulics Laboratory staff based in Sydney. A few months of operation revealed that operating costs were high due to the labour intensive nature of site support, data management and data transcription services.

1.1   Staffing Aspects

The pilot site showed that field operations staffing would be a major problem, and that the labour costs associated with wave data management at the laboratory would have to be reduced by the extensive use computers, if the 8 to 15 sites planned were to be suitably supported within budget by available staff.

The Public Works Department is decentralised and district offices exist to manage works in each region. Staffing would consist of district personnel, untrained in the use of electronic systems, allocated on a very part time basis to wave recording site support. High mobility of district staff would mean that operator training would have to be considered to be a continuing function over the span of the wave recording project.

Operations at the laboratory would use one junior staff member to carry out day to day logistic support and data management tasks with involvement from time to time by senior staff who would be responsible for overall quality control and system modification aspects. Installation and maintenance of systems would employ one civil engineering and one electronics technical officer, on a full time basis, supervised by engineering staff when required.

2.   FIRST GENERATION EQUIPMENT

Over the period 1975 to 1983 a series of offshore wave buoy and inshore wave pole stations were set up and operated. (Refs. 1,2) The experience gained in this way provided the basis for the selection and design of the second generation network. A review of the practical aspects of the work follows here to provide a general understanding of the type of operational problems that were encountered and show how the present system evolved.

2.1   Hardware

Offshore stations used the Datawell Waverider buoy and WAREP receiver as per the pilot site. Analogue data from the receiver was recorded by a mains powered digital cassette data logger (built by the Sydney firm Digital Electronics) which was triggered by the clock in the WAREP four times per day and captured just under 20 minutes of data each time. In cases of logger failure the WAREP provided backup paper chart records. The logger had an internal clock and so was able to record time and sequence numbers on the data cassette in addition to the wave data. This aspect was felt to be important so that tapes could later be identified back at the laboratory in Sydney even if paperwork that should normally accompany them was lost. Users at the sites were given on-site training and provided with a specially written manual aimed at those having no previous experience of electronic recording equipment.

The constant risk of damage to buoys moored close to ports and traffic lanes was known of from other users. Experience was that early detection of buoy drift was the best way of minimising operational losses and to this end all operators were provided with direction finding radio sets and instructions on how to make regular checks of buoy positions.

Wave poles used were of the Zwarts type, and consisted of a pair of concentric copper tubes mounted vertically into the water surface, providing a coaxial transmission line acting as a radio frequency resonator whose oscillatory period was linearly related to water level. Some of these sites, in order to allow battery operation, used a low power cassette logger (Datel LPS-16) and timer rather than the mains powered logger, however all cassettes could be read on the same cassette reader into the laboratory computer.

2.2   Data Management

The public postal system was used to transport data cassettes to and from the sites. Data storage at the laboratory was based on a newly installed HP 21MX minicomputer, which was shared with other laboratory projects. Support software was prepared using a version of FORTRAN to enable data entry from cassette tapes, checking of data, and some data reduction.

The 72 minutes of raw data obtained daily from each site led to a storage need on the central computer of about 10MB per site per year. Any gaps in cassette data could be partially documented by study of the WAREP chart records. These were used to provide a manual quality control check of correct station operation and zero crossings analysis done by hand on a regular basis to provide results as a check against the computer output.

2.3   Lessons From These Deployments

A major disadvantage was the inability of the recorders to take extra sets of readings at times of major storms. Manual override facilities enabled an informed operator to trigger the recorder, and so capture storm peaks, but in practice such events were often not fully captured. The automatic measurement of wave direction also remained an intractable problem as no affordable equipment could be located and so spasmodic manual observations by skilled persons remained the only source of this data.

Overall data recovery from sites via cassette tape was around 80 per cent over the periods for which buoys were installed. There were times at which sites could not be kept operational, following buoy losses, due to capital expenditure limits and logistic limitations. Purchase and delivery of a buoy could take as long as a year to achieve.

2.3.1 Buoy Components

Constant review and staff retraining on buoy and mooring deployment and maintenance were shown to be investments that paid dividends in the long term. The largest capital losses were due to mooring failure, often due to external factors, such as impact with ships or entanglement with fishing gear. For deployments to date, out of a total of 40 buoy hulls purchased, 17 have been lost or damaged beyond repair. The direction finding systems have been used on several occasions to assist recovery of drifting buoys, however recovery, unless possible within a few hours of mooring failure, becomes difficult and expensive. (Since losses are highest at times of severe sea conditions rapid recovery is often not an option. In practice an operator may have no indication of problems until the buoy has drifted close to the limit of the radio telemetry link. If the drift is onshore then total loss onto rocks often occurs as radio contact is lost. If drift is offshore then the expense of mounting a long distance recovery attempt may be greater than the value of the buoy.)

The buoy itself proved to be a most reliable and stable device, the only disadvantage being the well known fragility of the accelerometer system. Battery life of six months proved to be a good match to the deployment limit set by fouling build up on the hull. System calibration was carried out before and after each 6 month deployment.

2.3.2 Shore Components

The pen recorder fitted to the WAREP radio receiving system proved to be the part of the system most at risk from poorly trained operators. However, the utility of having a visual recording of wave data at the site was such that costs associated with its use were seen as justified. The data encoding method used by the Waverider system consisted of an audio subcarrier frequency modulated by the wave profile data. This subcarrier forms a most effective 'health monitor' of the buoy and data link and was available to the operator, via a small loudspeaker. Radio data link problems such as CB radio interference at some urban sites were very quickly identified in this way.

Digital data cassette tapes do not achieve the level of reliability normally expected from computer readable media. Although care was taken to only use certified data cassettes, data was at times lost. The fact that the data logger tape cassette transport only had the ability to write to tape proved to be a weakness. If an operator failed to insert a tape properly, the fact could not be indicated in any way and data was thus lost.

It had been hoped that it would prove possible to extend the role of the logger system to include additional functions, such as automatic warning of problems, on-line data reduction, and adaptive data recording. As the logger contained a general purpose single board computer, based on a 6502 microprocessor, such extensions were theoretically possible. However, owing to the problems of getting custom assembly code programs prepared no extensions were in fact made. The supplier was not able to provide support services on firmware and no contractor could be located to undertake such work.

Although the planned life of the logger system was 6 years two factors caused increased maintenance costs before this. Firstly the tape transports began to show mechanical wear after about 4 years. Secondly, the high operating temperatures of some parts of the logger caused a larger number of integrated circuit failures than expected, the A to D convertors and program EPROMs having the highest rates of failure. Some remedial mechanical work was done to improve the flow of cooling air to the hottest parts of the logger.

2.3.3 Computing Aspects

The area of data reduction software was one in which limited experience existed at the laboratory when the project started. In the mid 1970's the software tools available on minicomputers were very limited. For example, the Laboratory's minicomputer had rather limited file management facilities and minimal FORTRAN debugging aids. Support programs to allow the level of automation originally planned could not be prepared within the time scale hoped for. The problems of checking and storing large quantities of raw data became the overriding consideration and clearly demonstrated the limitations of both the hardware and systems software as follows:

i.   Raw data files were typically about 200k bytes in length, however program and variables space was limited to about 27k bytes, the computer having a 32k byte address space. Thus programs had to make extensive use of record by record access to data on disc rather than reading data sets into arrays. This had two disadvantages, firstly the amount of FORTRAN code was probably at least doubled as a result of the constant calls to machine dependent disc access subroutines, and secondly the code was thus rendered non portable.

ii.   Disc files were not time or date stamped and names were only allowed to be 6 characters long. Thus additional programs had to be written to aid indexing and tracing of data.

iii.   Disc storage of 50MB was fitted to the computer but only a portion
         of this could be devoted to the wave data project, thus archiving of
         data to tape was needed at intervals. As the project proceeded the
         time needed to support tape backups and retrieval of data started to
         rise greatly.

iv.   An extensive background in the use of the computer was needed to
         allow searching and retrieval of data so that in practice this could
         only be carried out by one skilled machine operator, and not as
         originally hoped, by persons from other projects who needed the
         processed data.

3.   IMPLEMENTATION OF THE SECOND GENERATION NETWORK

By late 1970's wave data had become the largest single data management task
facing the laboratory and pressures were building for a major upgrading of
computing resources to meet the needs of the next 5 to 10 years. Efficient
software development was seen as of paramount importance in containing
costs by allowing greater automation of many phases of the wave data
project. The following steps were seen as central to the implementation of
a viable network:

i.   Installation of a central computer allowing use of packaged data
         management tools, greatly expanded storage space, and low cost
         program development.

ii.   Use of computers at the field sites, so that on-site data reduction
         and adaptive logging could be used and local operator tasks de-
         skilled. These computers should, as far as was practical, offer
         users and programmers the same working environment as the central
         site computer.

iii.   The continued use of the well proven buoys and radio receivers.

iv.   Provision for later integration of wave direction data derived from
         shore mounted X-band radar sites. This would involve computational
         support for image processing of the radar data. (Pilot results,
         from a site installed in 1980 at North Head in Sydney, had
         established the feasibility of using this technique, but only manual
         analysis of photographic images from the radar screen had as yet
         been achieved. (Development work on the automatic analysis of radar
         images is continuing as a joint project with James Cook University,
         Queensland, Australia.)

v.   Provision for the integration of data from bottom mounted
         directional current meters into the wave data base as this could be
         used to infer wave direction and give a means of estimating
         directional spectra. Current meters (Marsh McBirney types 551 and
         ARC585) had been used in connection with specific projects needing
         sediment transport data at several sites.

vi.   Provision for later connection of sites via modem links to the
         central computer, so that all manual aspects of data capture into a
         central data base would be eliminated. This would also allow
         hardware support and maintenance to operate efficiently on an actual
         needs basis, rather than by routine preventive scheduled visits plus
         emergency repair calls.

3.1   Field Installation Changes

The major change was the replacement of the logger by a much more capable
computer able to carry out data compression, in the form of time series
analysis, thus allowing more data sets to be obtained in a given time while
still keeping the total volume of data within manageable limits. Selection
of the computing equipment for field sites was carried out with other

laboratory projects in mind. The cost of using a number of minicomputer-like systems would be high and thus good use of the systems on various projects well into the future should be possible. The aim was to achieve standardisation and re-use of the equipment for as long as possible. (Ref. 3)

It was felt, in the light of the support problems encountered with a number of other computer based instrumentation systems, that building a custom system from a well established board set would be the preferable path. This would allow use of low cost, mass produced electronic assemblies, so keeping costs within reasonable limits, while giving full control over aspects such as software, and the suitability of the equipment for the environmental conditions expected. Essential qualities were seen to be as follows:

i.   The board set should have reasonable market share and expected availability into the 1990's.

ii.   The supplier should have engineering staff in Sydney and be able to provide board level exchange and repair support as the laboratory would not be able to set up in-house repair support.

iii.   Extensive information should exist about the use of the board set and its environmental limits.

iv.   Configuration of custom systems should be easy by virtue of being able to purchase support hardware and fully wired backplanes and cables.

v.   Packaged software and a real time operating system should be available, together with contract support and local programming specialists able to undertake any work needed to configure the software to the particular hardware used.

vi.   The operating system would have to be able to run from tape as well as from disc, as the environment for some sites was too harsh to allow the use of floppy discs. A tape cartridge was also judged to be the best media to ship back to the laboratory by post.

vii.   Total power needs should be modest as some future applications could need battery operation and all wave recording sites were subject to loss of mains power from time to time. (Ref. 4)

The need for a tape based operating system turned out to be the determining factor. The combination selected was the RT-11 operating system running on a Q-Bus LSI-11/21, supported by a dual drive TU-58 tape subsystem and a user keyboard console unit fitted with a single line liquid crystal display. Spare serial ports were fitted to allow addition of an auto-answer modem and printer or full VDU terminal.

Some custom hardware was designed as follows to give the specific qualities needed.

i.   Automatic over temperature detection hardware was added so that warning could be given to the user via the keyboard console unit if the CPU temperature reached 60 degrees Celsius and power cutout when the temperature reached 65 degrees Celsius.

ii.   All items were powered by an uninterruptable power supply, constructed around a high efficiency switching convertor and a low maintenance gel cell rechargeable battery. Status lines from the power supply enabled the computer to sense abnormal conditions and give an audible alarm if the power supply was close to an undervoltage condition.

iii.   A full calendar clock with internal battery backup was added, together with control logic to enable clock set up of the system at the laboratory prior to deployment and semiautomatic startup when on site.

iv.   An internal fan and heat exchanger allowed heat transfer out of the system without the need to draw air into the equipment case. This was important as some sites were subject to very high dust levels.

v.   A fully digital interface was constructed between the computer and the WAREP receiver by using direct counting techniques to measure the audio subcarrier frequency and so avoid the possible drift in the analogue frequency to voltage processing stages previously used.

Software run by the system was mainly written using FORTRAN, but FFT routines were coded in MACRO assembler to meet speed requirements. A user could converse via a menu and a number of simple commands, including "Help" commands, were provided. A user could also store comments to the data tape. Data capture was carried out on an hourly cycle. About 34 minutes of data was captured, then over the next 25 minutes computations were carried out to produce zero crossing and one dimensional spectral statistics which were then filed to tape. Raw data was also stored to tape twice per day and automatically more often at times of significant sea conditions.

3.2   Central Site Details

Since the data cartridges could be read directly into a VAX 750 system, no special hardware was needed to support the wave data stations. An intelligent modem with full autodial capabilities was the only extra hardware item installed to enable remote control and data transfer from sites fitted with modems. The VAX 750 computer system, used as the central computer, ran the VMS operating system and was equipped with a 800/1600 bpi high speed tape drive to allow archiving of data. A disc area of 100MB was reserved on a fixed disc to allow for the on-line storage of the statistical part of the wave data base. Raw data files that take up very large amounts of space were, in general, intended to be stored on tape, but some representative samples were also stored on- line. Software products added specifically to assist the development of the wave data system were as follows:

RTEM, an RT-11 emulation package that allowed a programmer to use the VAX under RT-11 as a software development system for the field site computers. Utilities within RTEM were also used as part of the production wave data system to convert data files from the field computers to the formats needed by VMS.

DATATRIEVE, a data retrieval package with many features that would assist untrained persons in the use of the other software resident on the VAX.

VMS, plus RT-11 emulation proved to be a productive environment to work within and the wave data system was implemented using:

i.   DCL (the command language within VMS), to provide the upper levels of the application suite such as menu control file housekeeping and garbage collection.

ii.   FORTRAN 77 to do the actual calculation work which included data conditioning, spectral analysis, and computation of display output, such as graphs and tables of statistics.

iii.   DATATRIEVE to build and access the wave data base, and provide an audit trail of access and changes.

3.3   Data Base Access and Use

A software system - DWAVE - has been built to support a range of user types:

The DATA MANAGER is required to read and archive cartridges from the field loggers as they arrive at the Laboratory. A quick health check is made by examining automatically produced graphical summaries.

The DATA USER needs to access archived raw and analysed data. A wide variety of output formats are available (mostly graphical) and in all cases, screen previews can be made before hard copy. Operations that can be made on raw data include digital filtering, zero crossing and spectral analysis; and, for March McBirney/Zwarts combinations, directional spectral analysis.

The SYSTEM MANAGER keeps the system operating and adds new capabilities as the need arises.

A MENU driven environment enables all levels of users to operate what has become a large system accessing a very large data base.

In operating the system, data is addressed by SITE and DATESTAMP (ie: filenames need not be known). Full use is made of HELP, VALIDATION and MENU facilities.

The power of DATATRIEVE becomes apparent when making use of the record selection capabilities in order to window the data. For example, a typical user request would be:-

Do a linear regression on the significant wave heights from Eden and Port Kembla (two offshore Waverider sites) for 1984; but only include the data where the significant period at Port Kembla was between 8 and 10 seconds. The result of such a request is given as Fig. 1.

Other standard output formats available (Ref. 5) include:

Time histories of statistics, spectral plots, exceedence statistics, statistical summary tables.

4.   CONCLUSIONS AND CURRENT WORK

An efficient and powerful wave data network has been built for the N.S.W. coast. The network uses a variety of sensors (Waverider Buoys, Zwarts wave poles, and Marsh McBirney current meters), but a standard data logger and a central computer. Standard time series analysis takes place on-board the loggers so that more data can be stored in a compressed form. Management of the captured data on the central computer is via a menu driven software system. This allows ready access for casual users, as well as complex operations by sophisticated users.

Major developments in the near future include:

i.   Modifications to the logger operation in order to remove the need to post data cartridges. Instead, using modems at each site, data will be dumped automatically by telephone overnight, thus reducing the work load of both field and laboratory operators.

ii.   The development of automatic analysis of radar images for wave direction estimation.

5. REFERENCES

1. Public Works Dept. N.S.W.: "Wave Data Collection Systems at Manly Hydraulics Laboratory", Manly Hydraulics Laboratory, Report No. 311 April 1982.

2. Webb A.T.: "Wave Climate of the New South Wales Coast". Sixth Australian Conference on Coastal and Ocean Engineering, Gold Coast 13-15 July 1983, [poster paper].

3. Allan A.J.: "Adapting a Computer System for Use in Civil Engineering Projects", Proc. 6th Computers and Engineering Conference, Institution of Engineers Australia, Sydney 31 Aug. - 2 Sept. 1983, pp 44-48.

4. Allan A.J. and Todd M.: "A Mobile Computer System For Data Capture Tasks", Proc. DECUS Aust., Brisbane Aug. 1981, pp 1549-1555.

5. Public Works Dept. N.S.W.: "DWAVE Data Presentation", Manly Hydraulics Laboratory, Report No. 427, 2nd Edition Sept. 1985.

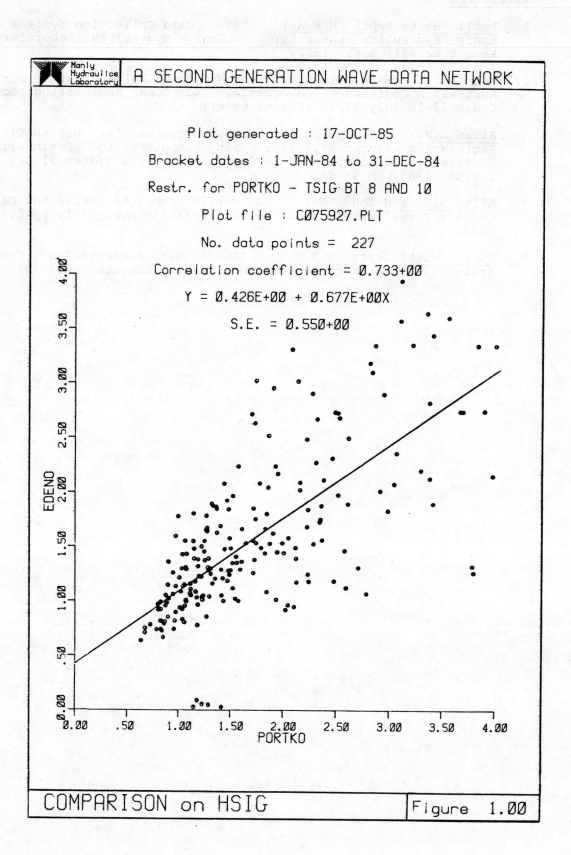

A SECOND GENERATION WAVE DATA NETWORK

Manly Hydraulics Laboratory

Plot generated : 17-OCT-85

Bracket dates : 1-JAN-84 to 31-DEC-84

Restr. for PORTKO - TSIG BT 8 AND 10

Plot file : C075927.PLT

No. data points = 227

Correlation coefficient = 0.733+00

Y = 0.426E+00 + 0.677E+00X

S.E. = 0.550+00

COMPARISON on HSIG

Figure 1.00

USE OF CORRECT OR NOMINAL
WATER VELOCITIES AND ACCELERATIONS IN MORISON'S EQUATION

G.Moberg, MSc, L.Bergdahl, PhD, B.Carlsson, Eng.
Chalmers University of Technology
S-412 96 Göteborg
Sweden

Summary

A laboratory experiment has been performed to evaluate the drag and inertia
coefficients in the Morison Equation. The total wave force, the pressure distribution
against a vertical circular cylinder and the wave kinematics of steep waves have been
measured.

The experimental arrangement is described and the data processing is sketched. The
technique for measuring velocity is described in detail.

It is shown that it is important to use the same type of velocity and acceleration in
design as is used in the experiments from which the coefficients are evaluated. It was
also found that the best choice of wave theory from five investigated theories for most
cases is the fifth order theory. For shallow water the best choice is the cnoidal
theory.

Held at Imperial College of Science and Technology, London. Organised and sponsored by BHRA, The
Fluid Engineering Centre. Co-sponsored by the American Society of Civil Engineers and the
International Association for Hydraulic Research.

NOMENCLATURE

A = characteristic area

C_D = drag coefficient

C_M = inertia coefficient

d = diameter of cylinder

F = force

F_D = drag force

F_M = inertia force

g = earth acceleration

k = roughness

KC = $u_{max}T/d$ = Keulegan-Carpenters number

p = pressure

Re = $u_{max}d/\varepsilon$ = Reynolds number

u = water particle velocity

u_{max} = maximum value of velocity during a wave period

u_s = free stream velocity

V = immersed volume of body

Δ = difference

ε = kinematic viscosity

ρ = density of water

$\partial/\partial t$ = derivative with respect to time

$|\ |$ = absolute value

1.   INTRODUCTION

In order to calculate forces on offshore structures with small diameters compared to the wave lengths the well known Morison Equation is often used.

$$F = F_D + F_M = C_D A\frac{1}{2} \rho\, u|u| + C_M V\rho\, \partial u/\partial t \qquad (1)$$

where   F = force

F_D = drag component of force

F_M = inertia component of force

C_D = drag coefficient

A = characteristic area

ρ = density of water

C_M = mass or inertia coefficient

u = water (particle) velocity

V = immersed volume

$|u|$ = absolute value of water velocity

$\partial u/\partial t$ = water (particle) acceleration

Since the introduction of this equation in 1950 by Morison et al (Ref 1) much work has been devoted to establishing values of its drag and inertia coefficients. Experiments have been made in oscillating u-tubes (Ref 2) in the real environment (Ref 3) and in wave tanks (Ref 4).

The coefficients have for fixed structures been found to be functions of the Reynolds number (Re), the structural roughness to diameter ratio (k/d), and the ratio between the diameter of the water particle path and the structural diameter (KC). For oscillating objects also the ratio between the double amplitude of oscillation and the diameter of the structure is important.

In this paper we will not discuss these functions or provide new values on the coefficients of the Morison Equation, but we will point to the fact that it is important to use the same type of velocity and acceleration in the design step as was used in the experiments from which the used coefficients were evaluated.

An example of the consequence of using the same experimental C_D and C_M values in different wave theories will be shown and then an experiment will be reported where undisturbed wave form, water velocity, and acceleration as well as the pressure distribution and total force against a vertical circular cylinder have been measured.

2.   DIFFERENT EXPERIMENTAL APPROACHES

There are basically two possible ways of making experiments to establish the drag and inertia coefficient.

One is to make experiments in an oscillating u-tube or possibly by oscillating a structural element in a closed conduit. This has the advantage that the water velocities and accelerations are easily determined and therefore nearly exact. Very high Reynolds numbers can also be realized. A disadvantage is that the undisturbed paths of the water particles are straight lines instead of ellipses. Such experiments are, however, generally good experiments.

Another is to make experiments in a wave tank, which has been done by us. The advantage is that the undisturbed paths and free surface effects are correctly modelled. Disadvantages are that we cannot reach realistic Reynolds numbers, and that the water velocity and acceleration must be measured separately or be evaluated from theories. In the latter case it must be noted that a wave theory that gives a fairly good approximation of wave form not necessarily gives a satisfactory approximation of velocity and acceleration.

3.   THE USE OF THE COEFFICIENTS IN DESIGN

If, in a design situation, the same coefficients of C_D and C_M are used together with different wave theories, some of which give good approximations of the wave form, substantial differences in calculated forces can be noted. As an illustration calculated forces per unit length against a vertical cylinder in transitional water depth are shown in Fig. 1 for four different wave theories using the same values of C_D and C_M. The experimentally measured force is also shown for comparison. Note that the second, third and fifth order theories are all good at approximating the wave form (Fig. 14a) but not velocities (Fig. 15a) and accelerations (Fig. 16a). Consequently they also give bad estimates of forces.

As a contrast, in Fig. 2 a comparison is made between the measured force and the force calculated with constant C_D and C_M values but using real measured velocities. C_D was here evaluated from the wave crest and C_M from the front slope of the wave in the experiment using measured velocities and accelerations.

4.   PERFORMED EXPERIMENTS

Experiments have been performed at Chalmers University of Technology in a wave tank measuring wave forces and pressure distributions against a vertical circular cylinder. The wave form, water particle velocity and acceleration were also measured. The experiments were subsequently analysed to give drag and inertia coefficients to be used in the Morison Equation. In Fig. 3 the used waves are plotted as points in a Le Méhauté diagram. The ranges of attained Keulegan-Carpenters numbers, KC, and Reynolds numbers (maximum at the still water level), R_e, were

$$1 < KC < 9 \qquad \text{and} \qquad 0.2 \ 10^5 < R_e < 2.3 \ 10^5$$

Further details on results in terms of C_M and C_D values will not be given here but will be reported in a thesis by G. Moberg to be published in 1986 (Ref. 5).

5.   EXPERIMENTAL ARRANGEMENTS

5.1   The wave tank

The used wave tank is 80 m long, 2.0 m wide and has a maximum water depth of 1.0 m (Fig. 4). The wave generator consists of a flap constricted to run parallel with the bottom in a sinusoidal motion. The amplitude of motion can be adjusted at the surface and bottom so as to approximate particle motions in a wave rather closely. Regular waves with wave lengths between 1 and 11 m can be generated. The downwave end of the wave tank has a gently sloping beach in order to minimize reflexion. In the experiments only six to ten fully developed waves were used in every run so that reflexions were almost completely avoided.

In the middle of the tank a vertical circular cylinder was placed.

5.2   Cylinder

The vertical circular cylinder had a diameter of 0.3 m and a hight of 1.0 m. It was made of steel and designed so that the resultant force from weight and buoyancy was small and directed downwards. The cylinder was hung in a cantilever. The bottom of the cylinder was hung as close as possible to the bottom of the wave tank without touching it.

The cylinder was equipped with pressure probes and a wave gauge in a vertical in its wall. The cantilever was equipped with strain gauges.

5.3   Gauge cantilever

The cantilever was equipped with strain gauges at two levels in two directions. See Fig. 5. The cantilever was clamped to a stiff heavy beam that was mounted across the top of the wave tank.

By measuring the moments at the two different levels of the strain gauges the magnitude and line of action of the force acting on the cylinder could be evaluated.

The cantilever was made of solid aluminium with a large diameter in order to avoid motions of the cylinder and to get large strains at the same time.

With the chosen cantilever the motions of the force and of the cylinder could not be visually observed in the experiments. For the static design load of +500 N at the end, the deflection of the cylinder was, however, measured to +2.5 mm. The maximum force in the experiment was around 150 N. Furthermore the natural frequency of the system can be calculated to approximately 10 Hz which is 10 times or more as great as the frequency of the used waves. The dynamic amplification could thus be considered to be negligibly small.

5.4   Pressure measurements

In order to measure the distribution of pressure around the cylinder, it was equipped with pressure transducers. The wish to measure the pressure above the level of the wave troughs demanded that the transducers should be able to work in both air and water and respond very fast to pressure changes. They should also be small in order not to disturb the flow and to give the pressure in a point.

Pressure transducers with a sensor diameter of 3.84 mm were found that met the requirements, although there were some problems with the transitions between water and air in the beginning of the experiments.

The limited number of channels in our measuring system and the price of the transducers led to equipping the cylinder with six transducers. They were mounted flush with the cylinder surface in a vertical line. See Fig. 5. The uppermost transducer was situated 0.05 m or 0.15 m above the stillwater level depending on the water depth being 0.8 m or 0.7 m. One of the transducers was in both cases at the still water level.

To obtain the pressure for the whole circumference of the cylinder with just one vertical line of transducers, the whole cylinder was rotated by 22.5 degrees increments. See Fig. 6. 16 separate experiments were consequently needed in order to obtain the whole pressure-distribution around the cylinder.

5.5   Wave probes

Two ordinary resistive wave gauges were used for measuring the water level. One was placed 2.5 m in front of the cylinder to get an approximate knowledge of the undisturbed wave. PRO2 in Fig. 6. The other, PRO3, was placed just in line with the front of the cylinder and was used to synchronize the different experiments, see paragraph 6.1 below.

A special wave gauge was built to measure the water level at the cylinder surface close to the vertical line of pressure transducers. This wave gauge was also resistive but was made of four thin (ϕ 0.2 mm) wires of stainless steel.

This wave gauge made it possible to record the water level at the cylinder wall and thus to approximate the pressure above the uppermost submerged cylinder. It was also used to handle problems with signal offset when the pressure transducers went from air to water or vice versa.

5.6   Velocity measurements

During the evaluation of the forces on and the pressure distribution around the cylinder it was suspected that available wave theories gave too poor estimates of water particle speed and acceleration. Therefore it was decided to measure these in the used laboratory waves which were easily reproduced. As the undisturbed velocity should be used in the Morison Equation the velocity and acceleration was measured without the cylinder. In a later experiment velocities in points around the cylinder were also measured for a MSc thesis (Ref. 6).

Available equipment for this type of measurements were found to be either too expensive or too poor, so we had to develop our own equipment. The basis for this was a Kent micro-propeller (Fig. 7) which has been found to have good hydrodynamic properties

(Ref. 7 and 8). As we intended to measure the water velocity also above the level of the wave trough it was important that the propeller stopped instantaneously when emerging from the water. This was seen to be the case and was caused by surface tension.

The propeller has five blades. Each time a propeller blade passes a point on the mounting frame a puls is created. In most equipment these pulses are counted and converted to an analogue signal which then must be roughly smoothed out. We chose, however, to measure the lapse of time between each pulse and transform this lapse of time to an analogue value in a digital to analogue converter, which thus gave a stepwise constant value out inversely proportional to the water velocity. The used propeller was calibrated from 0.04 m/s up to 0.85 m/s and was found to give an error about 2 mm/s. See also Ref. 8.

In Fig. 8 a scheme for the velocity recorder is shown in principle. It works as follows. An oscillator, C, is continuously delivering pulses to a 12-bit binary counter, CTR, that is successively filled. The terminals of the counter are connected, to a storage register, (D and E) that contains the current state of the counter. When the propeller, P, delivers a pulse the monostable flip-flop, A, is trigged and gives a signal to the storage register to transfer its information to the D/A converter. This information is saved on the terminals while a new count is made. The flip-flop, A, also trigs flip-flop, B, which gives a reset signal to the counter and the register. When the counter is reset it is immediately receiving new pulses from the oscillator and its status is again saved in the register until a new propeller pulse is delivered. In order to get the right order between clear and reset signals the flip-flop, B, is trigged by the negative-going edge of the flip-flop, A. This causes that data is transferred safely before reset is done. The digital value (number of oscillations) and consequently the analogue output is a measure of the lapse of time between two consecutive propeller pulses. The developed equipment only gives the magnitude of the velocity but not its direction. Later in the processing the direction was received indirectly from the wave recorder assuming positive velocity in the crest and negative velocity in the trough. In Fig. 9 an example of measured velocity is shown. As can be seen, velocities below 0.04 m/s could not be measured, and measured values close to this velocity were obscured by the changing of direction of rotation.

Calibrations at constant velocities showed that the propeller moves approximately 6 mm between two pulses. To get a good accuracy at high velocities the clock pulse must be chosen rather high. 10,000 Hz was chosen. At 0.85 m/s thus 70 or 71 clock pulses are counted depending on the phase difference between propeller pulses and clock pulses. This error in the digital/analogue converter is +0.5 bit. ("True" value in this example is 70.58 "pulses"). Another error is caused by the fact that the reset-pulse can be given at the same moment as a clock pulse. This clock-pulse will then be counted in the next lapse of time.

In Table 1 the estimated errors in velocity for a continuously submerged propeller at constant velocity are given for the used range of velocities. It should be pointed out that these faults are very much smaller than what can be experienced if the propeller is maltreated or dirty. The propeller must therefore be recalibrated often and be safegarded against damage and dirt.

In paragraph 7.2 comparisons between water particle velocity and stagnation pressure is performed as a cross check on velocity.

6.   MEASURING SYSTEM AND DATA PROCESSING

6.1   Measuring system

The experimental variables were recorded with a Compucorp 625M-II desk computer. The recorded data was subsequently transferred to an IBM3081 computer for processing. In parallel with the desk computer a paper chart recorder was used for supervision of some variables. A graphic terminal connected to the IBM-computer was also placed in the laboratory. See Fig. 10.

The arrangements made it possible to check the measured and transferred data immediately using the graphic display, to store large amounts of measured data and to retain flexibility during the evaluation process.

Thirteen channels were used in the experiments with the cylinder: six for the pressure transducers, three for the wave gauges, and four for the straingauges on the gauge cantilever.

6.2   Averaging technique

All the data in the experiments was averaged over six to ten waves. This was necessary because the individual waves in a wave train varied slightly, and because the circumferencial pressure distribution had to be calculated from different but almost equal experiments.

In the experiments good correspondence between the different runs was secured by starting each measurement at the "same" wave measured from the beginning of the wave train and using the same number of waves in corresponding runs.

In Fig. 11 the velocity from Fig. 9 has been averaged to a single representative cycle, which is shown by the dashed graph in Fig. 11. The direction of the velocity in the trough has then been assigned a negative value. Furthermore a smoothing from positive to negative velocity has been performed by fitting a smooth curve through the reliable velocity data on both sides of the zero crossing point. The result is shown by the dotted graph. For pressure, water level and strain the resulting graphs were calculated by averaging only as these parameters were well defined also for values close to zero and for negative values. Checks on the technique is referred in Chapter 7.

6.3   Calculation of acceleration

The acceleration was calculated from the averaged measured velocity by calculating the numerical derivative. This is dubious because the highest accelerations are often attained around the zero crossing where we have smoothed the velocity by fitting a curve, and later these high accelerations are used for evaluating the inertia coefficient. Nevertheless the results indicate that this was permissible.

7.   CHECKS ON MEASURING TECHNIQUE

7.1   Comparison between measured force and integrated pressure

A good test on the averaging technique was a comparison between the force obtained from the integrated pressure with the force obtained from the strain gauge cantilever. The obtained forces fitted well to each other and it was calculated that the total difference in maximum forces were less than 0.5%. In Fig.12 an example of such a comparison is shown.

7.2   Comparison between measured velocity and stagnation pressure

By assuming potential unidirectional flow in the wave crest and wave trough, the maximum measured water velocity in the undisturbed wave could be compared with the pressure difference, Δp, between the stagnation point on the front (back) of the cylinder and the points of 90° or 270°. The velocity in the latter points should be $2u_s$, where u_s is the free stream velocity. Thus

$$(2\,u_s)^2/2g = \Delta p/\rho g \tag{2}$$

or

$$u_s = \sqrt{\Delta p/2\rho} \tag{3}$$

For the shorter waves a correction was made for wave surface curvature. In Fig. 13 the measured velocity and the velocity according to Eq. (2) are shown for the crest and trough at different depths in two waves. The correspondence is rather good between the two velocities in most cases, while, as shown, the theories do not fit measured values well in all situations.

8.1 Evaluation of wave theories from the experiments

The desire to use a wave theory for the calculation of wave forces led to an investigation of the existing theories. In this respect we have compared wave form, course of velocity and acceleration as well as maximum and minimum values. The result of this investigation is shown in Tables 2 and 3. It was found that the fifth order theory was the best theory to be used, except for shallow water where the cnoidal theory is preferable.

8.2 Wave profile

In Fig. 14a,b the theoretical and experimental wave profiles are compared for transitional and shallow water. For both cases only the first order theory gives a poor fit for the presented theories. In the second case, realistic solutions from only the fifth order and cnoidal theories were obtained.

8.3 Horizontal particle velocity

In Fig. 15a,b the horizontal particle velocities at 0.10 and 0.20 m below the still water level are shown for the different theories and experiments. For transitional water, fifth and second order theories give the best results for the wave crest. In the trough, the first order wave theory gives the best approximation for the observed minimum velocity. It was also observed that the theories overpredict the velocity above the still water level for waves on shallow water. An example is shown in Fig. 13b where a decrease, instead of an exponential increase, for the velocity was found.

8.4 Horizontal particle acceleration

In Fig. 16a,b the horizontal accelerations at the same depths are shown. From Fig. 16a it is seen that second and fifth order theories are closest for both the crest and the trough situation. From Fig. 16b it is seen that the cnoidal theory overpredicts the acceleration, especially near the still water level. As described above, the acceleration was obtained from the measured velocity by calculating the numerical derivative, thus the criteria of closeness to wave profile and velocity was given greater weight.

9. EVALUATION OF C_D AND C_M COEFFICIENTS

For the different waves, C_M and C_D coefficients were evaluated using the measured velocities and accelerations and using the different wave theories. The coefficients were evaluated as functions of wave phase. An example is shown in Fig. 17.

10. CONCLUSIONS

The inexpensive technique for measuring instantaneous velocities with a micro-propeller, was found to work well for oscillating flow.

The performed experiments have shown that it is important to use a relevant wave theory for evaluating the C_D and C_M coefficients in the Morison equation, and that coefficient values obtained from experiments using one wave theory cannot be used when designing with the help of another wave theory.

Therefore it is recommended to use the Le Méhauté diagram, Fig. 3, for classifying the wave. If a shallow wave are to be studied, the cnoidal theory should be used. Otherwise the fifth order theory is recommended. Also if only the order of magnitude is needed, the linear theory can be used to obtain fairly good results.

REFERENCES

1.   Morison, J.R., O´Brien, M.P., Johnson, J.W. and Schaaf, S.A.: The force exerted by surface waves on piles. 1950. Petroleum Trans. 189, 149-154.

2.   Sarpkaya, T.: Vortex shedding and resistance in harmonic flow about smooth and rough circular cylinders at high Reynolds numbers. Naval Postgraduate School. 1976. Technical Report No NPS-59SL76021, Monterey, CA.

3.   Bishop, J.R.: R.M.S. Force coefficients derived from Christchurch Bay wave force data. 1979. National Maritime Institute Report NMI-R-62.

4.   Bearman, P.W., Chaplin, J.R., Graham, J.M.R., Kostense, J.K., Hall, P.F. and Klopman, G.: The loading on a cylinder in post-critical flow beneath periodic and random waves. 1985. Proc. 4th Int. Conf. on the Behaviour of Off-Shore Structures, Delft, July, 1985.

5.   Moberg, G.: Wave forces and pressure distribution on a vertical cylinder in shallow water. (Tentative title). 1986.
Thesis, Department of Hydraulics, Chalmers University of Technology, Gothenburg, Sweden.

6.   Nilsson, K., Nilsson, A.: An experimental and theoretical study of the velocity distribution around a vertical cylinder. 1985. MSc Thesis, In Swedish, Department of Hydraulics, Chalmers University of Technology, Gothenburg, Sweden.

7.   Basco, D., Svendsen, I., Christensen, J.: Measurement with a bi-directional micro-propeller current meter. 1982. Progress report NO.57, ISVA, Techn. Univ. Denmark.

8.   Wiuff, R.: Experiments on the surface buoyant jet. 1977. Series Paper No 16, ISVA, Techn. Univ., Denmark.

Table 1 Total error in velocity measurement due to propeller (constant $\pm$ 2 mm/s) and velocity meter (D/A $\pm$ 1.5 bit). Constant velocity.

Velocity (m/s)	D/A %	Propeller %	Total %
1.00	$\pm$2.5	$\pm$0.2	$\pm$2.7
0.50	$\pm$1.3	$\pm$0.4	$\pm$1.7
0.25	$\pm$0.6	$\pm$0.8	$\pm$1.4
0.10	$\pm$0.3	$\pm$2.0	$\pm$2.3
0.04	$\pm$0.1	$\pm$5.0	$\pm$5.1

Table 2 Ranking of theories according to closeness to waveprofile, horizontal velocity and horizontal acceleration

Wave characteristics				CLOSENESS TO EXPERIMENTAL RESULTS		
				wave profile	Horizontal part velocity	Horizontal part acceleration
H m	T s	d m	L m	1234	1234	1234
0.186	2.50	0.8	6.54	DC,BA	DB,AC	DB,AC
0.082	1.70	0.8	3.90	DCB,A	DBA,C	D,BA,C
0.223	4.20	0.7	10.70	EDA	DE,A	ED,E
0.146	3.36	0.8	9.15	DCBA	DBAC	BDAC
0.118	4.20	0.8	11.64	DCBA	DBCA	BD,CA
0.038	4.20	0.8	11.43	DCB,A	DBA,C	DB,AC
0.048	3.36	0.8	8.98	DCB,A	DBA,C	DB,AC
0.062	2.53	0.8	6.51	DCB,A	DBA,C	DBA,C
0.080	0.86	0.8	1.21	DCB,A	BA,CD	BADC
0.190	1.70	0.8	3.98	DCB,A	DABC	DBAC
0.198	1.30	0.8	2.67	DBC,A	BA,DC	DBAC
0.274	3.36	0.7	9.00	ED,A	DE,A	EDA
0.350	2.53	0.7	6.52	ED,A	ED,A	DEA
0.334	4.20	0.7	12.02	E	E	E

A. First order theory
B. Second order theory
C. Third order theory

D. Fifth order theory
E. Cnoidal wave theory

Table 3 Mean percentage deviation from measured maximum velocity for waves included in the experimental program. Only values from the still water level and further down is incorporated.

Mean percentage deviation					
Fifth order		First order		Cnoidal theory	
crest	trough	crest	trough	crest	trough
-13	-36	-28	-21	-14	-46

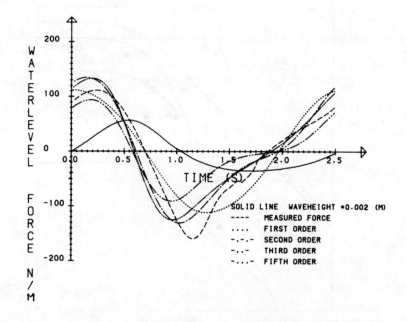

FORCE LE=-0.10 M

Fig. 1 Force against a unit length of the vertical
cylinder as a function of time for constant
C_M = 1.9 and C_D = 1.0. Wave profile in Fig. 14 a.

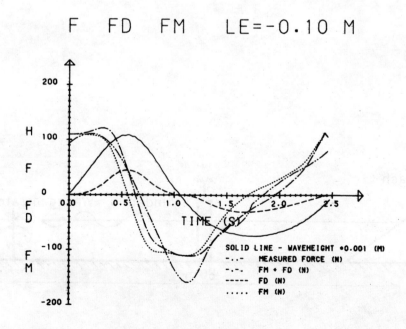

F FD FM LE=-0.10 M

Fig. 2 Measured force and force calculated with
C_M = 1.4 and C_D = 1.3 and real measured
velocities and accelerations. Wave profile
in Fig. 14 a.

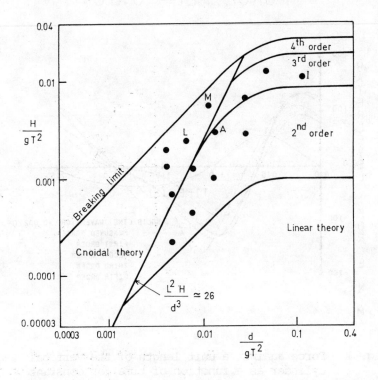

Fig. 3   Waves included in the experimental programme
          plotted in a Le Méhauté diagram.
          A.   Figures 1, 2, 9, 11, 12, 13a, 14a. 15a,
                16a and 17a
          L.   Figure 13b
          M.   Figures 14b, 15b and 16b

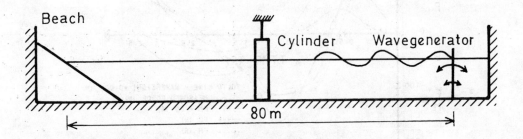

Fig. 4   Elevation of the wave tank with beach, ex-
          perimental cylinder and wave generator.
          (Not to scale).

Experimental arrangement

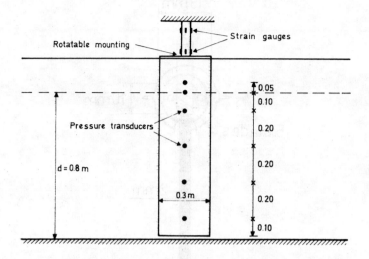

Fig. 5   The cylinder with its probe.

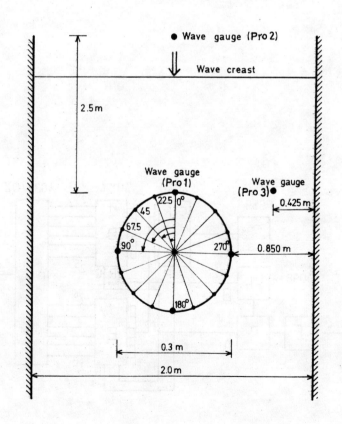

Fig. 6   Rotation of cylinder and positions
of wave gauges.

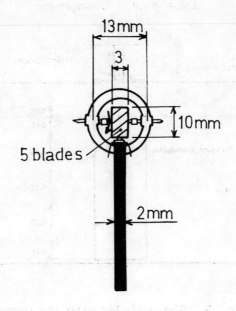

Fig. 7   Kent micro-propeller

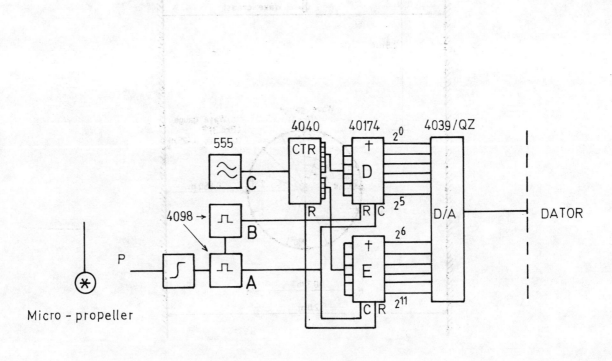

Fig. 8   Scheme of the velocity meter.

WAVE PROFILE-VELOCITY

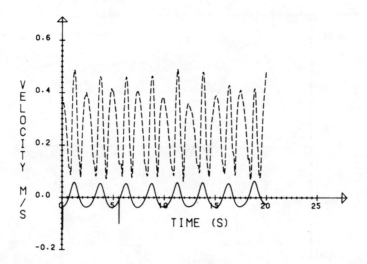

Fig. 9   Measured velocity at 0.10 m below still water level.

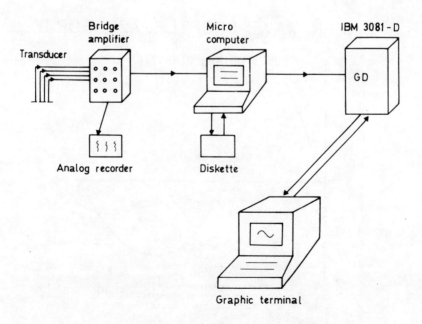

Fig. 10 Diagram of the data processing system.

MEAN VELOCITY AND HEIGHT

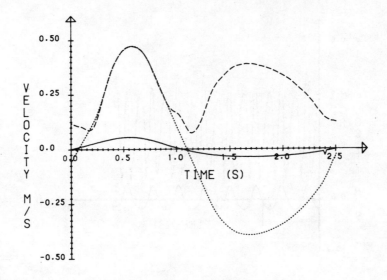

Fig. 11 Averaged water velocity at 0.10 m below
still water level.
--- measured and averaged over six wave periods.
··· corrected for direction and smoothed
around the zero crossing.

MEAN WAVE PROFILE AND FORCE

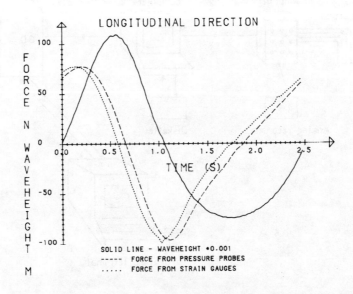

Fig. 12 Comparison between measured force and
integrated pressure.

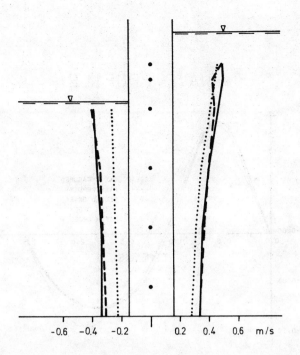

Fig. 13a Maximum horizontal particle velocity for
the crest and trough. Measured velocity,
velocity calculated from stagnation
pressure and from fifth order theory.

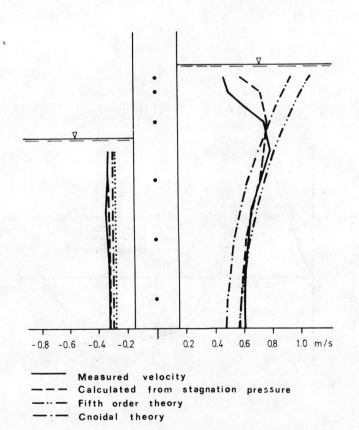

Fig. 13b Maximum horizontal particle velocity for
the crest and trough. Measured velocity,
velocity calculated from stagnation
pressure and from fifth and cnoidal theories.

WAVE PROFILE

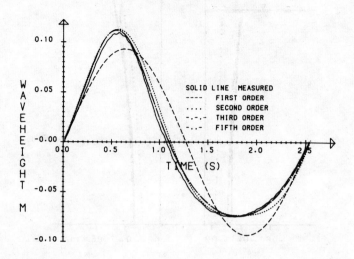

Fig. 14a Wave profile. Transitional depth.
Wave height 0.186 m. Wave period T = 2.50 s.

WAVE PROFILE

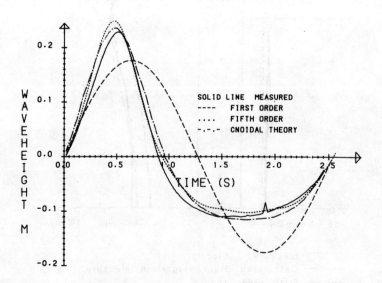

Fig. 14b Wave profile. Shallow water. Wave height 0.350 m.
Wave period T = 2.53 s.

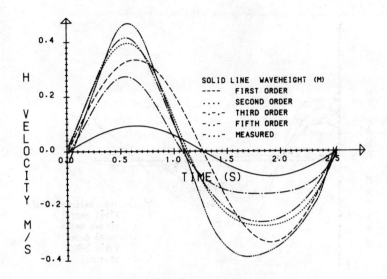

Fig. 15a Velocities at 0.10 m below the still water
level. Calculated and according to theories.

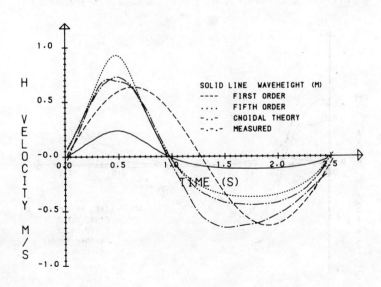

Fig. 15b Velocities at 0.20 m below the still water
level. Calculated and according to theories.

ACCELERATION   LE=-0.10 M

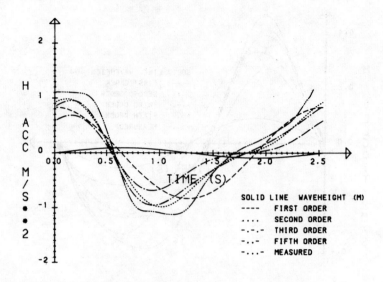

Fig. 16a Accelerations at 0.10 m below the still water
level. Calculated and according to theories.

ACCELERATION   LE=-0.20 M

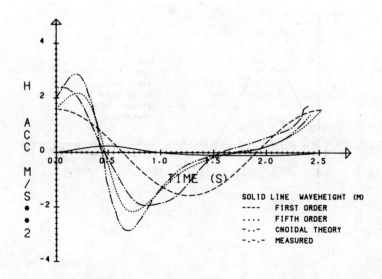

Fig. 16b Accelerations at 0.20 m below the stillwater
level. Calculated and according to theories.

FORCE COEFFICIENTS LE=-0.10 M

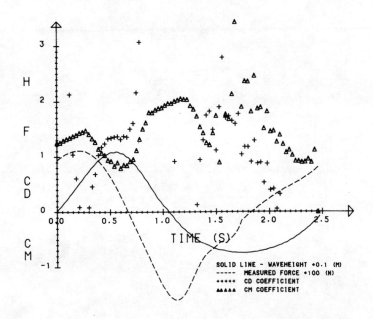

Fig. 17 Force coefficients at 0.10 m below the stillwater level, calculated.

MEASUREMENT OF WATER PARTICLE VELOCITIES USING PERFORATED BALL PROBES

by

J R Bishop   B.Eng., C.Eng., M.I.Mech.E.

and

J C Shipway   B.Sc.Eng.(Hons)

British Maritime Technology Ltd.
Hythe
Southampton.

SUMMARY:

The perforated ball probe was first used to measure atmospheric
turbulence. A new version of the instrument was developed to
measure water particle motion in wave force experiments at the
Christchurch Bay Tower. A special method of data processing has
been developed where both the particle velocity and accelerations
are derived, using both drag and inertia force calibrations for
the ball. Instrument zeros are determined from readings due to
tidal current when the wave activity is low. The probe has produced
very consistent results in extensive analyses of the wave force data.

Held at Imperial College of Science and Technology, London. Organised and sponsored by BHRA, The
Fluid Engineering Centre. Co-sponsored by the American Society of Civil Engineers and the
International Association for Hydraulic Research.
©BHRA, The Fluid Engineering Centre, Cranfield, Bedford MK43 0AJ, England 1986.

1. INTRODUCTION:

The first known use of a perforated ball instrument was in 1939 by L F G Simmons of the Aerodynamics Department, National Physical Laboratory. The purpose then was to measure atmospheric gusts and turbulence, and data were recorded at a number of altitudes, with the instruments suspended from a barrage balloon. The specific interest was the dispersion of poison gas. Further work on the instrument was carried out by Rosenbrock and Tagg (1), by R I Harris (see Morison (2)) and by Hopley and Tunstall (3). The attractive feature of the perforated ball device was its fast response and its ability to deal with omni-directional flows - both of these are important in unsteady flow.

The Christchurch Bay Tower was designed to study wave forces in real sea conditions at a scale giving Reynolds numbers in the post-critical plateau. The 3 dimensional nature of sea waves posed the same problems as atmospheric turbulence in their unsteady and omni-directional nature. Fortunately Ray Gould and John Wills of the National Physical Laboratory were acquainted with the "aerodynamic" perforated ball instrument and made us aware of the potential for the wave force studies. We quickly built and tested a water version and perforated ball probes were installed for the first experiments at the Tower, alongside electromagnetic instruments that had then had a limited use in sea conditions. It was clear at that time that the measurement of the wave particle motion was the most difficult of all the measurements at the Tower, hence the use of the two types of instrument in a "belt and braces" fashion. We soon found that the data from the electromagnetics was very poor so we took them off the Tower and concentrated on developing the perforated ball probe for the wave force project.

2. CHARACTERISTICS OF THE PERFORATED BALL:

The basic principle of the instrument is that the drag force on the ball depends on the square of the flow velocity. The use of a sphere rather than any other shape had an obvious logic when dealing with omni-directional flow. An unperforated ball would shed vortices in a frequency range which would make it useless as a measuring instrument (e.g. at a frequency of about 2Hz for a 100mm diameter ball in a flow velocity of 1 m/s). However the perforated surface, allowing flow in and out of the holes, changes the vortex shedding dramatically; major vortices are not shed, although there are small scale vortices shed at the holes. These latter are at a high frequency and their effects are not additive over the complete ball. Consequently the ball behaves as a simple "drag device" and has a constant drag coefficient over a very large Reynolds number range. An illustration of the ball now being used is shown in fig.1.

In oscillating flow, fluid inertia forces are also present and need to be taken into account if the detailed time history of the velocity is needed. If the fluid acceleration is also needed it is even more important to allow for the inertia force. An illustration of the error caused by neglect of the inertia force is shown in fig.2. This is a theoretical calculation based on known drag and inertia force coefficients and using a sinusoidal flow velocity with an amplitude and period which is reasonably representative of frequently occurring waves at the Tower. It is seen that:-

    i)   neglect of the inertia force causes negligible error in the <u>velocity peak</u>

   ii)   neglect of the inertia force causes a phase shift and distortion of the velocity wave form

  iii)   neglect of the inertia force causes substantial errors in the flow acceleration.

Because of the foregoing, two types of data processing are adopted

    a)   a simple drag-based derivation of velocity when only peak values are needed.

    b)   a drag and inertia based derivation of velocity and acceleration if velocity time histories or acceleration are needed.

The combined drag/inertia derivation is accomplished in an iterative process at suitable time steps. The calculation is of the form

$$F = Au^2 + B\frac{\Delta u}{\Delta t}$$

where F is the measured force on the ball

A contains the drag coefficient

B   "   " inertia   "

Δu is the velocity change in a time interval Δt.

In the first step Δu is calculated using a u value equal to that at the start of the interval. In subsequent steps the mean velocity for the interval is used. The process works very well and is easy to use as a routine with modern computers. Having outlined the basis of the data processing routine, it should be mentioned that we have had a number of queries regarding the soundness of it. For instance, when an experiment is studying wave forces on cylinders, with investigation of the adequecy of the Morison formulation, is it valid to assume the Morison equation for the perforated ball? The answer is that there are only a very few instruments that would claim to produce sensible results in waves and our experience points to the perforated ball as being amongst the most suitable. An instrument that responds to velocity alone will always have problems when the velocity signal is differentiated to obtain acceleration. The opposite procedure, of measuring acceleration and integrating to obtain velocity is a better system for oscillating flow provided the mean value can be obtained by some method. However, in sea waves a device that would have dominant inertia forces (i.e. responding to the fluid acceleration) would be excessively large, so this approach is not feasible. But it is clear that the presence of significant inertia forces, as with the perforated ball instrument, is advantageous, and leads to more stable results for both velocity and acceleration than would otherwise be possible.

A further question has been concerned with the square law characteristic. This leads to two problems.

a)   insensitivity at low velocity readings

b)   a need for data processing that is more complicated than is the case with a linear instrument.

The answer to b) is that this is already accommodated in the processing routine described above. Regarding a) this is certainly a disadvantage if accurate data are needed near the bottom of the measurement range. It becomes necessary to choose the instrument range with care to avoid this problem. The range can easily be adjusted by changing the ball diameter or the cantilever cross-section. A further aspect is that the square law gives higher sensitivity at the higher velocity readings, where it is usually more important. An example is in our wave force project. There are obviously many engineering situations where important parameters depend on velocity squared and it can be concluded that a square law instrument has advantages.

With a square-law instrument it is imperative to have well established zeros, particularly if data are needed low down in the velocity range. In the laboratory one can turn everything off to take zeros in still conditions. But in the real environment such measures are not possible, and we have developed special procedures for zero setting - these will be described later.

3.   CALIBRATION:

a)   In steady flow

A large number of calibrations have been carried out in towing tanks to determine Cd for the ball. At an early stage we got somewhat different results in different facilities, but this turned out to be due to the test

techniques, and we now get very consistent calibrations. We used to
calibrate all the perforated balls produced, but found that the Cd
variations from one ball to another were small enough to be statistically
insignificant. So we now only calibrate samples, if for instance the
manufacturing technique has been altered. A calibration is shown in fig.3.
The line corresponds to a Cd of 0.73.

b) In oscillating flow

The perforated ball has been tested in Sarpkaya's U tube oscillator, over
a Keulegan Carpenter number range of 4 to 40. The purpose was to determine
the inertia coefficient, Cm, and to see if Cd was consistent with steady
flow values. The results, fig.4, show that Cd and Cm are nearly constant
for Kc values down to about 10, which corresponds roughly to a waveheight
of 0.6m. If currents are present, these lead to a higher Kc value, with
consequent lowering of the waveheight below which the coefficients begin to
vary.

4.   SOME SAMPLE RESULTS:

All of the results presented will be from the Christchurch Bay project.

a) Measurement of tidal current and zero setting

As mentioned earlier the real sea never offers a quiet environment and so
different methods from those used in the laboratory for setting zeros have
to be adopted. Some special tests were set up to investigate the effects
of tidal current and develop a method of determining the zeros. This
involved a one-day test in conditions with very low wave activity (4). The
tidal current was measured by two rotor type instruments, which are accurate
at low current speeds (but unsuitable for measurements in waves). Data from
all the Tower sensors were recorded. Fig.5 shows indicated current speeds
and corresponding directions from one rotor instrument together with
perforated ball results. It was found that the tidal current values, from
the rotor instruments, possessed symmetry of the readings in the two halves
of the tidal cycle, and this property has been used for setting the
perforated ball zeros. The PBP data in fig.5. are the results obtained
after setting the zeros in this manner. The figures also show readings from
the cylinder force sensors on the small and large columns using chosen Cd
values. The good agreement between all sensors confirms that the instrument
sensitivities are good, particularly as the current-induced forces are very
low indeed. Figs. 6 and 7 show samples of current vectors in adjacent parts
of the tidal cycle. The fact that these have all swung round with the tide
in a consistent manner is a further indication of good results at low levels
of force.

b) Comparison of PBP velocity data with predictions from wave theory

Unidirectional linear random wave theory has been used to transform the
measured surface elevation data into time histories of particle velocity
and acceleration. The results have been compared with corresponding data
from the PBP. Fig.8 shows a sample result. The agreement is seen to be
good. It is intriguing how the two sources of data agree in waves which are
very complicated. We found that in other work where attempts were made to
fit results for discrete waves the results were very disappointing; the
real sea is sufficiently random to make Fourier-based treatments vital - as
in the linear random wave theory.

Spectra of the perforated ball velocity and acceleration are compared with
wave theory predictions in figs. 9 and 10. These illustrate a general
finding that measurement and prediction agree well at the spectral peak but
the theory over-predicts at higher frequencies. The over-prediction is
more marked with particle acceleration than with velocity. The justification
for saying that the PBP data are better than the surface elevation-based

data is open to question. At this stage there is not an irrefutable answer
to this, but there are many pointers from analyses carried out using
extensive amounts of data. It is difficult to utilise more convincing
methods because there is no reference instrument that can be used in sea
waves, with their random orbital flow. At this stage the best laboratory-
based confidence in the perforated ball comes from the U tube oscillator
tests. One reason for this is that the mean velocity at the test section
is obtained from positional information on the height of the water in the
limbs of the U tube, which is accurate. But this type of test does not
represent the orbital nature of wave flows.

c) Application of PBP data to wave force analyses

The particle velocity and acceleration data from perforated ball
instruments has been used extensively in the analyses of wave forces on the
two columns of the Christchurch Bay Tower. The large column has inertia
dominated forces, so particle acceleration is relevant, whilst the small
column has comparable levels of drag and inertia forces and so both
velocity and acceleration are important. Fig.11 shows a measured force
spectrum for the large column together with a PB based prediction using a
constant Cm value. It is the shape of the spectrum which is of interest
here because the general level can be adjusted by the Cm value. The shapes
do agree well. Similar data for the small column are shown in fig.12.
Again there is good agreement between measurement and prediction,
particularly when it is remembered that differences may be expected due to
vortex shedding from the cylinder.

We have found that the force spectra for the columns are less-well
predicted when using particle kinematics from wave theory. This is a
tentative indication that measured data from the perforated ball is of high
quality. This information is rather "circumstantial" because it is the
forces on the cylinders which is the subject of study. Nevertheless, the
good agreement is unlikely to arise from the independent sources of data
departing equally from the truth to give good observed agreement.

5. CONCLUDING REMARKS:

The perforated ball instrument has been used extensively at the Christchurch Bay
Tower to measure wave particle kinematics, with very encouraging results. The
omni-directional aspects of the wave orbital flow creates severe measurement problems
and there are only very few types of instruments that claim to be capable of producing
sensible results in these conditions. The data processing for the perforated ball
converts the measured force on the ball into particle velocity and acceleration using
calibrated drag and inertia force coefficients. Special methods have been developed
for setting the instrument zeros in the sea environment. These utilise known
behaviour in conditions of low wave activity where current effects are predominant.

Considerable effort has been put into laboratory calibrations for Cd and Cm and very
consistent results have been found. However, when going to the orbital flow
conditions of waves there is no "standard" instrument that can be used to evaluate
the perforated ball. One problem is that the mounting arms of any instrument will
give rise to wakes which will contaminate the flow in the test region, by being swept
back and forth in the oscillating flow. Nevertheless the perforated ball has
produced very good results in studies of wave forces and there is a lot of
circumstantial evidence that the data are reliable and accurate enough for such
purposes.

ACKNOWLEDGEMENTS

The Christchurch Bay project and the development of the perforated ball
instrument for the project has been funded by the Department of Energy.

REFERENCES

1   ROSENBROCK H M and TAGG J R,   1951.
     Wind and Gust-measuring Instruments developed for a Wind Power Survey.
     The Institution of Electrical Engineers, Symposium on Electrical
     Meteorological Instruments,   January 1951.

2   MORISON J G,   1968.
     The Development of a Miniature Gust Anemometer. Paper 30, Symposium on
     Wind Effects on Buildings and Structures. Loughborough University of
     Technology,   April 1968.

3   HOPLEY C E and TUNSTALL M J. (CERL, Leatherhead).
     A Fast Response Anemometer for Measuring the Turbulence Characteristics
     of the Natural Wind.
     Journal of Physics E: Scientific Instruments, 1971 Vol 4.

4   SHIPWAY J C,   1984.
     An Investigation into Tidal Current, Current Induced Loadings and
     Zero Readings at the Christchurch Bay Tower.
     NMI Report R181, also OT-O-82101 Part 4.

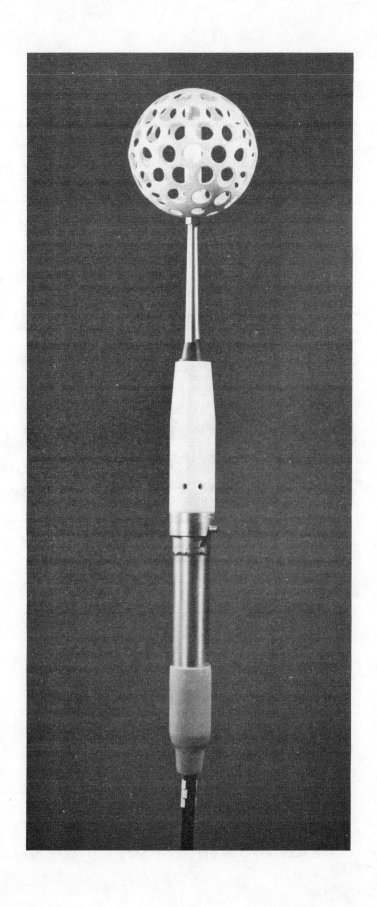

Fig. 1   B.M.T. Perforated Ball Probe used at Christchurch Bay Tower

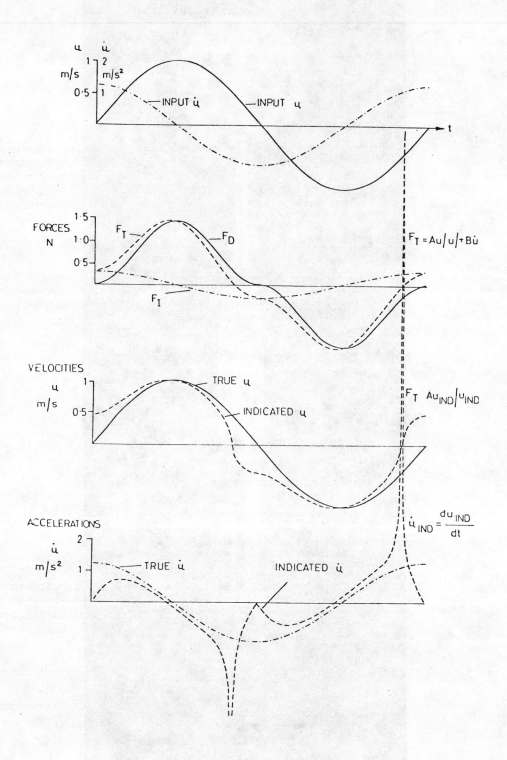

Fig.2   Velocity and Acceleration errors in Simple ($F=k_u/_u$) Perforated Ball Data
Analysis

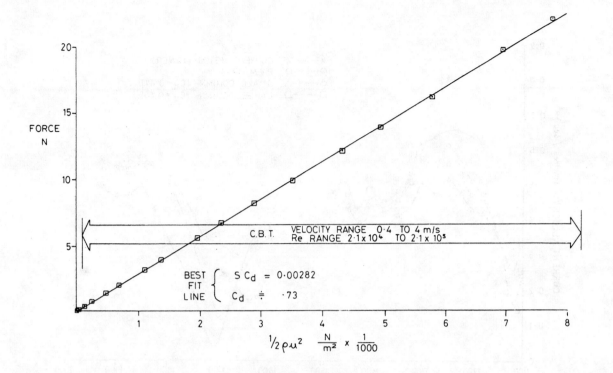

Fig.3   70mm Ball Steady Flow Calibration

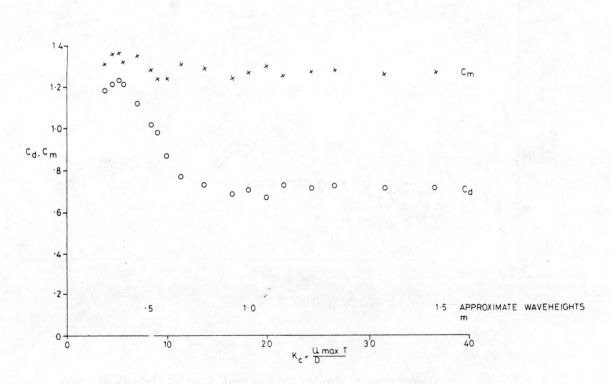

Fig.4   'U Tube' Calibration of 70mm Ball by Sarpkava

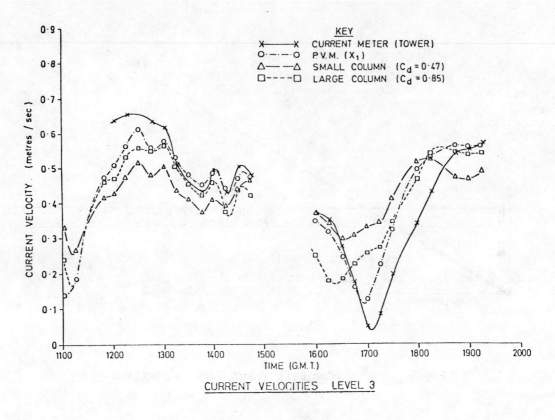

CURRENT VELOCITIES LEVEL 3

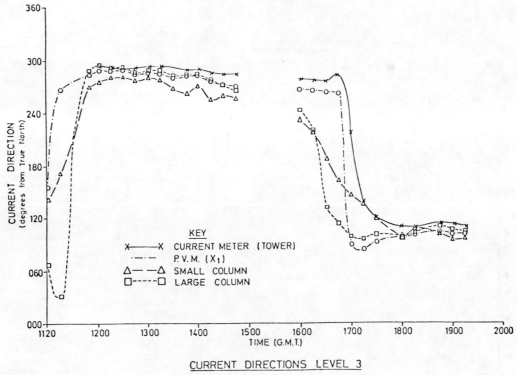

CURRENT DIRECTIONS LEVEL 3

Fig.5   Comparison of Indicated Current from Different Sources

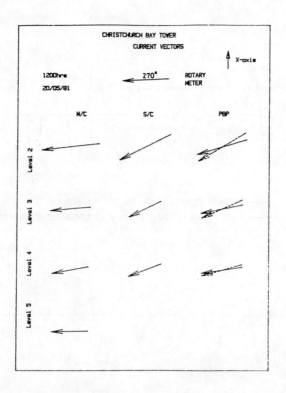

Fig.6   Current Vectors 1200Hrs (20/05/81) Derived From Force Sleeves &
         Perforated Ball Results

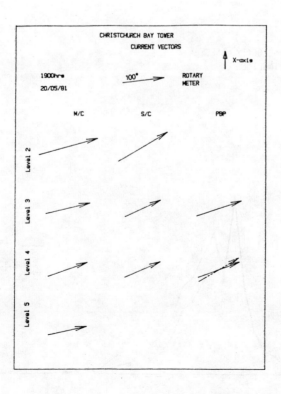

Fig.7   Current Vectors 1900Hrs (20/05/81) Derived From Force Sleeves &
         Perforated Ball Results

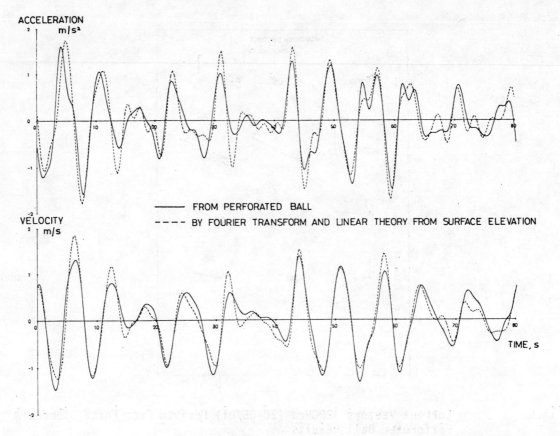

Fig.8   Horizontal Particle Velocity and Acceleration with 0.5H$_z$ Filter

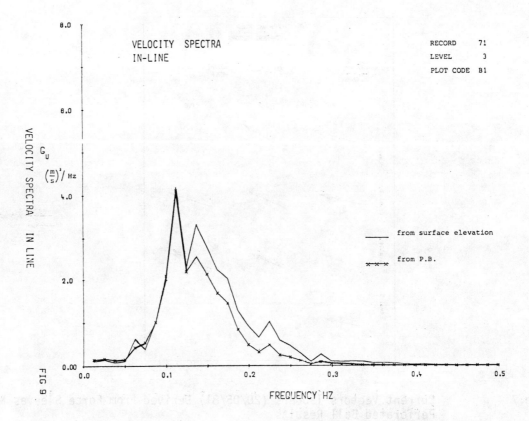

Fig.9   Velocity Spectra in-Line

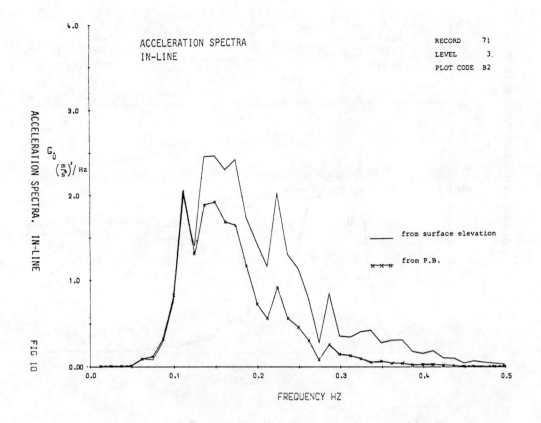

Fig.10   Acceleration Spectra in-Line

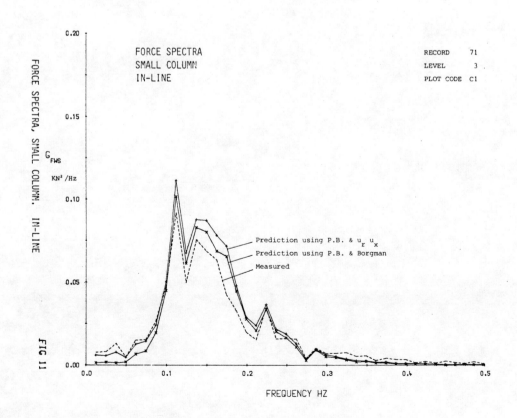

Fig.11   Force Spectra small column in-Line

119

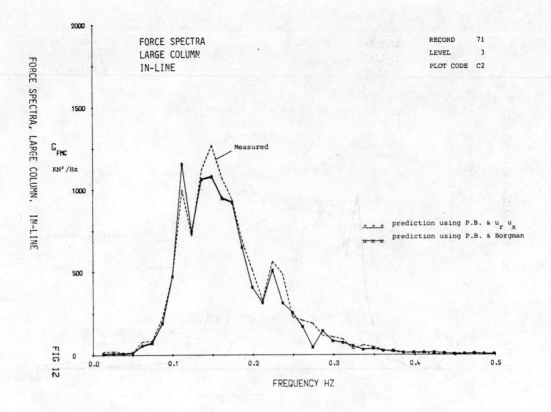

Fig.12   Force Spectra Large Column in-Line

SPECIAL PORE WATER PRESSURE MEASURING SYSTEM
INSTALLED IN THE SEABED FOR THE CONSTRUCTION
OF THE NEW OUTER HARBOUR AT ZEEBRUGGE, BELGIUM.

R. Carpentier and W. Verdonck

Belgian Geotechnical Institute
B 9710 Zwijnaarde
Belgium

Summary

A new outer harbour is under construction at Zeebrugge, situated on the Belgian coast. The breakwaters are of the rubble mound type and are constructed over large lengths on a dumped sand foundation.

In situ pore water pressure measurements have been made by means of electronic pressure measuring devices in order to gather information about the influence of tides and waves on the pore water pressures. Pressure transducers were built at seven different levels into a cone shaped shaft resulting in a multiple piezometer probe. This multiple piezometer probe was jacked into the sandy and clayey seabed. The measuring signals from the pressure transducers were transmitted to a real time computer by a sophisticated telemetry system.

Held at Imperial College of Science and Technology, London. Organised and sponsored by BHRA, The Fluid Engineering Centre. Co-sponsored by the American Society of Civil Engineers and the International Association for Hydraulic Research.

1. INTRODUCTION

A general lay-out of the new outer harbour under construction at Zeebrugge (Belgium) is given in Fig. 1. The main breakwaters ABC and FED protrude about 3.5 km outside the coast line and surround the old Leopold-II-dam.

The main breakwaters are of the rubble mound type with the crest at level Z+10 or Z+12.5 (Z being a reference level which is 0,08 m above the MLLWS condition at Zeebrugge). A typical cross-section of the north-eastern breakwater (α - α in Fig. 1) is given in Fig. 2.

Stability of the breakwaters is strongly related to the shear strength of the used building materials and the foundation layers. To know the shear strength of the foundation layers, the pore water pressures in these layers have to be known. As little is known about the pore pressure distribution and pore pressure changes beneath the sea-bottom due to tides and waves, it was decided to gather some information by field measurements.

The whole test set-up to measure the pore pressure distribution in the foundation layers and to measure the response of the pore pressures to the tide and wave action has been realized in co-operation between different partners : the Coastal Service of the Ministry of Public Works, the contractor Tijdelijke Vereniging Zeebouw-Zeezand, the consultant Haecon-Zeebrugge, the Belgian Geotechnical Institute and the Laboratory for Automatic Control of the State University of Ghent.

The location of the field measurements was chosen near the spot of a previously performed CPT test nr. NOD-GZ/D19 (Fig. 3) (ISSMFE, 1977). Based on the results of this test, the levels of the pore pressure measuring devices were selected. From Fig. 3 it can be deduced that from the sea-bottom, situated at the level Z-5, one has successively a dumped sand layer up to about Z-12, a stiff clay layer up to about Z-13, a clayey sand layer up to about Z-14.5 and a stiff clay layer continuing up to much greater depths than could be deduced from the general site investigation. The levels where the pore pressure measuring devices should be situated were fixed at the level of the sea-bottom and respectively 2, 4, 6, 8, 10 an 13 m under the sea-bottom, i.e. at the levels Z-5, Z-7, Z-9, Z-11, Z-13, Z-15 and Z-18.

At the test site, water depth and wave conditions can be characterized by the following data : (i) MLWS at level Z+0.32 ; (ii) MHWS at level Z+4.62 ; (iii) a water depth of about 5.5 m at MLWS and (iv) a significant wave height up to 6.1 m.

2. THE MEASURING SYSTEM

2.1 General lay-out

The measuring system (Fig. 4) consisted of four main components :

(i) a multiple piezometer probe (MPP) jacked into the seabed and having pore pressure measuring devices at seven different levels starting from the sea-bottom to a maximum depth of 13 m underneath the sea-bottom.

(ii) conditioning electronics connected to the multiple piezometer probe by a cable bundle. This conditioning electronics were installed on a platform near the multiple piezometer probe. A transmitter was also installed on the platform.

(iii) a complex telemetry system to send the data from the test site to a real time computer. This telemetry system consisted of a UHF link to bring the signals to the coast and subsequently a telephone link to bring the signals to the computer.

(iv) a real time computer to sample the signals and to store the sampled data on a suitable storage medium.

2.2 The multiple piezometer probe (MPP) (Fig. 5)

(a) Pressure transducers

Selection of the measuring devices depends on the actual situation : i.e. the objectives of the measuring campaign, the soil characteristics, the location and so on (Vaughan, 1973). In the case of measuring the influence of tides and waves on the pore pressures, one has to chose a measuring device with a very fast response. Moreover, it will be necessary to take a vast number of readings in order to have sufficient data to study the tides and waves influence. So manual data collection will be impossible. Automatic data collection can be done in the most convenient way when the input signals are electric signals. These two demands, fast response and electric signals lead to electronic transducers as measuring devices.

Among the different types of electric and electronic pressure transducers, strain gage transducers were chosen. They have the advantages (i) to have a very small volume displacement and hence a fast response, (ii) to consume very few energy and (iii) not to contain any moving part.

Transducers of the sealed gage type were selected to avoid problems with the reference pressure. The body of the selected transducers was made of titanium to withstand the severe marine environment. They had an integral sealed cable to eliminate contact problems that can occur in the case of fitting the cable by a connector. The strain gages of the integrated silicon type were arranged to get a full bridge. Temperature compensation was built into the transducers. The used transducers had a full range of 0.5 MPa (50 m water column), a full scale output of 100 mV and a non-linearity and hysteresis of 0.1 % BSL.

(b) Concept of the pore pressure measuring devices

Pore pressure measuring devices or piezometers can be constructed by building the pressure transducers into a convenient housing. The actual lay-out of the housing depends on the way the piezometers have to be installed into the ground. One can distinguish two major types : piezometers to be installed in bore holes or push-in piezometers.

When piezometers have to be installed at several levels, the piezometers can be installed in different bore holes. Installation of several piezometers in the same bore hole is also possible but it leads to severe sealing problems. On the other hand, installation in different bore holes results in a series of piezometers that are not exactly at the same vertical. Another problem with several piezometers installed in bore holes is the fact that it is hard to know the exact level differences.

With the push-in type piezometers the sealing problem is less severe. When installing several push-in piezometers, the problems of not being on the same vertical and the uncertainty about the level differences remain. So it seemed to be desirable to build a rigid multiple piezometer probe (MPP) with all the pore pressure measuring devices brought together (Massarsch, 1975). This solution had the following advantages : (i) all the piezometers are exactly located on the same vertical, (ii) the distance between the piezometers is perfectly known and (iii) all the piezometers can be installed during one operation. A possible disadvantage of the chosen solution is thought to be that the MPP has to be jacked into the sea-bottom and thus can constitute a kind of "hard point" which possibly can influence the actual pore pressures to some degree.

Pore pressure transducers have been installed at seven levels of the MPP. The seven levels are indicated on Fig.5. For reliability reasons, the measuring devices have been doubled at all the levels except at the two deepest ones. Here doubling was impossible due to space limitation.

Each pressure transducer is mounted in a housing (Fig. 5). The volume in front of the sensitive element of the transducer is connected to the outside of the MPP by a number of inclined channels. At the end of each channel, an inlet cone has been provided. Some preceding experiments were done to fix the inclination and the diameter of the channels and the form of the inlet cone. A very rough filter of perforated nylon disks was foreseen in front of the sensitive element of the transducer.

At the time of assembling the MPP, the piezometer channels have been filled with de-aired water. At the outside of the MPP, the channels were covered and sealed. These sealings have been removed during installation after immersion. In this way a good saturation of the piezometers could be obtained.

(c) Construction of the MPP

The MPP itself consisted of six steel pipes linked by stainless steel nipples (Fig. 5). The length of the steel pipes was about 2 m, except for the last one whose length was about 3 m. This resulted in a total length of the MPP of more than 13 m. At the nipples, the diameter of the MPP decreased downwards. In this way the MPP seemed to have a taper needle like shape. This particular shape has been chosen to prevent a direct contact between the different piezometer levels along the shaft of the MPP after installation into the seabed. The diameter of the MPP varied from 60 mm at the tip to 193 mm at the top. At the top a 250 mm diameter flange was foreseen for installation purposes. The diameters have been chosen based on construction, installation and operation considerations. From the construction point of view, the diameter had to be great enough to get the room for the transducers and to have a sufficiently stiff MPP to prevent buckling during installation. On the other hand, the diameter had to be small enough to limit the needed jacking force and not to disturb the soil too much around the probe by compaction.

Each pore pressure measuring device had its own cable. At the end of each cable, a connector was fitted to plug the device to the conditioning unit. At the top of the MPP, the cables were brought together into a cable bundle.

The MPP was fully assembled in the laboratory and was transported by truck in a suitable steel tube of 800 mm diameter and 12 m length over about 60 km from the laboratory to the harbour site. During transportation of the MPP, the cable ends with the connectors were fixed in a sealed box to eliminate damage and corrosion.

2.2 The measuring platform and equipment (Fig. 6)

The conditioning electronics for the pressure transducers of the piezometer probe, a transmitter and the related equipment had to be installed at the test site. This equipment could be installed on the sea-bottom near the MPP or could be placed on a platform above the high water level. The latter solution was chosen because it is easier to perform service on the equipment. Also difficult sealing, fixing and corrosion problems could be avoided.

(a) The measuring platform

The platform with all the necessary equipment was placed on a separate pile, the so-called measuring-station pile (MSP). This pile was driven into the seabed to a depth of 11 m. The platform was situated at the level Z+13 to avoid storm damage. Because of its length, the MSP was subjected to severe wind and storm attack. It was thought that the repeated horizontal loading and unloading of the pile could result in some kind of pumping action in the seabed which could influence the pore pressures in the soil layers in the immediate vicinity of the pile. So the MSP had to be kept at a sufficient distance from the MPP. This distance was fixed at 20 m.

For the MSP, a steel tube having a length of 29 m, an outer diameter of 813 mm and a wall thickness of 20 mm has been used. The platform itself has been built on two HEB sleepers. It had a floor surface of 2.26 m by 2.26 m. A railing of 1 m high surrounded the platform. A steel tube with a length of 2 m and an inner diameter somewhat larger than the outer diameter of the MSP was fitted under the platform. This steel tube slipped over the MSP. Steps were fitted to this tube and to the MSP to give the possibility to access the platform through a trap-door in the platform floor.

(b) Equipment on the platform

All the electronic equipment on the platform, including the conditioning electronics for the transducers and the transmitter of the telemetry system were housed in a hermetic box. A small continuous flow of dry nitrogen was fed to this box as an inert gas. A pressure relief valve evacuated the excess gas. In this way an inert atmosphere could be maintained in the hermetic box to protect the delicate electronics against the aggressive and corrosive sea atmosphere. The nitrogen flow, delivered by a bottle at high pressure, was controlled by a precision flow regulation system. One bottle of nitrogen did last for a period of 6 months.

Power supply for the electronic system was delivered by solar cells connected to rechargeable batteries. To guarantee an uninterrupted power supply throughout the year, photovoltaic solar cells having a peak power of 200 W and a battery pack having a total capacity of 500 Ah by 24 V have been installed. The batteries were housed in two vented boxes. The solar cells having a total surface of about 2 square meter were mounted on the railing of the platform. They were oriented to the south and had an inclination of 60 degrees in order to deliver a maximum of energy during winter. The UHF aerial of the transmitter was also fitted to the railing of the platform together with a lightning conductor.

Conditioning electronics for the transducers consisted of excitation circuits and amplifiers (Sheingold, 1980). The excitation circuits were constant current sources.

These were selected instead of constant voltage sources because errors due to resistance changes in the cables and at the connectors were minimized. For each transducer, a separate and independent current source was present in order to avoid any interaction between the different channels. This was very important in the case of failure of one or more channels. Such failures might not influence the other channels. The amplifier modules brought the 0 to 100 mV output signal of the transducers to a 0 to 2 V level. Gain and offset correction were possible. The amplifier modules were also completely independent of each other to avoid problems with the other channels when one channel would fail.

2.3. The telemetry system

The MPP was installed at about 2 km out of the shore line. The computer chosen to store and process the data was installed in a laboratory at about 60 km from the site of investigation. So a complex system had to be realized to bring the data to the computer.

A direct cable link between the electronics on the platform and the coast was hard to operate in a safe way due to the works in progress between the platform and the coast. As a direct link seemed to be inapplicable, two solutions were possible. A first possibility consisted of the installation of a data logger on the platform. Sampled data could be stored on an external medium, e.g. a data cassette. When needed, one had to go to the platform to get the stored data and bring them to the computer for processing. At this time a new data cassette could be loaded into the data logger. For tide measurements one could go to the platform and start the sampling process. This procedure however was not possible for wave analysis as the measurements were wanted during storm conditions. Unfortunately, in those circumstances it would have been impossible to go to the platform and start the sampling process. This problem could be solved by adding radio control to the data logger. In this way, start and stop the sampling process also became possible during storms.

As an alternative, a telemetry system could be installed. As the power consumption of the electronics on the platform including a transmitter could be kept at a sufficiently low level, this solution was selected. The use of directional antennae permitted to keep the necessary RF-signal power of the transmitter at only 20 mW to get a reliable data transfer. All the other electronics on the platform, namely the conditioning electronics for the transducers, the multiplexer, the A/D converter and the encoder were also designed to consume very few energy. The total energy consumption could be kept below 12 W. The telemetry system remained in operation all the time, so the output signals of the transducers were always present for sampling and storing whenever wanted.

The frequency of the carrier wave was fixed by the Post Office at 459.67 MHz. The sampled data were transmitted using the Pulse Code Modulation technique. Transmission speed was 6 kBit/s. In this way about 300 samples were transmitted per second, this means almost 20 samples per second for each channel. Each sample, a 10 bit digital value, was contained in a data frame that also included additional information for channel recognition and data integrity checking.

At the coast, a corresponding reciever was installed in the Meteo Station Shed of the Coastal Service. By this reciever, the transmitted data were transformed again into their original analog form. Now the data had still to be brought to the computer at 60 km away from the coast. Again two solutions were possible : store the signals by a data logger installed in the Meteo Shed and bring them afterwards to the computer or send them immediately to the computer. The last solution has been chosen because it made possible to have the signals all the time available in the computer room. Data acquisition could be performed by the computer itself. The advantages of this data acquisition process was decisive in the selection of the latter solution (Verdonck, 1980).

Bringing the signals to the computer had to be done by the services of the Post Office. Leased telephone lines and 1200 baud modems were offered by the Post Office. So an interface had to be built to convert the signals a second time to send them by the modem to the computer where a second modem was installed to bring the signals to the computer. The signals were multiplexed, translated to a digital form and imbedded in a data frame together with the needed channel information and control bits. About two samples per channel could be transmitted per second. Resolution again was 10 bit.

2.4 The real time computer

For final data collection and storage, a very popular 16 bit real time minicomputer could be used. As the data flow, two samples per second on each channel, was rather high and because one had to store the data from long sampling periods, a hard disk was needed to store the incoming data for later processing.

Two software packages were developed to run the data acquisition and processing system. A first package consisted of a few sampling programs. The second package contained the different processing programs. The software was written in Fortran, except for one single routine which was written in assembler. This routine had to be written in assembler because of the speed demand and because the incoming data frames had to be translated to a standard data format used in the computer.

A first sampling program was written to watch the transducer readings during installation. During the jacking operation of the MPP it was necessary to closely follow the readings of the transducers in order to stop the jacking operation when one or more readings did go beyond predetermined safety limits.

A second sampling program was written to be run at regular intervals (e.g. once a week) to check the working condition of the system. The charging state of the batteries on the platform was also checked. This could be done as the voltage of the batteries was transmitted by the telemetry system on a spare channel.

Next a sampling program for real data sampling was developed in order to study wave action. This program read the incoming data from the telemetry system, did some validity tests (e.g. parity check, overrun, lost data...) and stored all the data on disk for post-processing.

Finely a sampling program for real data sampling in order to study tide action was developed. This time the sampling process had to go on for days. It is clear that it was impossible to store all the incoming data. At a rate of two readings per second and two bytes per reading a data flow of 24 bytes per second for the 12 channels leads to a data volume of about 2 Mbytes per day. This would have consumed too much disk space and it would have disturbed too much the normal routine work performed on the real time minicomputer. So data reduction had to be done before storing the data on the disk. Therefore, a time averaging procedure was imbedded in the program. The sampling program computed the mean value on every channel for time periods of 15 minutes. Only the mean values were stored onto the disk. So an enormous disk storage reduction could be realized.

The sampling programs were supervised by a watchdog program. This watchdog program enabled the operator to schedule sampling operations several days in advance. In this way a completely unmanned and automated operation became possible. When a storm condition was forecasted, the watchdog program was informed to activate the appropriate sampling program at predetermined times when data acquisition seemed to be desirable.

The post-processing program package consisted of some Fortran written routines to list and plot the processed data. A first step in the data processing procedure was the transformation of the readings to pressures. The sampled data series $M_i(t)$ had values in the range 0 to 1023. This range corresponded to a voltage range of 0 to 2000 mV. As the gain of the amplifiers in the conditioning modules of the transducers was 20 and the offset was zero, the output voltages of the pressure transducers could be computed as :

$$x_i(t) = M_i(t) \quad \frac{2000}{1023 \times 20} \text{ mV}$$

Prior to the installation of the MPP a calibration of the transducers was performed. From this calibration, a sensitivity m_i (MPa/mV) and an offset n_i (MPa) were calculated by linear regression. Taking into account this calibration parameters, the pore pressures could be calculated as :

$$u_i(t) = m_i \cdot x_i(t) + n_i$$

With the gravitation constant being 9.81 m/s^2 and with an estimated density of 1025 kg/m^3 for the sea water, the pressures expressed in MPa could be transformed to pressures expressed in meter sea water column.

3. INSTALLATION AND REMOVE

The installation of the equipment has been performed in two stages. The MSP and the platform have been installed in the summer and the MPP in the fall of 1981. In the summer of 1983 the equipment had to be removed because the construction of the breakwaters could not be slowed down.

3.1 Installation of MSP and platform

The MSP and the platform with all the equipment, except the conditioning electronics for the pressure transducers and the transmitter of the telemetry system, have been installed in July 1981. Installation was performed from a jack-up platform. The MSP was vibrated into the seabed to a depth of about 11 m. With the sea-bottom at level Z-5, the base of the pile was situated at level Z-16 and the top at level Z+13. After installation of the MSP, the completely mounted platform itself was put over the MSP by a crane. During this operation the solar cells were covered by protection panels to prevent damage.

3.2 Installation of the MPP

The MPP and the electronic system were installed in October and November 1981. The jack-up platform used to install the MSP and platform was also used to do this job. The MPP was jacked into the sea-bottom through an opening in the jack-up platform. To push the MPP, a 1 MN hydraulic jack was used. The reaction force was delivered by two watertanks having a capacity of 56 m^3 each.

Due to the length, the weight and the slenderness of the MPP and taking into account the required penetration force, special measures had to be taken to assure a safe installation. To keep the MPP into position and to prevent buckling, a supporting casing having an inner diameter of 268 mm was installed first. At the bottom this supporting casing was enlarged over a length of 3 m to an inner diameter of 800 mm. Two openings were made in the enlarged part to enable a diver to remove the piezometer sealings when they passed the openings during the jacking operation. This operation had to be done at the turn of the tides. The anchorage of the supporting casing into the seabed was provided by a large supporting plate with a ring of bedding pins.

After installation of the supporting casing, a CPT test was performed at the spot ot the MPP. This test was stopped at a depth of 12.5 m in order not to disturb the soil under the MPP tip. Next the MPP was withdrawn from its protection casing (Fig. 7). Due to the slenderness of the MPP this had to be done in a vertical position. After withdrawal, the MPP was lowered into the supporting casing. To get good centering of the MPP into the supporting casing, wooden guiding blocks were provided at 3 levels of the MPP (Fig. 7). These guiding blocks also had to be removed by a diver through the openings in the enlarged part of the supporting casing during the jacking operation. Pushing tubes successively were slipped over the cable bundle and fitted to the MPP as the jacking operation progressed.

During the jacking operation, the penetration resistance was recorded and the readings of the pressure transducers were taken. At a certain time, the readings of some pressure transducers showed unexpected values so the jacking procedure had to be modified.

After complete penetration of the MPP, the jacking tubes and supporting casing were removed. The cable bundle was fixed to a steel beam installed on the sea-bottom between the MPP and the MSP and also to the MSP itself. On the platform, the cable bundle was fitted to the hermetic box. The electronic system including the conditioning electronics for the pressure transducers and the transmitter of the telemetry system were installed in the hermetic box, connections were made and the whole system was brought into operation.

3.3 Remove of the equipment

Because the construction of the breakwaters had not to be delayed, the equipment had to be removed in June 1983. The same jack-up platform was used again to do this operation. The cable bundle was released from the steel beam on the sea-bottom and from the MSP. Next the MPP was pulled out of the seabed by a crane. It was slipped back in a suitable protection casing for transportation purposes. After the remove of the MPP, the platform and the MSP were removed. All the equipment was brought on shore and the MPP was brought to the laboratory for inspection and calibration. From this inspection it could be concluded that some cables were broken. Post-calibration was possible only for 5 transducers. For these transducers, sensitivity and offset changes were very limited.

4. DATA SAMPLING AND PROCESSING

The data sampling procedure can be seen as a time and event driven process. One had to start the sampling programs on regular times in order to check the working condition of the system. On the other hand, real data sampling was mainly dictated by weather forecasts. The data processing procedure was a user driven process as complete processing was performed for the data that seemed to be the most interesting ones.

4.1 Data sampling

Considering the data sampling, one has to distinguish the real data sampling and the data sampling done for performance guarding and check purposes. A first data sampling for check purposes was performed during the jacking operation. During this operation, the readings were continuously taken and checked against limit values to detect overflow conditions during the jacking process. Whenever these limit values were passed, an alarm signal was generated in order to take appropriate measures.

After installation, at least once a week data sampling was started to check the working condition of the whole system. Some of the stored data were just printed without any more sophisticated processing to see if the system worked well. At the same time, the charging condition of the batteries was inspected. Whenever a problem with the system was detected, a fault finding and repair procedure was started.

Real data sampling was performed on the basis of user demand in the case of data sampling for tides analysis. Data sampling for waves analysis was performed in accordance to a scheme worked out by the users. In this scheme, weather forecast was the dominant parameter. Whenever a forecast for important wave heights arrived from the Meteo Station at Zeebrugge, a watchdog program in the computer was activated to start data sampling at the expected tide related times. Whenever started, the sampling program went on storing the incoming data for a period of 30 minutes.

4.2 Data processing

Data processing for the sampled data to check the working conditions of the system only consisted of the production of a listing of the stored data. Data processing for the data stored during the real data sampling process was more complicated.

In the case of wave height analysis, a first routine screened the data : all individual calculated pressures $u_1(t)$ of the top level piezometer (level A, i=1 ; Fig. 8) were plotted. The waves thereby were referred to a hypothetical reference line corresponding to the mean value $u_{1,m,30}$ of all data of piezometer 1 inside the observation window of 30 minutes. Furthermore on the same diagram, a short term tide pressure curve $u_{f,1}(t)$ was plotted. This tide curve was computed by performing a digital low pass filtering on the samples with a cutt-off frequency of $(1/8 \text{ min}^{-1})$. Thereby all frequency components with a frequency greater than 0.002 Hz are heavily damped. Within the bandwidth of interest (frequency < 0.002 Hz) no phase shift occured.

Next the user screened the plotted data. He made a choice and selected one or more areas of important wave height. These particular areas of interest were then stretched out by the computer. The pressures $u_i(t)$ at all piezometer levels were plotted in function of time (Fig. 9).

The long term tide analysis operated in a similar way. However the stored data already were mean values calculated from the incoming data with an aperture time of 15 minutes. After screening of the whole observation period (e.g. two weeks : Fig. 10) for the top piezometer, an area of interest was selected and afterwards plotted throughout all the levels (Fig. 11).

5. CONCLUSIONS

At Zeebrugge, situated on the Belgian coast, a new outer harbour is under construction. The main breakwaters which are of the rubble mound type, are constructed over large lengths on a dumped sand foundation.

In order to gather information about the pore pressure distribution in the dumped sand layer and in the top zone of the natural stiff clay layer, and about the influence of tides and waves upon these pore pressures, field measurements were made by means of a multiple piezometer probe. At seven levels, pore pressure measuring devices were built in the probe. Pressure transducers with integrated silicon strain gages were used. They were mounted in the nipples of the probe. The volumes in front of the sensitive elements of the pressure transducers were connected to the outside of the probe by a number of small slightly inclined channels.

Output signals of the pressure transducers were brought to the shore by an UHF radio link. The transmitter together with the conditioning electronics for the transducers were installed on a platform in the vicinity of the multiple piezometer probe. Power supply for this equipment was delivered by solar cells connected to a battery pack.

At the shore, the signals were captured and put on telephone lines to bring them to a suitable real time computer. Data acquisition and processing was performed by this computer.

Complete analysis and evaluation of the sampled data is not yet finished. A first discussion of the obtained results and some preliminary conclusions can be found in De Wolf et al., 1983.

6. REFERENCES

De Wolf, P., Carpentier, R., Verdonck, W., Boullart, L., De Rouck, J., de Saint Aubain, T., De Voghel, J., 1983, In situ pore water pressure measurements for the construction of the breakwaters of the new outer harbour at Zeebrugge. Proc. 8th Int. Harbour Conf., Antwerp, pp. 1.203 - 1.215.

ISSMFE, 1977, Standard for Penetration Testing in Europe. Proc. Ninth Intern. Conf. on Soil Mech. and Found. Eng., Tokyo, Vol.3, pp. 95-135.

Massarsch, K.R., Broms, B.B., Sundquist, O., 1975, Pore pressure determination with multiple piezometer. Proc. Conf. In Situ Measurement of Soil Properties – North Carolina State Univ., Raleigh, Vol. 1 pp. 260-265, ASCE, New York.

Sheingold, D.H., 1980, Transducer Interfacing Handbook. A Guide to Analog Signal Conditioning. Analog Devices, Inc. Norwood, Massachusetts, USA.

Vaughan, P.R., 1973, The Measurement of pore pressures with piezometers. Field Instrumentation in Geotechnical Engineering. A Symposium organized by the British Geotechnical Society. Part 1 pp. 411-422. Butterworth, London.

Verdonck, W., 1980, A Data Acquisition and Processing System for Engine Tests. SAE Paper 800411 and SAE Transactions Vol. 89. Society of Automotive Engineers, Warrendale, USA.

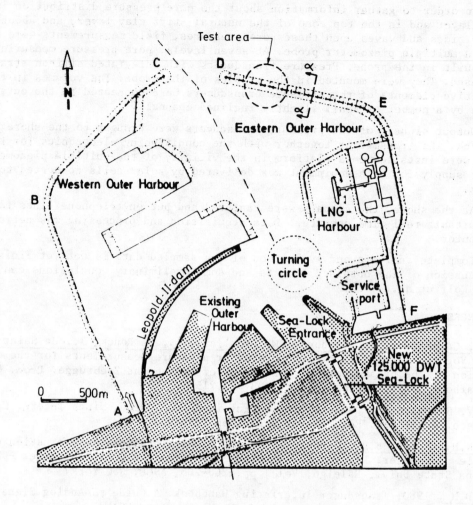

Fig.1 – Lay-out of the new outer harbour
at Zeebrugge, Belgium.

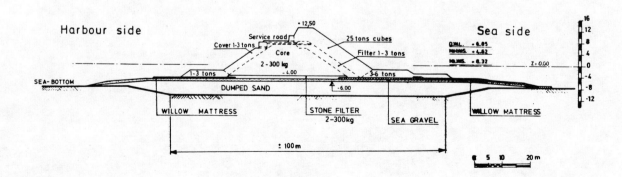

Fig.2 – Cross-section (α-α in Fig.1) of the north-eastern breakwater.

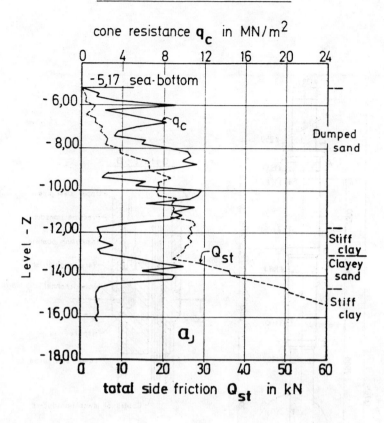

CPT test NOD-GZ/**D19**

Fig.3 – Results of CPT test n° NOD–GZ/D19.

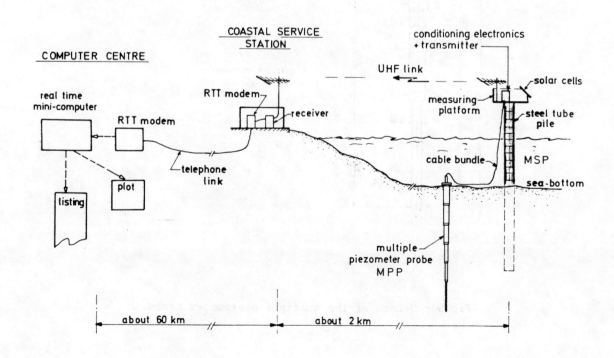

Fig.4 – Lay-out of the measuring system.

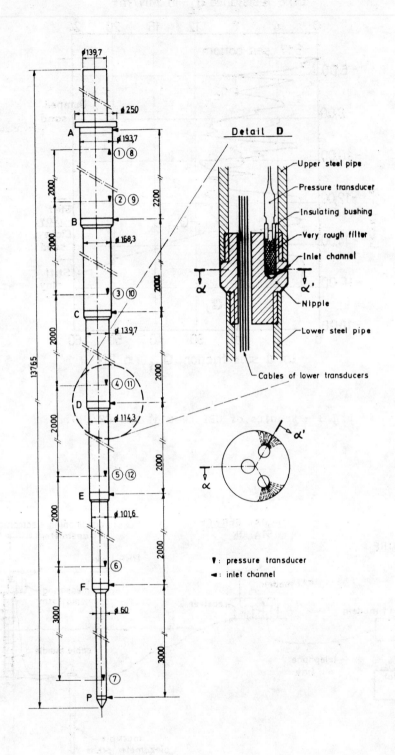

Fig.5 - Scheme of the multiple piezometer probe.

Fig.6 - View of the measuring platform
taken from the jack-up platform.

Fig.7 - The multiple piezometer probe
withdrawn from its protection casing.

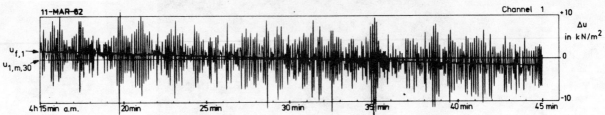

Fig.8 – Typical plot of pore pressure changes due to waves at piezometer 1 – level A.

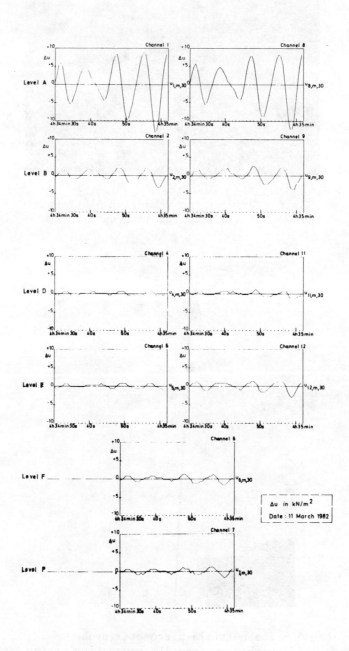

Fig.9 – Example of pore pressure changes due to waves at all piezometer levels.

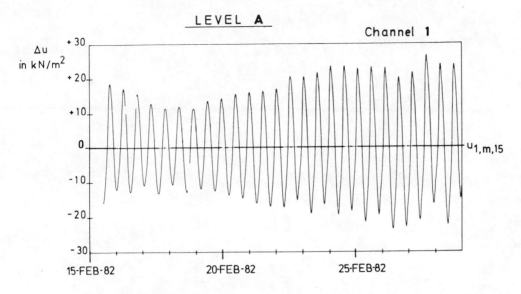

Fig.10 - Typical plot of pore pressure changes
due to tide at piezometer 1 - level A.

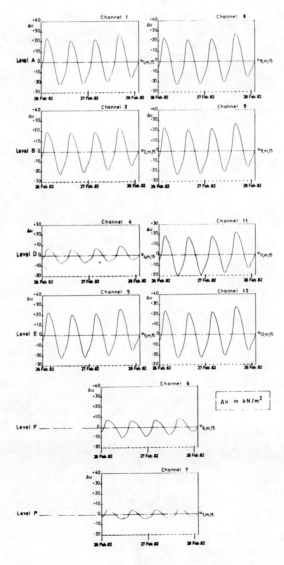

Fig.11 - Example of pore pressure changes due to tide at all piezometer levels.

135

WAVE INDUCED FORCES ON SUBMARINE PIPELINES

Assoc. Prof. N Jothi Shankar, B.E., M.Tech., Ph.D(Texas)

Assoc. Prof. Cheong Hin Fatt, B.Eng. M.Sc., Ph.D(Colorado State)

K. Subbiah[*], B.E. (Hons), M.Tech.

Department of Civil Engineering

National University of Singapore, Kent Ridge, Singapore 0511

SUMMARY

A measuring technique is outlined for the measurement of forces on submarine pipelines subjected to random wave action using strain gauge principle. A simulation method is described to evaluate the hydrodynamic coefficients in the Morison equation. An experimental investigation is carried out on a model pipeline placed at two different clearances above the simulated ocean floor and subjecting it to random waves (P-M spectra). The hydrodynamic coefficients of horizontal drag and inertia and coefficients of vertical lift and inertia have been evaluated using the simulation method.

[*] Research Scholar, Department of Civil Engineering, National University of Singapore

Held at Imperial College of Science and Technology, London. Organised and sponsored by BHRA, The Fluid Engineering Centre. Co-sponsored by the American Society of Civil Engineers and the International Association for Hydraulic Research.

NOMENCLATURE

C_D	drag coefficient
C_m	inertia Coefficient
C_L	lift coefficient
C_{mv}	vertical inertia coefficient
D	diameter of Cylinder
d	still water depth
e	vertical clearance between flume bed and bottom of cylinder
F_H	total inline force on the cylinder
f_D	drag force per unit length of cylinder
f_H	horizontal wave force per unit length of cylinder
f_I	inertial force per unit length of cylinder
f_L	lift force per unit length of cylinder
f_v	vertical force per unit length of cylinder
f_{VI}	vertical inertial force per unit length of cylinder
f	linear frequency
f_o	frequency at which the spectral density is maximum
g	gravitational constant
H_s	significant wave height
K	period paramter
k	wave number ($2\pi/L$)
L	wave length
l	length of cylinder
p	probability of occurrence of η in the class interval between η and $\eta + d\eta$
$S_r(f)$	spectral density estimate of realization r
t	time variable
u(t)	horizontal water particle velocity
$\dot{u}(t)$	horizontal water particle acceleration
$\dot{w}(t)$	vertical water particle acceleration
x,z	cartesian coordinates in the horizontal and vertical directions
η	water surface elevation
$\bar{\eta}$	mean value of η
ρ	mass density of fluid
ν	kinematic viscosity of fluid
ε	random phase
ε_s	spectral width parameter
σ_η	standard deviation of η

1. INTRODUCTION

Submarine pipelines are used to convey oil and natural gas from offshore oil rigs to onshore terminals, as ocean outfalls and for the protection of communication cables. These pipelines are either buried or laid on the sea floor or placed on saddles leaving a certain clearance between the pipe and the sea floor. The submarine pipelines are subjected to the action of waves, currents and other environmental loads. The design of these pipelines depends on the accurate assessment of wave induced forces and this can be achieved by proper determination of water particle kinematics and selection of proper force transfer coefficients. These coefficients have to be necessarily determined from laboratory and/or field experiments.

Laboratory studies on regular wave forces on submerged horizontal circular cylinders near a plane boundary have been carried out by many researchers, (Refs. 1, 2, 3, 4, 5, 6, 7). The forces on horizontal circular cylinders have also been investigated in U-tube water tunnels (Refs. 8, 9, 10, 11). Keulegan and Carpenter(Ref. 12) have placed their cylinder in standing waves but this cylinder was always below the nodal point of the waves. In both these cases, the flow is considered to be perfectly oscillatory and hence the vertical component of water particle velocity is 0. Laboratory experiments on horizontal cylinders in regular wave and current coexisting fields have also been conducted (Refs. 13, 14, 15, 16).

Laboratory studies on random wave forces on piles have been conducted and analysed (Refs. 17, 18). Field tests on vertical piles (random wave forces on piles) have been conducted by many U.S oil companies and the data were analysed (Refs. 19, 20, 21). The different methodologies like spectral density least square method and method of moments to evaluate the hydrodynamic coefficients for the piles subjected to random waves are given in the Technical Memorandum (Ref. 19). Grace and Nicinski (Ref. 22) conducted field experiments on 40.6 cm diameter pipeline in the Hawaiian waters at 0.076 m clearance above the bed in 11.28 m of water. The waves in the random wave records were treated as individual regular waves. These waves were superimposed by currents and hence their results on coefficients of drag, and lift showed considerable scatter.

This paper presents an experimental methodology for the measurement of inline and transverse forces due to random waves on a model pipeline placed at two different clearances above the simulated ocean floor and describes data acquisition and analysis of random data for evaluation of force transfer coefficients. A micro computer controlled data acquisition system is used for collection of random wave and force data. A computer program is developed based on the Fast Fourier Transform (FFT) algorithm to compute the spectral density estimates for two random records simultaneously. A simulation method is presented for the evaluation of force transfer coefficients for inline and transverse forces on the pipelines.

2. THEORY

2.1 General aspects

A horizontal circular cylinder of diameter D, placed at a clearance e, above the ocean floor in a water depth of d and parallel to wave crests is considered. The position of the centre of the cylinder is fixed at x = 0 and z is measured positive upward from still water level. The ocean floor is assumed to be horizontal so that no refraction effects be considered. A train of long crested and unidirectional random waves act upon the cylinder. Fig. 1 shows the definition sketch of the pipeline.

The depth of submergence of the cylinder considered is greater than 3 cylinder diameters so that the free surface effects can be neglected. The time history of the water surface elevation η, above the centre line of the pipe is described as a stationary Gaussian process.

$$\eta(t) = \int_{0}^{\infty} \sqrt{2S_\eta(f) \ df} \ . \ \cos(2\pi ft + \varepsilon) \tag{1}$$

in which $S_\eta(f)$ is the spectral density of η at frequency f. $S_\eta(f)$ decomposes the total variance of $\eta(t)$ into its various frequency components and ε is the random phase associated with the component having frequency f. The random phase ε is assumed to be distributed uniformly between 0 and 2π.

2.2 Inline forces

The spectra of horizontal water particle velocity, u and acceleration, $\dot{u}$, based on linear wave theroy are given in terms of spectral density of water surface elevation, $S_\eta(f)$, as

$$S_u(f) = \frac{(2\pi f)^2 \cosh^2 k(d+z)}{\sinh^2 kd} \cdot S_\eta(f) = R_H(f,z,d) \ S_\eta(f) \tag{2}$$

$$S_{\dot{u}}(f) = \frac{(2\pi f)^4 \cosh^2 k(d+z)}{\sinh^2 kd} \cdot S_\eta(f) = (2\pi f)^2 \ R_H(f,z,d) \ S_\eta(f) \tag{3}$$

$$R_H(f, z, d) = \frac{(2\pi f)^2 \cosh^2 k(d+z)}{\sinh^2 kd} \tag{4}$$

and $\qquad (2\pi f)^2 = gk \tanh kd \tag{5}$

where k is the wave number corresponding to the component having the frequency f, z is the depth of submergence of centre line of cylinder below still water level and g is the gravitational acceleration.

The time histories of horizontal water particle velocity, u(t) and acceleration, $\dot{u}(t)$ can be obtained from their corresponding spectra as:

$$u(t) = \int_0^\infty \sqrt{2S_u(f) \ df} \cdot \cos(2\pi ft + \varepsilon)$$

$$= \int_0^\infty \sqrt{2R_H(f,z,d) \ S_\eta(f) df} \ \cos(2\pi ft + \varepsilon) \tag{6}$$

and

$$\dot{u}(t) = \int_0^\infty - \sqrt{2S_{\dot{u}}(f) df} \ \sin(2\pi ft + \varepsilon)$$

$$= \int_0^\infty - \sqrt{(2\pi f)^2 R_H(f,z,d) S_\eta(f) df} \cdot \sin(2\pi ft + \varepsilon) \tag{7}$$

The horizontal acceleration is 90° out of phase with horizontal velocity and both are described by a stationary Gaussian process. From the estimated horizontal water particle velocity and acceleration spectra given by Eqns. (2) and (3) respectively, it is possible to generate their corresponding time histories using the Eqns. (6) and (7), after generating a set of random phases ε , associated with each of the frequency components.

The total horizontal wave force per unit length of a circular cylinder is given by Morison et al. (Ref. 23). The Morison equation as applicable to a horizontal cylinder is given by

$$f_H = f_D + f_I = 0.5 \ C_D \ \rho D \ u(t)|u(t)| + 0.25 \ C_m \ \rho\pi D^2 \ \dot{u}(t) \tag{8}$$

in which f_H is the horizontal force per unit length of cylinder and f_D and f_I are respectively the drag and inertial component forces, ρ is mass density of liquid, C_D

and C_m are horizontal hydrodynamic coefficients of drag and inertia respectively and other parameters are as defined earlier. From the measured time history of the total horizontal wave force and generated horizontal water particle velocity and acceleration time histories, it is possible to evaluate the coefficients of drag and inertia using least square method.

2.3 Vertical force

The approach to the evaluation of transverse hydrodynamic coefficients of lift and inertia is similar to that for inline force coefficients. The total vertical wave force per unit length of a circular cylinder, f_v is given as the summation of lift, f_L and vertical inertia force f_{VI} and is given by

$$f_v = f_L + f_{VI} = 0.5 \, C_L \, \rho D u^2(t) + 0.25 \, C_{mv} \, \rho \pi D^2 \, \dot{w}(t) \qquad (9)$$

in which $\dot{w}$ is vertical water particle acceleration and C_L and C_{mv} are hydrodynamic coefficients of lift and vertical inertia respectively. The time history of horizontal water particle velocity, $u(t)$ can be generated using Eqs. (2) and (6) and that of vertical acceleration, $\dot{w}(t)$ can be generated as follows:

The vertical acceleration spectrum based on linear wave theory in terms of $S_\eta(f)$ is given by

$$S_{\dot{w}}(f) = \frac{(2\pi f)^4 \sinh^2 k(d+z)}{\sinh^2 kd} \, S_\eta(f) = (2\pi f)^2 R_w(f,z,d) \, S_\eta(f) \qquad (10)$$

$$R_w(f,z,d) = \frac{(2\pi f)^2 \sinh^2 k(d+z)}{\sinh^2 kd} \qquad (11)$$

and the time history of vertical acceleration in terms of $S_\eta(f)$ is given as

$$\dot{w}(t) = \int_0^\infty \sqrt{(2\pi f)^2 R_w(f,z,d) \, S_\eta(f) df} \, . \, \cos(2\pi ft + \varepsilon) \qquad (12)$$

From the measured transverse force time history and generated horizontal water particle velocity and vertical particle acceleration time histories, the hydrodynamic coefficients of lift and vertical inertia can be evaluated using the least square method.

3. MEASURING TECHNIQUES

The total inline and transverse forces due to waves on submarine pipelines are measured using force dynamometers, which work on strain gauge principle. The total dynamic strains in the strain gauges are converted into measurable voltages by strainmeters.

The force dynamometer is designed as a cantilever beam supported at one end of the test section considering a minimum force for sensitivity criteria. From the design minimum force and assumed distance of the centre of the strain gauge beam from the end support and section modulus of the strain gauge beam, the strain at the strain gauge section can be determined. The section which gives a strain of about 60 units (1 unit = 1μ) is chosen as the best section for strain gauge fixing, considering the centre of the strain gauge beam from the support as an important criterion. The details of the strain gauge beam along with bulkhead are shown in Fig. 3. The strain gauges are selected based on the width of the beam and their capacity to measure maximum strains due to loading from the high waves that could be generated in the flume.

A pair of strain gauges is fixed in the vertical plane in order to measure total strains due to inline wave forces and another pair in the horizontal plane for measurement of strains induced by transverse wave forces. The strain gauges are properly water proofed. The force dynamometers thus designed are embedded inside the model pipeline to measure total inline and transverse forces due to waves. A resistance type wave gauge is used to measure the time history of water surface elevations.

Fig. 4 shows the block diagram of the data acquisition system. The strain gauges fixed on both sides of the vertical plane beam form a bridge and hence transmit total horizontal strain (inline force) and those in the horizontal plane form another bridge and transmit total vertical strain (transverse force). similarly the resistances in the wave gauge due to changes in water surface elevation are converted into voltages by a wave monitor.

A HP9816 micro computer is used for data acquisition. The analog voltages from strainmeters (inline and transverse forces) and wave monitor (water surface elevations) are converted into digital voltages by HP6942A Multiprogrammer. The data may be collected at any desired sampling interval of time between 300 μs and 65500 μs and 128 to 2048 samples ($N = 2^p$, p = 7 to 11) per realization. The collected data may be stored in diskettes for future analysis. The results of the analysis are plotted using HP7470A plotter and output printed using HP82906A line printer.

4   EXPERIMENTAL INVESTIGATION

An experimental investigtion was carried out on random wave forces on a 6 cm dia perspex model pipeline in a wave flume at the Hydraulic Engineering Laboratory, National University of Singapore. The flume is 2m in width, 1.3 m in height and 36 m long. A paddle type wave generator at one end of the flume generates random waves using inputs from a punched tape and a beach at the other end is used to absorb the waves so that reflection is negligible. The test section is chosen near the middle of the flume.

The force dynamometers were fixed to the bulkheads which inturn were embedded inside the pipeline and rigidly fixed to the pipe. The strain gauge beams were aligned in such a way that one beam was in the vertical plane to measure inline forces and other in the horizontal plane to measure transverse forces. The sides of the model were properly sealed to prevent entry of water into the model. The strain gauged beams were supported at the ends of the pipe by self aligning bearings. These bearings were housed in Aluminium channels, which can be raised or lowered to have any specific clearance between the flume bed and bottom of pipeline. The flume walls were streamlined for a length of 1.5 m on both upstream and downstream sides of the test section. A resistance type wave gauge was installed near middle of test section and above the centre line of the pipeline model to record time history of water surface elevation.

A punched tape with digitized wave data corresponding to P-M spectrum was read at a constant speed of 15 characters per second and fed to the wave generator as input signal to generate random waves. Time histories of water surface elevations, inline and transverse forces were collected simultaneously at the sampling interval of 15000 μs (1024 data samples per realization) and stored on the diskettes for further analysis. The pipe was placed at two different clearances of 6 mm and 30 mm giving relative clearance e/D of 0.1 and 0.5 respectively.

5.   RESULTS AND DISCUSSION

Statistical and spectral analysis of the measured time history of instantaneous water surface elevation η at the pipeline axis was carried out for purposes of comparison with theoretical probability density function, p(η) and the Pierson-Moskowitz energy spectrum. Random waves are generally assumed to follow the Gaussian or normal distribution given by:

$$p = p(\eta) \ d\eta = \frac{1}{\sigma_\eta \sqrt{2\pi}} \exp \ [- \frac{(\eta - \bar{\eta})^2}{2\sigma_\eta^2}] \ d\eta \tag{13}$$

in which p is the probability of occurrence of η in the class interval between η and $\eta + d\eta$, $p(\eta)$ is the probability density function of water level fluctuation η, $\bar{\eta}$ is the mean value of η and σ_η is the standard deviation of η. Figure 5 illustrates that the probability of occurrence of η normalized with respect to the peak probability p_o based on the measured time history of η follows closely the theoretical normal distribution.

There are two methods commonly used to compute the energy density spectrum; 1) Covariance method or the autocorrelation method, and 2) the Fast Fourier Transform method (FFT). A computer program was developed to compute Fast Fourier Transform coefficients for two discrete random processes of length $N(2^P)$ simultaneously. The complex operations in the FFT algorithm were effectively utilized by having a seperate complex vector of length N, in which one random sequence is stored in the real part and the other in the imaginary part. After transformation the complex valued sequence was decoded to obtain seperate transforms of the two original sequences. The computation time for the combined sequence is about half that required for transforming the two sequences seperately.

Using the above technique spectral density estimates for two water surface elevation records were computed simultaneously. In Fig. 6 the energy spectrum computed based on measured time history of η is compared with the theoretical Pierson-Moskowitz spectrum given by

$$S(f) = S(f_o) \left(\frac{f}{f_o}\right)^{-5} \exp\left[1.25 \left\{1 - \left(\frac{f}{f_o}\right)^4\right\}\right] \tag{14}$$

where $S(f)$ and $S(f_o)$ are respectively the spectral density estimates corresponding to the frequency f and peak frequency f_o. It is observed from Fig. 6 that the measured spectrum of η is in close agreement with the PM spectrum.

The results of the simulation method for the depth parameter d/D = 8 are given in Table 1 for two relative clearances of e/D = 0.1 and 0.5. In the table the significant wave height H_s, is defined in terms of zeroth spectral moment as

$$H_s = 4\sqrt{m_o} \tag{15}$$

The zero crossing period T_z, and spectral width parameter ε_s are defined in terms of higher spectral moments as,

$$T_z = \sqrt{m_o/m_2} \qquad \text{and} \tag{16}$$

$$\varepsilon_s = \left(1 - \frac{m_2^2}{m_o m_4}\right)^{1/2} \tag{17}$$

in which the n-th spectral moment is defined as

$$m_n = \int_o^\infty f^n S(f) \, df \tag{18}$$

The local period parameter $K(u_m T/D)$ is computed based on maximum horizontal water particle velocity u_m due to the significant wave with height H_s, period T_z and water depth d. Similarly the local Reynolds number R_e is defined as $u_m D/\nu$ where ν is the kinematic viscosity of water. The horizontal hydrodynamic coefficients of drag C_D and inertia C_m and vertical hydrodynamic coefficients of lift C_L and inertia C_{mv} are evaluated using the proposed simulation method.

The values of the hydrodynamic coefficients C_D, C_m, C_L and C_{mv} for random waves reported in Table 1. are comparable with the values reported for regular waves under similar period parameter values (Ref. 6). Furthermore from the results for the two relative clearances tested, it is observed that the inline force coefficients of drag C_D and inertia C_m, decrease with increase in the clearance from the bed.

In the case of transverse forces it is seen that the hydrodynamic coefficients of inertia C_{my} and lift C_L, decrease with increase in pipe clearance for a given depth parameter d/D. The bed boundary causes asymmetry of flow below and above the cylinder. This gives rise to corresponding change in the pressure distribution over the surface of the cylinder resulting in the transverse force in the vertical direction. The asymmetry in the flow decreases as the pipe is moved away from the boundary with consequent reduction in the vertical force.

Further research is underway at the National University of Singapore to establish correlation of the hydrodynamic coefficients associated with random wave forces with significant wave and physical parameters.

6. CONCLUSIONS

A two component force dynamometer utilizing strain gauge technique is developed for the measurement of horizontal and transverse forces on a model pipeline subjected to random waves. Spectral analysis of the random data involving two records of water surface elevations is accomplished effectively using an FFT algorithm. A simulation technique is proposed for the evaluation of the hydrodynamic coefficients associated with inline and transverse forces induced on pipelines by random waves. The hydrodynamic coefficients are of the order of magnitude obtained for regular waves.

7. ACKNOWLEDGEMENTS

The financial support of the National University of Singapore under NUS Grant RP 99/83 which has made this study possible is gratefully acknowledged.

8. REFERENCES

1. Garrison, C.J., Field, J.B., and May, M.D.: "Drag and inertia forces on a cylinder in periodic flow". Jl. of Waterway, Port Coastal and Ocean Div., ASCE, Vol. 103, No. WW2, May, 1977, pp. 193-204.

2. Holmes, P., and Chaplin, J.R.: "Wave loads on horizontal cylinders". Int. Conf. on Coastal Engg, 1978, pp. 2449-2459.

3. Jothi Shankar,N., and Sundar,V.: "Wave forces on offshore pipelines". Proc. 17th Conf. on Coastal Engg, 1980, pp. 1819-1828.

4. Maull, D.J., and Norman, S.G.: "A horizontal circular cylinder under waves". Proc. WIFCO '78 Symposium, Bristol, 1978, pp. 359-377.

5. Nath, J.H., Yamamoto, T., and Wright, J.C.: "Wave forces on pipes near the ocean bottom". 8th OTC, paper No. 2496, 1976, pp. 741-747.

6. Wright, J.C., and Yamamoto, T.: "Wave forces on cylinders near plane boundaries". Jl. of Waterway, Port, Coastal and Ocean Div., ASCE, WW1, 1979, pp. 1-13.

7. Yamamoto, T., Nath, J.H. and Slotta, L.S.: "Wave forces on cylinders near plane boundary". Jl. of Waterways, Harbors and Coastal Engg. Div., ASCE, WW4, 1974, pp. 345-359.

8. Maull, D.J., and Milliner, M.G.: "Sinusoidal flow past a circular cylinder". Int. Jl. of Coastal Engg. 2, 1978, pp. 149-168.

9. Sarpkaya, T.: "Forces on Cylinders and spheres in a sinusoidally oscillating fluid". Jl. of App. Mech. Trans., ASME, 1975, pp. 32-37.

10. Sarpkaya, T.: "Forces on Cylinders near a plane boundary in a Sinusoidally oscillating fluid". Jl. of Fluids Engg. Trans., ASME, Vol. 98, No. 3, 1976 pp.

499–505.

11. Yamamoto, T., and Nath., J.H.: "High Reynolds number oscillating flow by cylinders". 8th Int. Conf. on Coastal Engg. 1976, pp. 2321-2340.

12. Keulegan, G.H., and Carpenter, L.H.: "Forces on cylinders and plates in an oscillating fluid". Jl. of Res. of the National Bureau of Standards, Vol. 60, No. 5, 1958, pp. 423-440.

13. Bryndum, M.B.: "Hydrodynamic forces from waves and current loads on marine pipelines", 15th OTC, paper No. 4454, 1983, pp. 95-102.

14. Chandler, B.D. and Hinwood, J.B.: "The hydrodynamic forces on Submerged horizontal cylinders due to simultaneous wave and current action". 5th Australian Conf. on Coastal and Ocean Engg. 1981, pp. 6-7.

15. Iwagaki, Y., Asano, T., and Nagai, F.: "Hydrodynamic forces on a circular cylinder placed in wave – current coexisting fields". Reprinted from the memoirs of the Faculty of Engg, Kyoto Univ. Vol. X1V, Part 1, Jan. 1983.

16. Knoll, D. A., and Herbich, J.B.: "Wave and current forces on a Submerged offshore pipeline". 12th OTC, paper No. 3762, 1980, pp. 227-234.

17. Pierson, W.J., and Holmes, P.: "Irregular wave forces on a pile". Jl of Waterways and Harbors Div. ASCE, WW4, Nov. 1965, pp. 1-10.

18. White, J.K., and Carr, P.L.: "On the estimation of Morison coefficients in irregular wave". Mechanics of wave induced forces on Cylinders, Ed. by T.L. Shaw, Pitman, 1979, pp. 314-323.

19. Brown, L.J., and Borgman, L.E.: "Tables of the statistical distribution of ocean wave forces and methods of estimating drag and mass coefficients". U.S. Army Coastal Engg Res. Centre. Tech memo. No. 24, 1967.

20. Wiegel, R.L., Beebe, K.E., and Moon, J.: "Ocean wave forces on circular cylindrical piles". Jl. of Hydraulics Div., ASCE, Vol. 83, No. HY2, 1957, pp. 89-116.

21. Wilson, B.W.: "Analysis of wave forces on a 30-inch diameter pile under confused sea conditions". CERC, Tech. Memo. No. 15, 1965.

22. Grace, R.A., and Nicinski, S.A.: "Wave force coefficients from pipeline research in the ocean". 8th OTC, paper No. 2676, 1976, pp. 681-694.

23. Morison, J.R. et al: "The forces exerted by surface waves on piles". Petroleum Transactions, Vol. 189, TP2846, June 1950, pp. 149-154.

TABLE 1: RESULTS OF SIMULATION METHOD FOR DEPTH PARAMETER $d/D = 8$

Sl. No.	e/D	H_s cm	T_z sec	f_o Hz	ε_s	K	R_e	C_D	C_m	C_L	C_{mv}
1	0.1	28.1	1.922	0.391	0.707	16.78	31430	3.02	1.77	0.61	2.67
2	0.5	25.7	1.928	0.391	0.673	15.46	28870	0.85	1.67	0.25	2.07

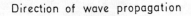

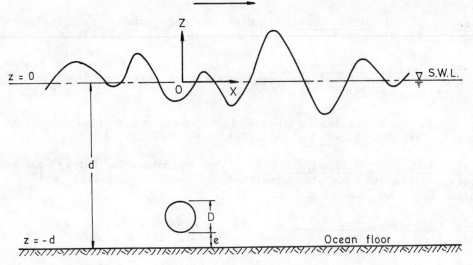

FIG. 1 DEFINITION SKETCH – Model pipeline

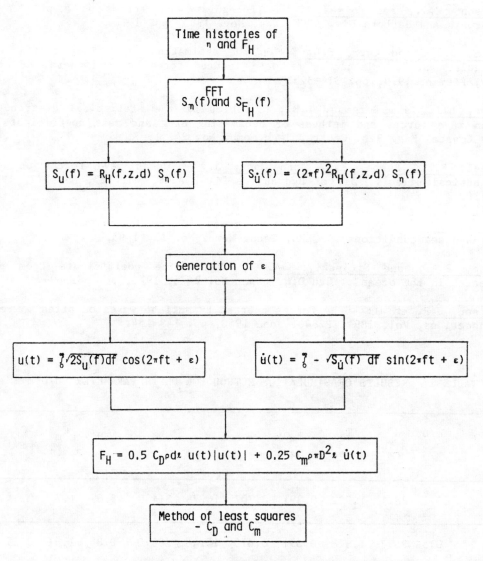

FIG. 2 FLOW CHART FOR THE DETERMINATION OF HORIZONTAL
HYDRODYNAMIC COEFFICIENTS

146

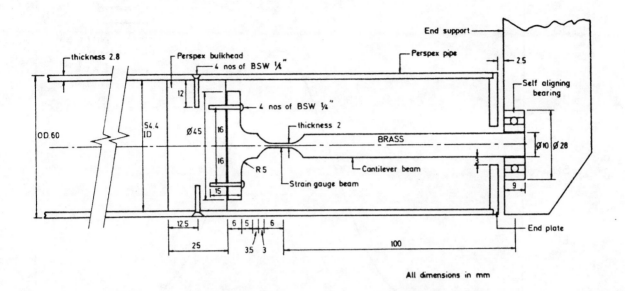

FIG. 3 DETAILS OF THE STRAIN GAUGE DYNAMOMETER AND PIPELINE MODEL

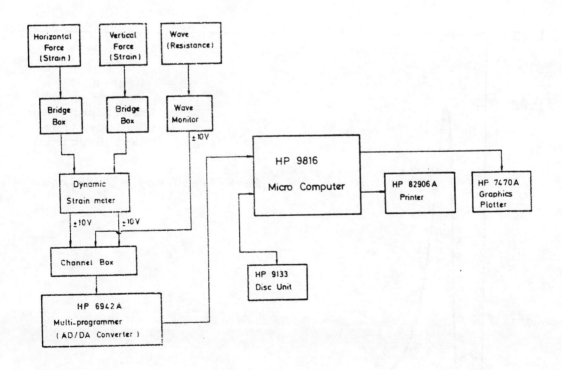

FIG. 4 BLOCK DIAGRAM OF THE DATA ACQUISITION SYSTEM

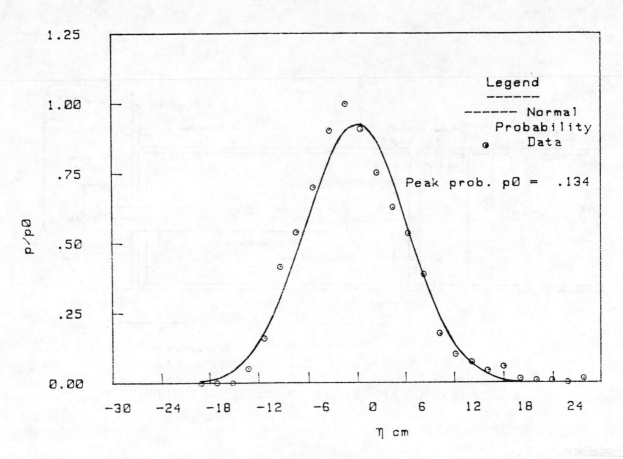

FIG. 5 THEORETICAL AND MEASURED PROBABILITY DISTRIBUTION OF η

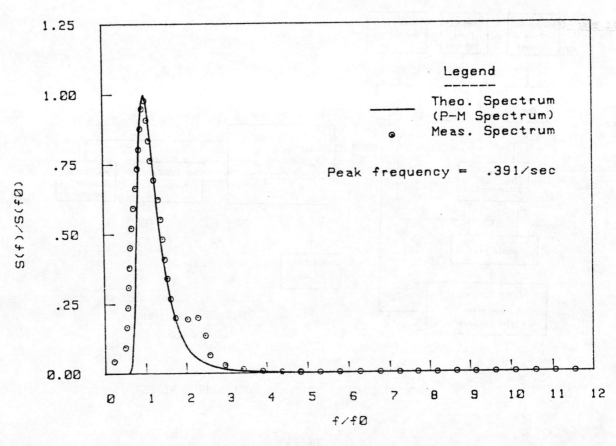

FIG. 6 COMPARISON OF MEASURED AND THEORETICAL SPECTRA

Measuring Techniques

of Hydraulics Phenomena in offshore, Coastal & Inland Waters

London, England: 9-11 April, 1986

A FORCE MEASURING DEVICE TO STUDY

THE IMPACT OF WAVES UPON SOLID STRUCTURES

Authors

Mr. H.B. Boyle - British Maritime Technology Limited

Dr. P.C. Barber - Ceemaid Limited

Held at Imperial College of Science and Technology, London. Organised and sponsored by BHRA, The Fluid Engineering Centre. Co-sponsored by the American Society of Civil Engineers and the International Association for Hydraulic Research.

1.   INTRODUCTION

This paper describes the development and testing of a Force Measuring device to study the impact of waves upon solid structures for engineering applications, with an initial brief review of theory as a background to the development of the instrument.

The sea wall unit described here, effectively integrates the total pressure normal to its face, which has a surface area of approximately $0.07m^2$. It is designed to have a low profile, to minimise any protrusion from a sea wall if surface mounted, or to minimise material removed if flush mounting is essential.

2.   THEORY

Many coastal and harbour structures subject to wave action are built with vertical, or nearly vertical, faces. Such structures will reflect a high proportion of the wave energy which reaches them. However, under many circumstances the possible disadvantages associated with wave reflection are unimportant or are outweighed by other considerations.

Waves breaking directly against vertical-face structures exert high, short duration, dynamic pressures that act near the region where the wave crests hit the structure. These impact or shock pressures have been studied in the laboratory (Bagnold (1), Denny (2), Ross (3), Nagai (4), Carr (5), Leendertse (6), Kamel (7), Weggel (8) and Weggel and Maxwell (9).) Some measurements on full-scale breakwaters have been made by de Rouville, et al., (10). Wave tank experiments by Bagnold (1) led to an explanation of the phenomenon. Bagnold found that impact pressures occur at the instant that the vertical, front face of a breaking wave hits the wall. Because of this critical dependence on wave geometry, high impact pressures are infrequent against prototype structures.

However, the possibility of high impact pressures must be recognised, and considered in design. The high impact pressures are of short duration (of the order of hundredths of a second), and their importance in the design of breakwaters against sliding or overturning is questionable. A good review of data available on impact pressures on sea walls is provided in Blackmore and Hewson (1984) (11). This work showed that only 0.04% of waves recorded produced impact pressures illustrating the sensitivity of the process to wave form and obliquity, and wall profile. It should be noted that measurements were made by pressure transducers of relatively small surface area and unstated response characteristics underwater. Following on from this recent work it was decided to develop an instrument for the measurement of wave forces including those due to impact pressures.

The theoretical approach to defining the wave force distribution from a sea state clearly influences the approach taken for instrument design.

Pressures from non-breaking waves

Non-breaking waves which approach perpendicular to a vertical wall are reflected and form a standing wave pattern. The pressure beneath the standing waves is given by linear wave theory as:-

$$p = \rho g[\eta_s \frac{\cosh k(d+z)}{\cosh kd} - z] \tag{1}$$

η_s is the water surface elevation of the combined system of incident and reflected waves. Assuming perfect reflection, $\eta_s = H_i$ when a wave crest is at the wall and the corresponding distribution of pressure is as shown in Figure 1(a). Figure 1(b) shows the pressure distribution for $\eta_s = H_i$ when a trough is at the wall. Note that in each case

151

linear wave theory predicts a small excess pressure at the water surface.

The total force at any instant on the seaward side of the structure may be determined by integrating equation (1) over the wall's wetted area. When a crest is at the wall the force per unit length, F_c, is:-

$$F_c = \int_{-d}^{H_i} p \, dz$$

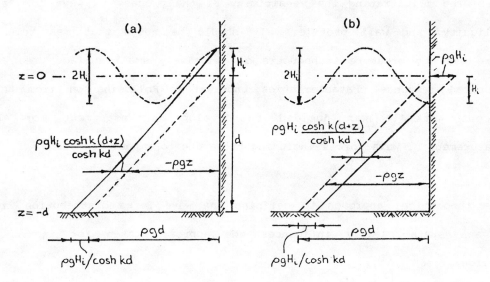

Figure 1   Pressure Distributions Beneath Small–Amplitude Standing Waves

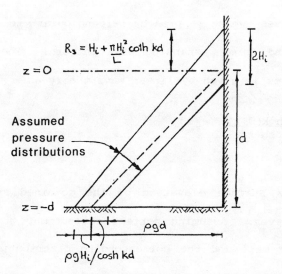

Figure 2   Assumed Pressure Distributions For Finite Amplitude Waves

$$= \rho g[\tfrac{1}{2}(d^2 - H_i^2) + \frac{H_i}{k}\frac{\sinh k(d+H_i)}{\cosh kd}] \qquad (2)$$

Similarly, when a trough is at the wall the force per unit length is F_t, given by:-

$$F_t = \int_{-d}^{-H_i} p\, dz$$

$$= \rho g[\tfrac{1}{2}(d^2 - H_i^2) - \frac{H_i}{k}\frac{\sinh k(d-H_i)}{\cosh kd}] \qquad (3)$$

The above results are based on small-amplitude wave theory. When finite amplitude waves are reflected from a vertical wall the maximum and minimum levels reached by the water surface on the wall are higher than the small-amplitude values. The run-up on a smooth vertical wall is given approximately by:-

$$R_s = H_i + \frac{\pi H_i^2}{L} \coth kd \qquad (4)$$

The minimum level reached by the water on the wall would be $(R_s - 2H_i)$ from still water level. Then, assuming that the pressure at sea-bed level $(z = -d)$ remains as given by equation (1) and that the pressure distribution reduces linearly from this value to zero at the water surface, a simple, generally conservative estimate may be obtained of the range of the standing wave forces and moments on the wall (see Figure 2).

In general there will also be pressures on the backs of the walls from water, soil and other sources. Furthermore, when considering the stability of such structures it is also necessary to take account of possible variations in uplift pressures on their bases, particularly when they are constructed upon permeable foundations.

153

<u>Pressures from Breaking Waves</u>

The most severe pressures on walls are caused by waves which plunge directly onto them. Measurements of the impacts generally show a pressure history like that shown in Figure 3. There is a rapid rise to pressure p_m as the breaking wave makes contact with the wall and then, following a pressure reduction, there is a more gentle rise as the bulk of the wave runs up the wall face. This second pressure peak is of the same order of magnitude as that associated with standing waves.

The magnitude of the initial impact pressure, p_m, depends upon many factors including the amount of air entrained in the water or trapped between the wall and the wave as the crest curls over. A trapped pocket of air tends to cushion the impact.

From the impulse momentum relationship and an assumption that horizontal water velocity within the breaking wave is reduced to zero it can be shown that:-

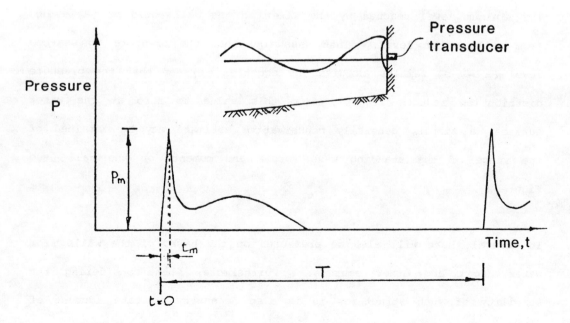

Figure 3   Typical Pressure History Produced By
Wave Breaking On A Wall

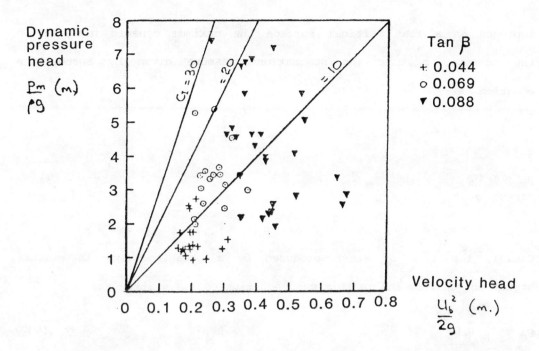

Figure 4   Measurements Of Impact Pressures
(After HAYASHI and HATTORI, 1958)

$$p_m = \frac{2\rho \, l_v \, U_b}{t_m} \qquad\qquad (5)$$

where t_m = rise time to maximum pressure.

ρ = water density.

U_b = horizontal water velocity before impact.

l_v = virtual length of water involved in impact.

Equation (5) predicts a hyperbolic relationship between p_m and t_m — the shorter the impact duration the greater the peak pressure. In practice the ratio t_m/T has been found to be of the order of 10^{-3}, where T is the wave period.

When a steady horizontal jet of water with uniform velocity U_b

155

impinges on a flat vertical surface the maximum dynamic pressure on the surface, p_m, is the stagnation pressure given by Bernoulli's equation as:-

$$p_m = \rho \frac{U_b^2}{2} \qquad (6)$$

Clearly the flow of water produced by a breaking wave is neither steady nor uniform and in this case equation (6) is written as:

$$p_m = C_I \frac{\rho U_b^2}{2} \qquad (7)$$

in which C_I is an impact coefficient which must be determined experimentally. Note that from equations (5) and (7):-

$$C_I = \frac{4 l_v}{U_b t_m}$$

The actual value of C_I (and, therefore, of p_m) will depend upon various factors including the height and period of the breaking wave, the slope of the sea-bed in front of the wall and the particular elevation on the wall which is being considered.

Figure 4 shows the results of some measurements of pressure p_m undertaken on model walls sited on beaches with slopes, $\tan \beta$, of 0.044, 0.069 and 0.088 (Hayashi and Hattori, 1958) (12). It may be seen that C_I had values of up to about 30 and this value might reasonably be adopted to predict p_{max}, the maximum value in the vertical distribution of p_m. Note that, theoretically, the maximum possible value of C_I, (C_I) lim, is given (Von Karman, 1929) (13) by:-

$$(C_l)_{\lim} = \frac{2\,c_w}{U_b} \qquad\qquad (9)$$

c_w is the speed of sound in water which is approximately 1450 m/s. However, this value may be reduced very considerably by the entrainment of a small percentage of air (see Figure 5 and experiments by Gibson (1970) (14)). Entrainment of air inevitably occurs during wave breaking and the strong influence which this entrainment has on the value of c_w may largely explain the wide variation in the measured values of p_{max}. Substituting equation (9) into equation (7) gives the upper limit of p_m, $p_{\lim}$, as:-

$$P_{\lim} = \rho\,c_w\,U_b \qquad\qquad (10)$$

This is the water-hammer pressure.

In using equation (7) to estimate wave impact pressures it is obviously necessary to assign a value to U_b, the initial horizontal water-particle velocity produced by the breaking wave. This value is usually assumed to be equal to the breaker celerity, C_b. For shallow water conditions, linear wave theory gives $C_b = \sqrt{g\,d_b}$. For conditions where waves are breaking however the waveform may be closer to description by solitary wave theory rather than linear in which event $C_b = \sqrt{g(d_b+H_b/2)}$.

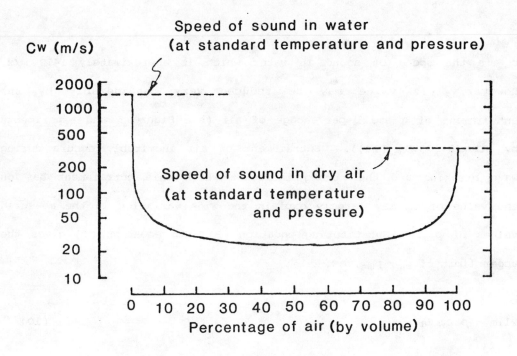

Figure 5   Effect Of Air Bubbles On The Speed

Of Sound In Water   (After GIBSON, 1970)

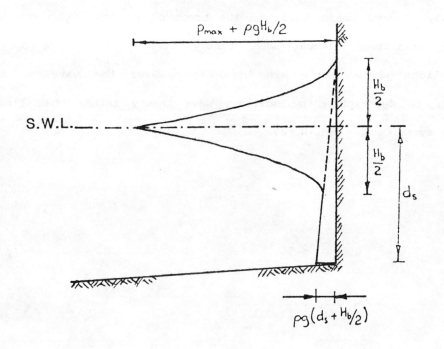

Figure 6   Pressures From Breaking Waves

The maximum value in the vertical distribution of p_m, p_{max}, often occurs at, or near to, still-water-level. However, its position can be influenced considerably by the presence of features such as a mound at the base of the wall as exists, for example, in a composite breakwater (a vertical-faced structure sited on a rubble mound). When the peak pressure is at still-water-level then there is usually a rapid decrease in the value of p_m above and below this level, approximately according to the equation:-

$$P_m = P_{max}\left(1 - \left|\frac{2z}{H_b}\right|^2\right) \qquad \text{for} \quad \frac{-H_b}{2} < z < \frac{H_b}{2} \qquad (11)$$

In this case the total dynamic force, F_m, acting on unit length of wall is given by:-

$$F_m = \int_{-\frac{H_b}{2}}^{\frac{H_b}{2}} P_m \, dz$$

$$F_m = 2 \int_{0}^{\frac{H_b}{2}} P_m \, dz = \frac{P_{max} H_b}{3}$$

Other possible pressure distributions are considered by Nagai (1973) (15) who also discusses a number of different equations for predicting p_{max}.

In addition to the dynamic force acting on the face of the wall there will also be a hydrostatic force. The hydrostatic pressure distribution may be assumed to act from a level of $H_b/2$ above still water-level at the instant of impact. Figure 6 shows the combined loading.

Although derived from more empirical considerations than adduced above the traditonal approach in the calculation of forces due to breaking wave is that developed by Minikin (1955, 1963) (16/17) who developed a design procedure based on observations of full-scale breakwaters and the results of Bagnold's study. Minikin's method can give wave forces that are extremely high, as much as 15 to 18 times those calculated for nonbreaking waves. Therefore, the following procedures should be used with caution, and only until a more accurate method of calculation is found.

The maximum dynamic pressure assumed to act at the still-water-level (SWL) is given by:-

$$P_{max} = 101 \, w \, \frac{H_b}{L_D} \frac{d_s}{D} (D + d_s) \qquad\qquad (12)$$

H_b is the breaker height, d_s is the depth at the toe of the wall, D is the depth one wavelength offshore of the wall, and L_D is the wavelength used to determine D. The distribution of dynamic pressure is shown in Figure 7. The pressure decreases parabolically from p_m at the SWL to zero at a distance of $H_b /2$ above and below the SWL.

The hydrostatic contribution to the force and overturning moment must be added to the results obtained from Minikin's formula to determine total force and overturning moment.

The Minikin formula was originally derived for composite breakwaters comprised of a concrete superstructure founded on a rubble substructure. Strictly, D and L_D in Equation 12 are the depth and wavelength at the toe of the substructure; d_s is the depth at the toe of the vertical wall (i.e. the distance from the SWL down to the crest of the rubble substructure). For caisson and other vertical structures where no substructures are present, the formula has been adapted by using the depth at the structure toe as d_s; D and L_D are the depth and wavelength a distance one wave-length seaward of the structure. Consequently, the depth D can be found from:-

$$D = d_s + L_d \, m,$$

where L_d is the wave length in a depth equal to d_s, and m is the near - shore slope. The forces and moments resulting from the hydrostatic pressure must be added to the dynamic force and moment computed above.

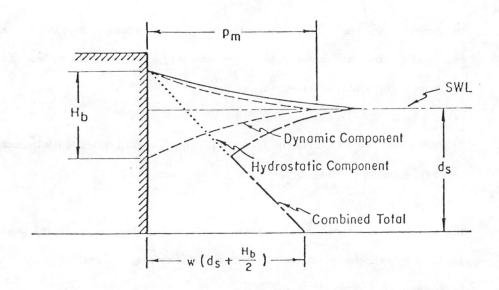

Figure 7   Minikin Wave Pressure Diagram

The triangular hydrostatic pressure distribution is shown in Figure 7; the pressure is zero at the breaker crest (taken at $H_b/2$ above the SWL), and increases linearly to $w(d_s + H_b/2)$ at the toe of the wall.

Pressures from Broken Waves

After a wave breaks it forms a highly turbulent bore of water which may interact strongly with the uprush/backrush cycle of the previous wave. The whole process is extremely complicated and amongst other factors the high degree of air entrainment which takes place would make it difficult to relate the pressures experienced by small scale models to the values likely to be encountered in the field. Consequently there is little information on the forces experienced by structures subjected to the action of broken waves. A small amount of field data (Miller et al., 1974) (18) suggest that if the impact pressures associated with bores are represented by equation (7) then C_I may have values of up to about 5.

3. REQUIREMENTS

(a)   To provide an instrument to measure total pressure loading on a sea wall, on an area of approximately $0.07m^2$ in order to monitor a representative area.

(b)   To provide an instrument casing which will cause minimum interference with the wave action on the sea wall.

(c)   To provide an instrument capable of measuring very high peak loads of extremely short duration, dictating a sensor with a high natural frequency.

(d)   To provide adequate waterproofing.

(e)   To provide means of attachment to the sea wall.

It is tempting in any pressure measurement to use existing pressure transducers since there is no development to undertake and their performance is well defined from manufacturer's literature. However, experience shows that the integration of a number of essentially point sources to provide a true area integration of pressure on a surface is inaccurate unless the total number of transducers is high, which in turn is expensive and adds additional data analysis. Thus an alternative was sought in the form of a large area pressure transducer which effectively provides from its single output the integrated value. It was this philosophy which led to the final design configuration.

4.   DESCRIPTION OF THE SEA WALL INSTRUMENT

The instrument consists essentially of a body for containment of load cells and electrical connections, fitted with external attachments for ease of mounting on sea walls, (Fig. 8).

A load plate, fitted with a rubber diaphragm seal completes the casing. Loads applied to the load plate are transmitted to the body through three shear load cells, spaced radially at 120°. The three shear load cells are interconnected, to provide a combined output proportional to the total load applied to the load plate. Power in and signal out, are taken to measuring instruments by a single multi-core cable, glanded at the casing.

Shear cells using electrical resistance bonded strain gauges were selected for their sturdiness, and their ability to respond to high shock loads.

The body and loading plate are both made from high quality carbon steel, to reduce the overall size of the unit, compatible with strength requirements. The internal space is filled with an electrical insulating liquid, and small air sacs are placed within the liquid to allow small deflections of the load plate relative to the body to be achieved without hydraulic locking, whilst allowing a high frequency response.

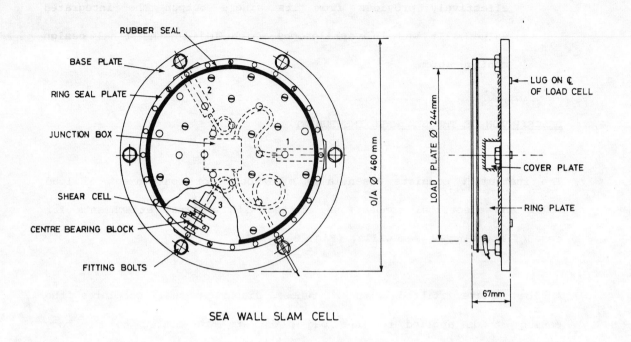

SEA WALL SLAM CELL

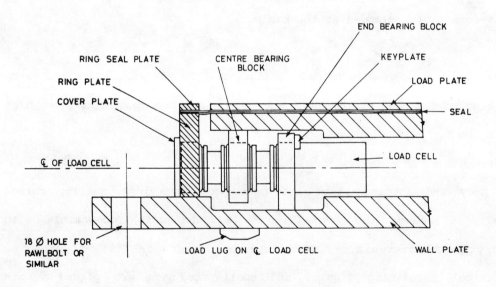

G.A. OF LOAD CELL CARRIER

FIGURE 8

5. TEST PROGRAMME

(a) To load the unit, statically up to a limit of 2 tonnes, across diameters, ensuring that each shear load cell was loaded in turn.

(b) To subject the complete assembly to slam loads, by dropping face down in water from pre-determined heights.

The tests described in (a) are to act as a calibration, or span check of the complete assembly, and to note if there is any variation on measured load dependant on position of the centre of pressure of the applied load. Tests described in (b) are intended to reproduce slamming conditions in a controlled manner. This is to indicate the steepness of the load response, and to provide information on the natural frequency and damping of the system.

6. TEST RESULTS AND DISCUSSION

(a) Fig. 9 is obtained by loading the load plate across diameters. It will be seen that providing the centre of pressure remains within the triangle bounded by the sensitive sections of the 3 shear load cells, then the unit is tolerant of centre of pressure. If the centre of pressure lies outside the triangle, (an unlikely occurrence), then the load measured shows a variation of 4%. The most likely explanation for this is that whilst the centre of pressure (C.P.) remains within the triangle, all shear cells are subject to a 'compressive' load, but if the C.P. is outside the triangle, two shear cells will normally see 'compression', the other seeing 'tension'.

(b) Figs. 10 and 11 show the output from the unit resulting from dropping flat into water from a pre-determined height. For this test, the unit was connected to a storage oscilliscope, fitted with a threshold level trigger.

The unit was dropped, freely into the water, and the storage oscilliscope automatically triggered and stored the data, for replay over a longer period using a standard BMT Ltd beach data acquisition and analysis system. Immediately obvious from the trace is the very steep rise and decay of the initial water impact, and the natural frequency of the system in water.

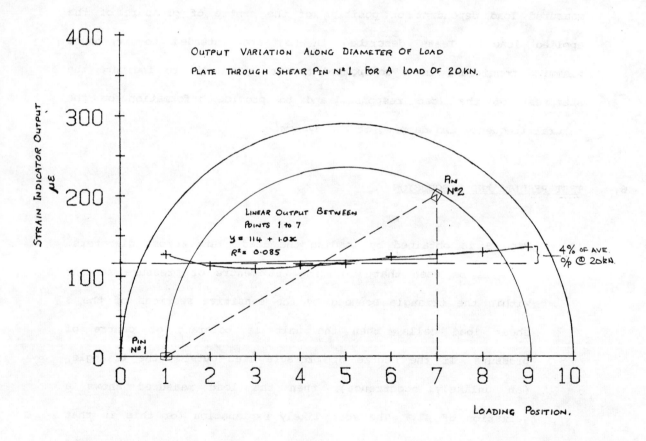

FIGURE 9

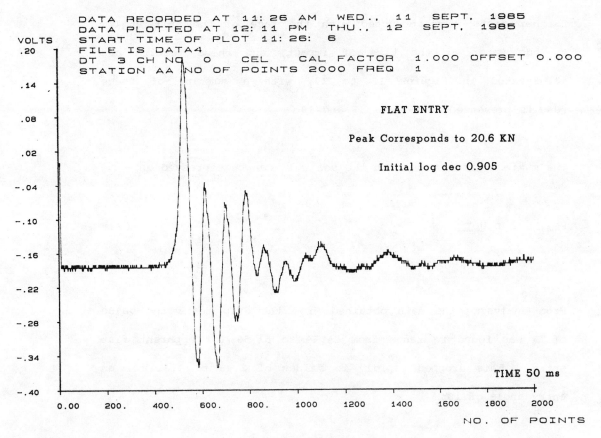

FLAT ENTRY

Peak Corresponds to 20.6 KN

Initial log dec 0.905

TIME 50 ms

FIGURE 10

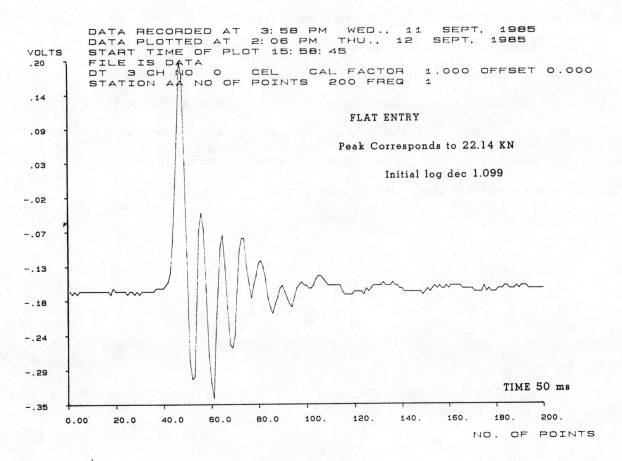

FLAT ENTRY

Peak Corresponds to 22.14 KN

Initial log dec 1.099

TIME 50 ms

FIGURE 11

167

Further tests were conducted with the face of the unit angled to the water at the time of impact, and these results are illustrated in Figures 12 to 17, with a summary of these results presented in Figures 18 and 19.

The maximum pressure P_{max} on the sea wall can be expressed as:-

$$P_{max} = C_l \ \rho \ \frac{Ub^2}{2} \qquad\qquad (7)$$

From analysing the data obtained from the drop tests the value of C_l was found to range from 24.44 to 51.56 for apparent flat entry. This dropped rapidly to values of 2.96 to 4.31 for an entry angle of 10°.

In practice it was not possible to guarantee the squareness of the angle of entry. On the tests where the angle of entry was controlled, there was still some 'whip' as the cell penetrated the surface. The indicated slam loads dropped considerably with increased entry angle. This is due to the fact that when the cell lands square the rate of load plate area contact with the water surface is greater than the speed of sound in water, and a compressive shock wave is generated.

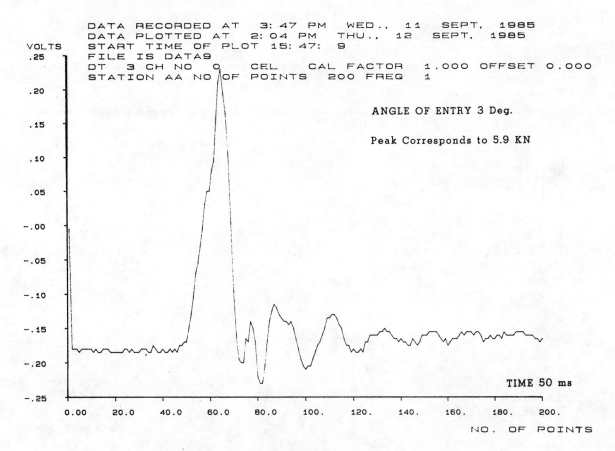

FIGURE 12

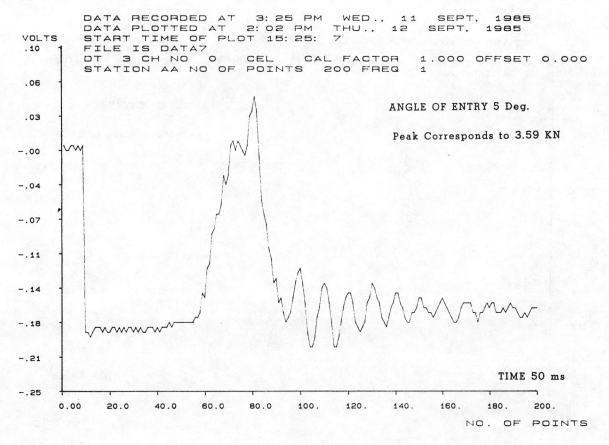

FIGURE 13

169

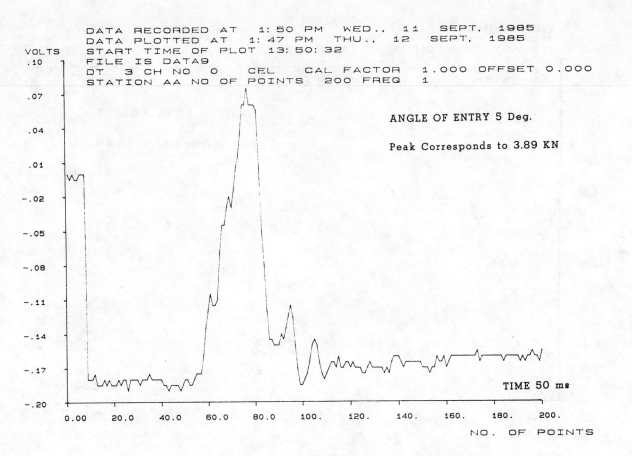

FIGURE 14

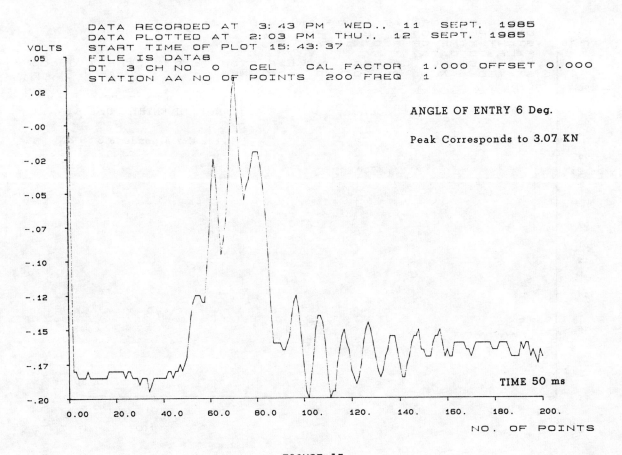

FIGURE 15

170

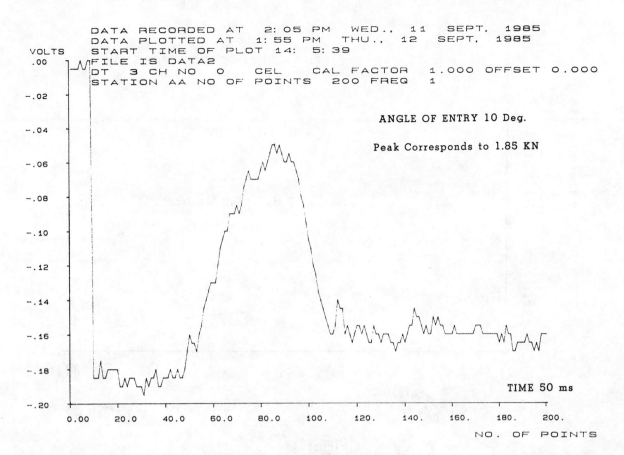

DATA RECORDED AT 2: 05 PM WED., 11 SEPT. 1985
DATA PLOTTED AT 1: 55 PM THU., 12 SEPT. 1985
START TIME OF PLOT 14: 5: 39
FILE IS DATA2
DT 3 CH NO 0 CEL CAL FACTOR 1.000 OFFSET 0.000
STATION AA NO OF POINTS 200 FREQ 1

ANGLE OF ENTRY 10 Deg.

Peak Corresponds to 1.85 KN

TIME 50 ms

NO. OF POINTS

FIGURE 16

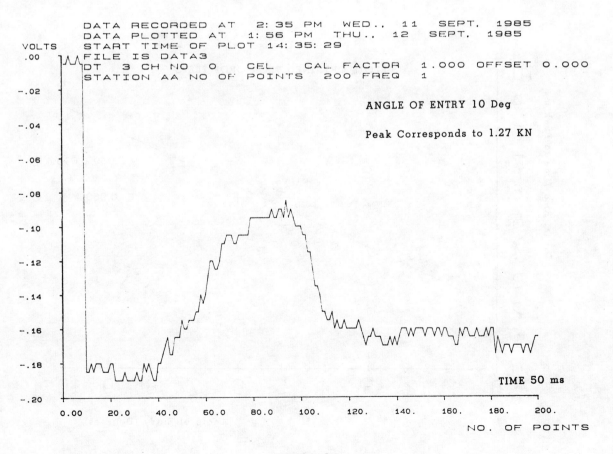

DATA RECORDED AT 2: 35 PM WED., 11 SEPT. 1985
DATA PLOTTED AT 1: 56 PM THU., 12 SEPT. 1985
START TIME OF PLOT 14: 35: 29
FILE IS DATA3
DT 3 CH NO 0 CEL CAL FACTOR 1.000 OFFSET 0.000
STATION AA NO OF POINTS 200 FREQ 1

ANGLE OF ENTRY 10 Deg

Peak Corresponds to 1.27 KN

TIME 50 ms

NO. OF POINTS

FIGURE 17

171

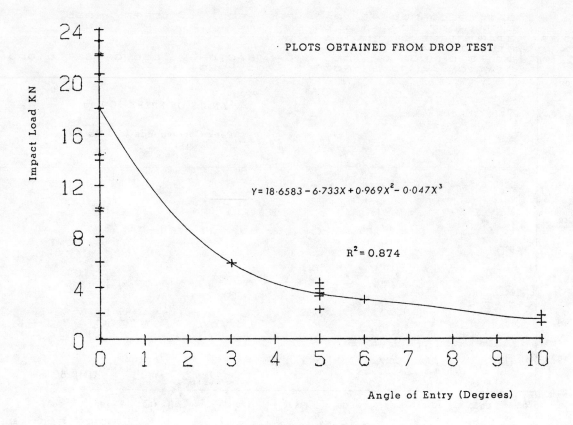

PLOTS OBTAINED FROM DROP TEST

$Y = 18.6583 - 6.733X + 0.969X^2 - 0.047X^3$

$R^2 = 0.874$

Impact Load KN

Angle of Entry (Degrees)

FIGURE 18

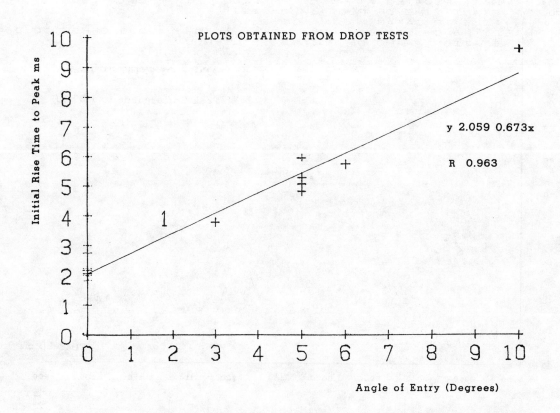

PLOTS OBTAINED FROM DROP TESTS

y 2.059 0.673x

R 0.963

Initial Rise Time to Peak ms

Angle of Entry (Degrees)

FIGURE 19

7. CONCLUSIONS

Irrespective of angle of entry, the load cell can integrate the total pressure normal to the load plate, and hence give an accurate reading of the load upon the structure being monitored. The natural frequency of the cell in water, 460 Hz observed in Figs. 10-17, is greater than the initial rise time to peak under the highest expected slam conditions, and the damping coefficient enables the cell to be restored within 25ms.

Results obtained from the British Maritime Technology Limited data collecting station in Christchurch Bay show that sea surface elevation spectra can be expected to peak from 0.1 to 0.3Hz. Therefore, the slam cell is well able to stabilize itself before the next wave impact.

From Hayashi and Hattori method the expected C_l values do not exceed 30 (Fig. 4). The maximum value calculated from the drop test results was 51.56, against a theoretically possible value of 888. The variation of the maximum value was 24.44 to 51.56 at flat entry.

Although these calculated values are much lower than the theoretical maximum, they are generally higher then the expected values on sea walls.

Because of the nature of the drop test the air entrainment is very low compared to actual wave impact. This would give rise to the higher values of C_l. The inability of the test rig to guarantee perfect flat entry, and the fact that a shock wave is also generated in the body of the cell itself causes the C_l value to be lower than $(C_l)_{lim}$, and would also account for the variation in C_l.

The previous work on wave forces against sea walls has identified the significance of air entrainment within the wave upon the wave loading. The tests described here upon simulated impact at different angles of entry suggest that this also is of critical importance in determining the wave load upon the sea wall.

As the results clearly demonstrate the sea wall cell is clearly capable of measuring wave impact loads with C_l values greater than the expected maximum value of 30.

It is proposed to deploy this sensor upon a vertical sea wall for full-scale trials with the incident directional wave spectrum measured concurrently. It is concluded from work already undertaken that without such definition of incident sea state no meaningful relationship between wave conditions and their loading upon a sea wall is possible.

8. **ACKNOWLEDGEMENT**

The authors wish to acknowledge the invaluable assistance of Mr. T. Hedges in the preparation of this paper with regard to the development of the theory and Mr. G.W. Pearce for his advice and assistance with regard to the instrument design.

REFERENCES

1. BAGNOLD, R.A. Interim Report on wave Pressure Research.
   Journal of the Institution of Civil Engineers, Vol. 12, London, 1939.

2. DENNY, D.F. Further Experiments on wave pressures.
   Journal of the Institution of Civil Engineers, Vol. 35, London, 1951.

3. ROSS, C.W. Laboratory study of Shock Pressures of Breaking Wave.
   TM 59, U.S. Army, Corps of Engineers, Beach Erosion Board, Washington D.C., February 1955.

4. NAGAI, S. Experimental studies of specially shaped concrete blocks for Absorbing wave Energy, Proceeding of the seventh conference on Coastal Engineering, ASCE, Council on Wave Research, The Engineering Foundation, 1961a.

5. CARR, J.H. Breaking Wave Forces on Plane Barriers. Report No. E 11.3 Hydrodynamics Laboratory, California Institute of technology, Pasadena, California, 1954.

6. LEENDERTSE, J.J. Forces Induced by Breaking Water Waves on a vertical wall. Technical report 092, U.S. Naval Civil Engineering Laboratory, 1961.

7. KAMEL, A.M. Water Wave Pressures on Seawalls and Breakwaters. R.R. No. 2 - 10, U.S. Army, Corps of Engineers, Waterways Expeiment Station, Vicksbury, Miss., 1968.

8. WEGGEL, J.R.   "The Impact Pressures of Breaking Water Waves, Thesis presented to the University of Illinois, Urbana, Illinois, in partial fulfillment of the requirement for the degree of Doctor of Philosophy, (unpublished, available through University Microfilms, Ann Arbor, Michigan).

9. WEGGEL, J.R., and MAXWELL, W.H.C.   Experimental Study of Breaking Wave Pressures, "Preprint volume of the Offshore Technology Conference, paper No. OTC 1244, 1970b.

10. ROUVILLE, A., de, BESSON, P. and PETRY, P.   Etat Actuel de Etudes Internationales sur les Efforts dus aux Lames, Annals des Ponts et Chaussees, Paris, Vol. 108, No. 2., 1938.

11. BLACKMORE and HEWSON, P.J.   1984 Experiments on Full Scale Wave Impact Pressures Coastal Eng., 8 : 331 – 346.

12. HAYASHI and HATTORI (see page 176)

13. VON KARMAN (see page 176)

14. GIBSON, 1970 (see page 176)

15. NAGAI (1973) (see page 176)

16. MINIKIN R.R.   Breaking Waves, A Comment on the Genoa Breakwater, "Dock and Harbour Authority, London, 1955, pp 164 – 165.

17. MINIKIN R.R.   Winds, Waves and Maritime Structures, studies in Harbour Making and in the protection of coasts, 2nd rev. ed. Griffin, London.

18. MILLER et al. 1974 (see page 176)

KORNHAUSER, M.

Structural Effects of Impact.

U.S. ARMY COASTAL ENGINEERING RESEARCH CENTRE.

Shore Protection Manual Vol. II.

GIBSON, F.W.

(1970), Measurement of the effect of air bubbles on the speed of sound in water.

J. Acoustical Society of America, Vol. 48, No. 5, pp 1195 – 1197.

HAYASHI, T. and HATTORI, M.

(1958), Pressure of the breaker against a vertical wall.

Coastal Engineering in Japan, Vol. 1, pp 25 – 37.

MILLER, R.L., LEVERETTE, S., O'SULLIVAN, J. TOCHKO, J., and THERIAULT, K.

(1974), Field Measurement of impact pressures in surf.

Proc. 14th Coastal Engineering Conference Am.Soc.Civ. Engineers pp 1761 – 1777.

NAGAI, S.

(1973), Wave forces on structures.

Advances in Hydroscience, Vol. 9, pp 253 – 324.

VON KARMAN, TH.

(1929), The impack of seaplane floats during landing, Tech. Note No. 321, National Advisory Committee for Aeronautics, Washington.

A SYSTEM FOR THE MEASUREMENT OF THE STRUCTURAL RESPONSE
OF DOLOS ARMOR UNITS IN THE PROTOTYPE

Gary Howell

Coastal Engineering Research Center
Waterways Experiment Station
U.S. Army Corps of Engineers
Vicksburg, Ms. 39180-0631
U.S.A.

Summary

Twenty new dolosse to be installed as part of repairs at Crescent City Harbor, California, will be instrumented to determine stresses within the dolosse due to impact loads and wave forces in the prototype. Offshore incident wave conditions will be simultaneously measured to allow correlation of wave forces and impact energy with sea state. Velocity of six of the dolosse will be measured with six degrees of freedom in order to determine actual motion of the dolosse and calculations of impact energy.

Implementation of the program required development of specialized instrumentation, data acquisition systems, and environmental protection. All dolos instrumentation is permanently sealed within the concrete armor unit. An internal microcomputer system for each dolos digitizes stress and motion data at rates sufficient to characterize dynamic stress response. Data are returned to the top of the breakwater by a power and data cable system protected with a modified anchor chain assembly. Data from all instrumented dolosse are cabled to a shore based computer system. A hierarchical, multi-processor architecture is used to implement the shore computer system. Preliminary data reduction is performed in real time to reduce the data quantity required for storage to magnetic tape.

Results of the measurements will be used together with finite element structural analysis and physical models to develop design criteria for the structural strength of dolos armor units.

Held at Imperial College of Science and Technology, London. Organised and sponsored by BHRA, The Fluid Engineering Centre. Co-sponsored by the American Society of Civil Engineers and the International Association for Hydraulic Research.
© BHRA, The Fluid Engineering Centre, Cranfield, Bedford MK43 0AJ, England 1986.

1. INTRODUCTION

Future advances in hydraulic and coastal engineering are becoming more dependent on improvements in prototype scale measurement techniques. The lack of adequate prototype data has delayed the successful application of standard engineering analysis techniques such as physical and numerical modeling to the solution of many current problems. In the area of breakwater engineering, assessing the structural strength requirements of concrete armor units is such a problem. Progress requires the modeling of complex interactions of physical processes, and because of the severe environment of the prototype breakwater, very little prototype scale data is available.

The economic benefits of concrete armor units have been widely recognized during the past two decades. As breakwaters are designed for more severe wave conditions, sufficiently large natural stone is rarely available. Hydraulic models of breakwaters using concrete units for armor layers are routinely used for most designs. However, contemporary hydraulic scale models consider only stability and assume that all armor units have sufficient structural strength, if stability conditions are maintained. Experiences with breakage of concrete armor units have shown that armor unit structural strength must now be considered in the breakwater design process. Unfortunately, there is not presently a design methodology capable of predicting the forces, loads, and stresses required to establish structural strength requirements.

As part of an effort to develop a structural design procedure, the U.S. Army Corps of Engineers, Waterways Experiment Station (WES) has begun a program to acquire data on the forces and structural response of dolos armor units in the prototype. The goal of the program is to provide data from armor units in an actual prototype breakwater exposed to storm wave conditions, along with supporting measurements sufficient to provide calibration and verification data for physical and numerical models of stuctural response. The site for the study is Crescent City, California, where a previously constructed breakwater was scheduled for rehabilitation with new layers of 42 tonne dolos. Because of the severe wave conditions at Crescent City and the desire to acquire dynamic structural response data, new measurement technology had to be developed.

2. MEASUREMENT PLAN

The Crescent City prototype study is a comprehensive monitoring program to characterize the 42 tonne dolos unit from casting through transport, placement, nesting, and exposure to storm wave conditions (Ref. 1). The rehabilitation of the breakwater requires the placement of 500 new, metal fiber reinforced dolosse over existing unreinforced dolosse along a 100 m trunk section of the jetty. Twenty of the new dolosse will contain instrumentation cast inside for measurement of strains at the shank-fluke interface. Six of the twenty instrumented dolosse will be fitted with a motion measurement package in addition to the strain instruments. Wave measurements will be made offshore and near the breakwater. The structural measurements will be supplemented with photographic and bathymetric measurements. This paper will describe the dolos instrumentation and the data acquisition system developed to acquire data from the units placed in the breakwater matrix.

The measurement plan for the study was the result of efforts to compromise the most desirable experimental data with constraints imposed by cost and technical feasibility. It was developed after consideration of many alternative plans and several approaches to the problems of instrument survivability and data acquisition. A key element in the measurement plan was the decision to define the fundamental measured parameters as the two moments and the torque taken about the shank-fluke interface. This was adopted over, perhaps, more traditional techniques of surface mounted concrete strain gages, because it was a measurement that could be performed by internally mounted gages, thus providing protection for the instruments. An additional advantage was that the measurement of the moments and torques could be accomplished with a minimum number of gages, thus keeping the data rates per dolos lower than would be required with sufficient surface strain gages. It should be emphasized that one of the original goals of the study was to obtain dynamic stuctural strains such as might be associated with the fundamental modes of the dolos when excited by an impact load. Measurement of dynamic responses requires data rates significantly higher than measurement of static strains and those strains with periods on the order of magnitude of ocean waves. Preliminary experiments with finite element (FE) numerical grids indicated that, given measurements of moments and torques, the FE models could yield stress distributions throughout the areas of interest in the dolos.

Another assumption of the measurement plan is that interpretation of the data will be, of necessity, stochastic. In order to have reasonable confidence in the results, a large sampling of instrumented dolosse would be required. The original proposals for the study recommended thirty to forty instrumented dolosse, however, cost limitations forced reductions. Rather than limit the number of instrumented units to an unacceptably low number, an additional compromise was made. Strain instrumentation is provided only at one shank-fluke interface. The selection of the verticle tending shank-fluke interface for instrumentation was based on field experience with breakage, and consideration of potential stress concentrations predicted by simple beam element models of dolos.

Given that the structural response of the instrumented dolosse are measured via the two moments and torque taken about the shank-fluke interface, it remains to quantify the forcing function from which the response arises. The actual stress can be considered as the summation of the static stresses, stresses due to wave forces, and impact stresses. The static stresses arise from the dolos' own weight as well as the loads of other dolosse resting on it in the matrix. Wave force stresses are drag related forces caused by wave induced water velocity components. Impact stresses are excited when a dolos moves and collides with another dolos. Due to the potential importance of impact stresses in dolos breakage, it is important to quantify motions of dolos in the matrix. While it would be desirable to measure the motions of all of the instrumented dolosse, cost constraints again forced a compromise. Six of the dolosse are instrumented with accelerometers to measure motion with six degrees of freedom.

The placement of the instrumented dolos in the breakwater is a critical part of the measurement plan. Again, many alternatives were considered, including equal dispersement along the breakwater, random placement, and grouping in three areas roughly corresponding to low, moderate, and severe wave energy. The plan chosen was based on concern that interpretation of the results depended on the ability to have useful statistical comparisons between units exposed to the same wave energy. Therefore, it was decided to locate all of the instrumented dolos in a single matrix centered approximately around the mean water level. Fig. 1 shows the planned placement area superimposed on a photograph of the hydraulic model of the post rehabilitation breakwater. Fig. 2 is a side view of the placement area.

3.   SENSORS

Limiting structural measurements to the moments and torque about the shank-fluke interface permits the sensors to be arranged in a configuration which facilitates installation and protection of the gages and interconnecting wiring. Fig. 3 shows a view of the the arrangement of the sensors inside the dolos. The moments and torque are derived from strain gages placed on steel rebar members arranged in a rosette geometry. Four rosettes are cast in the dolos, one at each face. Fig. 4 is a photograph of one rosette. Moments are determined from the center bars on opposite sides. Two redundant measurements of torque are obtained. A torque measurement is derived from the algebraic combination of the strains from two opposite sets of rebars at 45 degree angles.

The accelerometers for the six dolosse instrumented for motion are mounted in a special cylinder contained in the trunk section of the dolos. Fig. 3 shows a steel cylinder sleeve cast into the dolos along with the strain gage assembly. The strain gage wiring terminates in a pressure balanced, oil filled cable assembly inside the steel cylinder sleeve. The accelerometer package is contained in a second cylinder installed inside the steel cylinder after the dolos has cured. The second cylinder is known as the dolos processor because it contains the digitizing electronics for the strain gages as, well as the accelerometers. Dolos without accelerometers contain dolos processor cylinders with empty space for the accelerometer assemblies. The accelerometer assembly is composed of three 1G linear servo accelerometers and three 100 rad per sec^2 angular servo accelerometers. Locating suitable angular accelerometers with the required accuracy and long term reliability for a feasible cost was quite difficult. A somewhat lower range would have been desirable for the angular accelerometers, however, they could not be obtained within the cost and schedule limitations of the project. The combination of the six accelerometers gives an inertial reference with six degrees of freedom, which assuming rigid body dynamics, allows the determination of the motion of any point on the dolos. The ranging of the dolos accelerometers is taken to provide measurements of dolos motion, but not to directly measure the deceleration due to an impact. Determination of an accurate velocity vector just prior to impact was preferred over direct measurement of impact, because impact energy can be calculated from the velocities.

4. DATA ACQUISITION

The design of the data acquisition system for the dolos instrumentation is heavily influenced by the requirements for environmental protection. Traditional techniques, such as locating an instrumentation van on the top of the breakwater were ruled out, because of the frequent overtopping of the breakwater during even moderate storm events. The decision was made to trade increased complexity in the internal dolos electronics for simple physical requirements at the breakwater site. By configuring a microcomputer controlled analog to digital converter system inside each dolos, data can be transmitted via standard digital, serial communications techniques to a safely located shore facility.

Fig. 5 shows a block diagram of the data acquisition system. The strain gages are digitized directly by the dolos processor. The signal from each bridge is band limited to 100 Hz by a Bessel filter and then sampled at 400 Hz. The choice of bandwidth for the system was made by estimating the frequencies of modal oscillation for the 42 tonne dolos. Using a simple beam element model, it was estimated that the first five modes could be identified with a 100 Hz bandwidth. The analog to digital (A-D) converter system contains an automatic programmable gain feature to allow the full dynamic range of the strain gages to be represented by a 12 bit A-D converter. The resulting combination allows 12 bit accuracy within a dynamic range of approximately 16 bits.

For the six dolosse with accelerometers, the analog signal is integrated once in the dolos processor to yield an analog velocity signal for each of the six degrees of freedom. The integrators incorporate a low frequency washout with a period greater than 200 seconds to eliminate accumulation of velocity errors due to accelerometer offsets. The high end frequency response is limited to 10 Hz and the velocity signal is digitized at a 40 Hz sample rate.

After digitization, the dolos processor formats all of the sampled data into packets which are communicated over an RS-422 serial link to the data concentrator, located on the breakwater cap. There are two data concentrators, each receiving cables from 10 instrumented dolosse. The data concentrator distributes power to each dolos and combines serial links from each dolos into one cable, maintaining a separate physical wire pair for each dolos. The shore cable provides one physical wire pair for the command serial link from the shore computer which is redistributed by the concentrator to each individual dolos. Thus each dolos processor receives all messages and commands from the shore computer. It must recognize and act on messages intended for it by decoding a dolos processor address contained in command and message packets. Command packets are not required in normal operation, and measured data is transmitted in a continuous stream from each dolos, through the concentrator, and back to the shore computer. The serial data rate from the dolos is 38.4 K Baud.

5. ENVIRONMENTAL PROTECTION

5.1 Protection requirements.

It is difficult to imagine a place less conducive to the reliable operation of electronic instruments than an ocean breakwater. For any hope of successful acquisition of data, the requirements of environmental protection must be considered in every phase of the measurement system design. The problem can be considered in three parts, consisting of the internal dolos instrumentation, the data cable from each dolos to the top of the breakwater, and the cables from each data concentrator to shore.

5.2 Internal instrument protection.

The internal dolos instruments rely on the dolos itself for structural protection. The main task is to maintain water tight protection for the sensors, cabling, and the dolos processor. The sensors on the rebar were installed and pressure tested at each stage of multiple levels of potting protection. All cabling was pressure tested for water tightness. After test, cables were installed in an oil filled conduit and pressure balanced tubing assembly that maintains an oil path from the strain gage back to the connector for the dolos processor. The dolos processor is then connected to the oil filled cable with a water proof connector and installed in the steel cylinder sleeve. A foam shock absorbing material fills the intercylinder volume. After installation in the dolos, the data cable is connected to the cylinder with a right angle underwater connector and the cable is passed out through a steel and polyurethane grommet assembly. The recessed space below the dolos processor (see Fig. 1) is then filled with a special water proof grout, making the surface of the dolos completely smooth. The only projections are the anchor bolt for the cable strain termination, the grommet assembly, and the cable itself.

5.3 Dolos data cable.

Perhaps the most difficult environmental protection problem is the method of retrieving data from an instrumented dolos. The feasibility of various methods of acquiring data from a dolos without a cable were explored. The high data rates eliminated from consideration most types of low bandwidth RF and acoustic telemetry systems. Estimates of total power consumption for a high data rate telemetry system yielded battery sizes that were impractical due to cost and size considerations. Use of a cable allowed both power to be supplied to the dolos instrumentation and retrieval of data at high rates. The use of a microcomputer inside each dolos was implemented to digitize data from the strain gages and accelerometers. This allowed the digital data to be formatted and sent on each cable as a serial data stream, thus limiting the number of required conductors in the cable to eight. By minimizing the number of conductors, a highly reliable cable was constructed consisting of the core cable, an extra thick polyurethane jacket, covered by a double layer of steel armor, with a final jacket of polyethylene, yielding a total outside diameter of 2.2 cm. Even though the cable is highly crush resistant, additional protection and anchoring is required.

A cable exposed to high velocity, oscillatory flows associated with the armor layer of a breakwater, will fail due to fatigue if it is allowed to oscillate with the flows. An additional armoring mechanism is required which will increase the density of the cable sufficientlty to prevent oscillatory motion. Precedents for this type of cable protection are rare. A system known as split pipe has been developed by the U.S. Navy Civil Engineering Laboratory (Ref. 2) for surf zone cable transitions. However, use of split pipe was ruled out due to the relatively small increase in density (about double cable alone), high cost, and time consuming installation procedures which would delay the dolosse placement operation.

After a period of design and field testing, a system employing a modified anchor chain was developed. The assembly shown in Fig. 6 is manufactured with used, 3 in. anchor chain, modified with a split sleeve assembly fabricated from structural steel pipe. The sleeve assembly accomplishes two objectives. First, it provides a tunnel for the cable along the chain eliminating the need for fasteners to hold the cable to the chain. Second, the sleeves are specially sized such that they limit the minimum bend radius of the chain to approximately 0.7 m. This requirement was found to be necessary after field tests demonstrated that it was very difficult to handle the chain without forcing a sharp bend in the cable below its minimum bend radius. The sleeves also afford the cable additional protection from sharp objects, while still relying on the chain itself to provide crush resistance for objects the size of a dolos. Fig. 7 shows a test section of the complete chain, indicating the minimum bend radius and the ability to easily pass the cable through the assembly.

For the prototype installation, the modified cable protection chains will be attached to the underside of the trunk of each instrumented dolos, and terminated at an anchor bolt on the edge of the breakwater cap. When the dolos is placed, the chain will be laid loosely up the slope. Since the breakwater will be constructed from the toe towards the cap, the cable assemblies will be covered by succeeding rows of dolos. From the point the chains are connected on the cap, the cables from each dolos will then traverse the cap in a buried conduit to the sheltered side, where they will terminate in the data concentrator assembly. There are two data concentrators, in water proof cylinders which are installed in horizontally drilled holes in the side of the cap. This configuration eliminates the need for any structure on the breakwater, using the breakwater itself as a protective structure.

5.4 Shore cable.

One double armored cable from each concentrator will carry power and data from the concentrators back to shore. The cable will be laid in the harbor for a distance of 1.6 km to the shore located data acquisition computer system. The shore based computer system will be located in a semi-trailer specially modified for the purpose.

6. DATA REDUCTION AND STORAGE SYSTEM

6.1 Data reduction algorithms.

Under normal operation the total data rate for all twenty instrumented dolosse will be approximately 60,000 bytes per sec, not counting overhead for packet synchronization and error checking. If all data were saved the data accumulation rate would be some 200 million bytes per hour. Even if it was possible to record all of this data for, say, one storm of three days duration, the total quantity of data for post analysis would be very difficult to efficiently manipulate.

The solution to this difficulty is to recognize that while the data rates are quite high, the required information is much less. Recall that the total stress can be considered the sum of static stress, impact stress, and stress due to wave forces. Let us adopt the terminology of Burcharth (Ref. 3) to refer to stress due to wave forces as pulsating stress. Because of the large time scale differences between the classes of stress, it is possible to separate the data by a process of time decimation and framing according to the type of observed stress. This can significantly reduce the total quantity of data and present it in a more directly useful form. If this process can be performed in real time, the data recording task becomes much more feasible.

Specifically, the real time algorithm is implemented as follows. For each dolos, take the moment and torque data and separate it into two types of processed packets. The first type of packet consists of a static/pulsating packet which is generated by digitally filtering the 400 Hz sampled data with a low pass filter set at 2.5 Hz and then time decimating the result to a sampled data rate of 10 Hz. Static/pulsating packets are generated for all received data. The second type of packet, the impact packet, is generated from a moving window which is programmed to recognize an impact event. The required length of the window can be estimated by taking the fundamental mode of the dolos, that is approximately 0.04 seconds for a 42 tonne dolos, and applying an approximate damping for concrete, to estimate that impact events can be represented by a time series record of 0.5 seconds. Using an event detection algorithm with variable thresholds to observe the data in the window, impact packets are created by taking a one second long frame of 400 Hz sampled data, and time stamping it with its position relative to the simultaneously generated static/pulsating packet. With this approach, the time resolution required for impact data is preserved, while minimizing the saving of data when impacts are not occuring.

6.2 Shore computer architecture.

To implement this strategy, a special architecture for the shore based computer system can be devised. Since all data will be formatted in packets and data can be processed independently for each dolos, a parallel architecture can be implemented which will make the still remaining problem of the high real time data rate more manageable. The parallel architecture takes advantage of the well known fact that many small computers are less expensive than a single large computer of equivalent power. For problems that can be treated in parallel, cost savings can be realized.

Fig. 8 shows the block diagram of the computer system used for the shore based data reduction and storage system. The system is a hierarchical parallel processor system with three levels. At the lowest level, there is a single board PDP-11 microcomputer for each dolos. This board known by the manufacturer's model number as the DEC KXT11-C is configured as a slave processor which can reside on the bus of an arbiter processor, in this case a PDP-11/23 processor. The KXT11-C can function as an independent processor but also has a section of its memory which is dual ported, that is the memory can be accessed by either the slave, KXT11-C, processor or the arbiter processor. This architecture facilitates control and data transfer from the slave processors by the arbiter.

Each KXT-11C handles communications with a single dolos processor, and performs all overhead processing necessary for packet synchronization and communications error checking. In addition the processor performs the real time reduction of the data from one dolos by reducing the incoming data to static/pulsating data packets and impact data packets. All packets contain header information sufficient to allow interpretation by post processed data analysis routines. In addition, the impact data packets have a relative energy indicator which is calculated as the ratio of the rms value of the impact data frame to the rms value of the current static/pulsating data packet. This relative energy indicator is used by the higher level processors as a priority indicator for deleting impact packets should queues start to overflow. The system is designed such that under normal conditions, queues will not fill, but to add robustness to the system under the conditions of erroneous data, the priority based packet deletion mechanism is provided.

Reduced data packets are posted by each KXT11-C to a PDP-11/23 arbiter processor. There is one arbiter for each concentrator, and therefore each group of ten instrumented dolos. Arbiter processors queue packets from all of the KXT11-C processors to the highest level processor, a DEC VAX 11/750. Communications between the two arbiters and the VAX computer are via an ETHERNET physical link using the DECNET software protocol. The VAX computer assembles packets arriving from both arbiters and writes them to one of a bank of four 500 MegaByte digital tape drives. The VAX computer automatically switches to a new drive when the previous drive is full. Simultaneously, the latest data packets are saved in a circularly organized disk file so that they can be accessed by real time monitoring, analysis, and graphics display software.

Control of the system may be accomplished by graphics display terminals locally, or via a remote communications link. Sufficient data analysis and graphic display software support is available on site to provide quality control and verify proper operation of the instruments. Final data analysis will be performed by transporting the recorded tapes back to WES for post processing.

7. CONCLUSION

Advances in hydraulic engineering have generally followed the development of field measurement capability. Further progress requires the development of measuring techniques which can obtain prototype data which has been previously unobtainable. New tools provided by advances in ocean engineering and computer technology represent one means to improve measuring techniques.

8. ACKNOWLEDGEMENTS

Many individuals contributed to the definition of the measurement problem and the development of the system described here. The author wishes to acknowledge his colleagues both in the Structures Laboratory and the Coastal Engineering Research Center of the Waterways Experiment Station.

Mr. Paul Mlakar suggested the concept of measurement of moments and torques and the geometry of the internal gages. Dr. Hans Burcharth made valuable contributions to the placement plan for the instrumented dolosse. The participants of the Workshop on Measurement and Analysis of Structural Response of Concrete Armor Units held at WES in January 1985 also helped solidify the measurement plan.

The work described in this paper was conducted as part of the Crescent City Prototype Dolosse Study sponsored by the San Francisco District, U.S. Army Corps of Engineers. Permission to publish this paper was granted by the Chief of Engineers.

9. REFERENCES

1. Howell, G.L. and Domurat, G.W.: "Measurement of Forces in Dolos Armor Units". Proceedings of the 41st Meeting of the Coastal Engineering Research Board, Coastal Engineering Research Center, Waterways Experiment Station, Vicksburg, MS., Aug. 1984, pp. 70-83.

2. Brackett, R.L., et al.: "Design and Installation of Nearshore Ocean Cable Protection Systems". U.S. Navy Civil Engineering Laboratory, Nov. 1979, Report FPO-1-78(3).

3. Burcharth, H.F.: "Fatigue In Breakwater Concrete Armor Units". University of Aalborg, Denmark, Oct. 1984.

Figure 1   Placement Area for Instrumented Dolos. The shaded area
            superimposed on the hydraulic model of the Crescent City
            Breakwater indicates the area where the instrumented dolosse
            will be placed.

Figure 2   Side View of the Placement Area. The shaded area indicates
            the instrumented dolosse placement area.

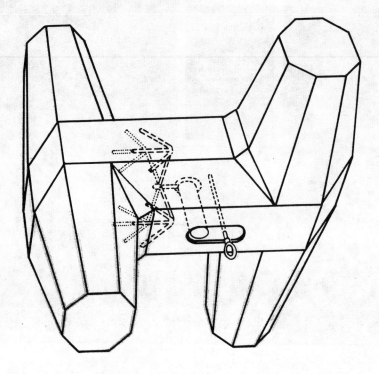

Figure 3   Internal Dolos Instrumentation. Strain gage rebar rosette
             assemblies, wiring conduit, and the cast steel sleeve for
             the dolos processor.

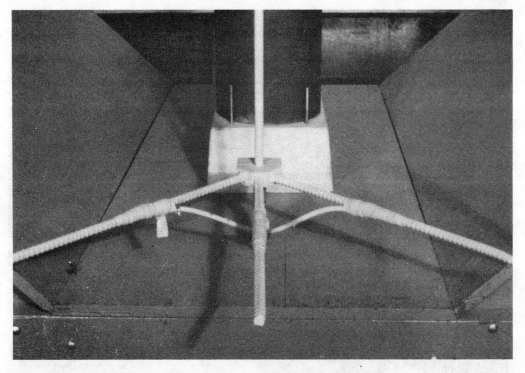

Figure 4   Rebar Rosette. View of one rebar rosette. Strain gages are
             applied to machined surfaces and protected with water proof
             potting. The conduit to gage wiring is in oil filled
             tubing.

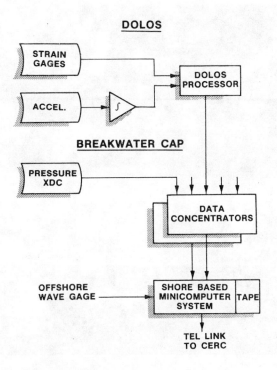

DOLOS

STRAIN GAGES → DOLOS PROCESSOR

ACCEL. → ∫ →

BREAKWATER CAP

PRESSURE XDC →

DATA CONCENTRATORS

OFFSHORE WAVE GAGE → SHORE BASED MINICOMPUTER SYSTEM | TAPE

TEL LINK TO CERC

Figure 5   Data Acquisition System Block Diagram. The major components of the dolos data acquisition system.

Figure 6   Dolos Cable Protection. Split steel sleeves are through bolted around used 3 in anchor chain to form a channel for the cable. The assembly serves as an anchor for the cable and provides added crush resistance.

187

Figure 7   Cable Protection Chain Assembly. The assembly is lifted at a single point, demonstrating the minimum bend radius. The cable can freely move in the assembly.

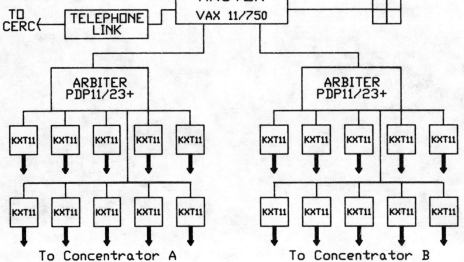

Figure 8   Shore Based Data Acquisition Computer System.

International Conference on

Measuring Techniques

of Hydraulics Phenomena in offshore, Coastal & Inland Waters

London, England: 9-11 April, 1986

EVALUATION OF ESTIMATION TECHNIQUES

C.A. Fleming

Sir William Halcrow & Partners
Burderop Park, Swindon, Wiltshire SN4 0QD
England

and

B.M. Pinchin

Keith Philpott Consulting Ltd
Thornhill, Ontario L3T 4A5
Canada

Summary

The Canadian Coastal Sediment Study included numerous activities related to improving the measurement and prediction of sand transport in the nearshore zone. Some of the measurement techniques are described together with the results obtained. The data was used to evaluate a number of different predictive techniques commonly used for coastal zone modelling, the ultimate aim being to accurately estimate alongshore sand transport rates. Results from the 1983 field programme at Pointe Sapin are used. The computed transport rates are compared with volumes of sand accumulated in the prototype sand trap formed by an offshore breakwater at the mouth of Pointe Sapin harbour. Models utilised for computation were a parametric wave hindcasting model, a spectral wave refraction model, nine alternative sediment transport predictors, seven of which rely on computed alongshore current distributions. Comparison of predicted and measured data of both offshore and nearshore wave climates was improved significantly by inclusion of energy saturation limits in the wave hindcasting and transformation models. This in turn improved the accuracy of the sediment transport predictions.

Held at Imperial College of Science and Technology, London. Organised and sponsored by BHRA, The Fluid Engineering Centre. Co-sponsored by the American Society of Civil Engineers and the International Association for Hydraulic Research.

1. INTRODUCTION

The Associate Committee for Research on Shoreline Erosion and Sedimentation of the National Research Council, Canada (ACROSES) organised the Canadian Coastal Sediment Study (C2S2), which is a multi-departmentally sponsored research study of alongshore sand transport in the nearshore coastal zone. The broad objectives of the programme described here were (i) to provide comprehensive analyses to develop an overall understanding of sand transport at the field sites and to provide baseline data during the field trials, (ii) to evaluate a number of different measurement techniques and (iii) to improve the present understanding of nearshore sand transport predictive techniques based on standard data sources and the site specific data gathered from the study and thus the evaluation of estimation techniques. The authors were directly involved in parts (i) and (iii) of the exercise which implicitly involved the extensive application of numerical modelling of nearshore processes including wave hindcasting, wave transformation in shallow water, generation of wave induced currents and prediction of both alongshore and onshore/offshore sediment transport.

A large number of studies have been carried out under the overall umbrella of C2S2 and those relating to field work by a number of investigations are referred to frequently. The first site to be studied was Pointe Sapin, New Brunswick which lies within the general confines of the Gulf of St Lawrence opposite Prince Edward Island. A second site was investigated at Stanhope Beach which lies on the north shore of Prince Edward Island. Those studies are not reported in this paper.

2. STUDY AREA

Pointe Sapin, New Brunswick is located at the northern entrance to the Northumberland Strait and that section of coastline generally faces east-southeast. The site is exposed to waves generated in the Gulf of St Lawrence (See Figure 1) and is sheltered approximately from east to south by Prince Edward Island. Thus, the wave climate is dominated by waves approaching from the northeast quadrant.

The northeast coast of New Brunswick is characterised by outcrops of erodible, carboniferous sandstone and shales. The rocks dip seaward and have been folded along an axis which runs perpendicular to the coast. Post-glacial submergence of this topography has resulted in a series of headlands and bays. The beach at Pointe Sapin is a wedge shaped body of sand between 2 and 3 metres thick on the backshore thinning to nothing between 60 and 75 metres offshore. Thus, the beach overlies a wide rock shelf which is composed of reddish-brown siltstone. The offshore bedrock is covered by a thin layer of gravelly sands. The sediments in the littoral zone are predominantly fine sand sized particles which are well sorted with a mean diameter between 0.18 and 0.25 mm. Local deposits of pebbles do occur.

There have been several investigations of littoral drift along the northeast shore of New Brunswick in general and a few specifically at Pointe Sapin. There is general agreement that the net sediment transport is dominantly to the southwest. Several investigators have estimated the net potential sediment transport rates past Pointe Sapin varying between 120 and 205 thousand cubic metres per year.

There is a small fishing harbour at Pointe Sapin which has suffered quite severe sedimentation problems in the past. As a result an offshore breakwater was constructed to act as a sand trap and this has been effective in significantly reducing the siltation problem. The offshore breakwater, the layout of which is shown in Figure 2, lies approximately 100 m off the main breakwater of the fishing harbour. It has been estimated that in an average year at least 60% of the littoral drift should be caught in the sand trap. The existence of the sand trap on the south side of the study site was one of the factors that led to the selection of this site. The mean tide range at the site is 0.9 metres and the tidal currents run reversing parallel to the coast with speeds generally less than 0.1 metre/sec.

3. FIELD MEASUREMENTS

The field programme at Pointe Sapin was carried out in the latter part of 1983. Instrumentation included, but was not limited to the following:

(i)   A buoy capable of measuring roll, pitch and heave deployed in 16.5 m depth of water 8.3 km east of the site. The buoy was provided by the National Water Research Institute and is called the CCIW Wave Direction Buoy. It uses a

Geodyne model A-92 toroidal buoy with an instrument package of Sea Data 1250 data logger, three Sunstrand servo accelerometers, a Humphry roll-pitch gyroscope, a digital magnetic compass and two cup anemometers. They buoy thus provided both directional wave information as well as overwater wind data (Skarfel, 1984).

(ii)   A conventional Waverider buoy was deployed by MEDS (Canada) in close proximity to the CCIW buoy.

(iii)   A Sea Data Corporation model 635-9 directional wave gauge was deployed in 6.15 m water depth by Woods Hole Oceanographic Institution. The instrument package consists of a Paroscientific 245-A pressure transducer, a two axis Marsh McBirney 512/OEM electromagnetic current sensor, a Digicourse 225 gimballed compass and a cassette tape drive (Aubrey and Spencer, 1984).

(iv)   Six Marsh McBirney 512/OEM two axis electromagnetic current metres with 4 cm probe.

(v)   Four 245A-002 Digiquartz pressure transducers.

(vi)   Two Sea-Tech Transmissometers.

(vii)   Two University of Washington Optical Backscatter Sensors.

(viii) Climatronics Weather Station.

Data relating to items (iv) to (viii) are summarised by Daniel, 1985.

In addition macro-scale experiments were carried out using a number of different types of bottom drifters as well as radioactive tracers. The sand trap was dredged clear immediately prior to the experiment and profile surveys of the sand trap area and adjacent beaches were carried out at regular intervals throughout the experiment period (Gillie, 1984 a).

3.1   Wave Measurements

The data from the CCIW wave buoy sensors were sampled at 2Hz per channel in 30 minute bursts, four times per day. Each 30 minute record was subsequently divided into three sub-samples each of about 8 minutes 32 secs duration. Spectra for each sample were calculated and the final results for any given sampling time were the average of the three sub-sets. The spectra were grouped into frequency bands 0.02Hz wide and therefore had 61 degrees of freedom per band.

The data from the Waverider buoy (MEDS) does not provide directional information. Standard 20 minutes bursts every three hours were collected at 2Hz and analysed in the standard format.

A comparison of the wave heights and periods from the two data sets is shown in Figure 3 generally exhibiting good agreement with the exception that the CCIW wave periods displayed a number of spikes that were not present in the MEDS data. The spikes were a result of two-peaked spectra where the lower frequency peaks were dominant.

The nearshore wave gauge (WHOI) was sampled at 1 Hz with 2048 second bursts every 6 hours. Spectral estimates were made using the full record length with 16 ensemble averages, yielding an ensemble length of 128 seconds, and a frequency resolution of 0.0078125 Hz. The resultant spectral estimates have 32 degrees of freedom with the expected spectral value within 0.65 and 1.76 of the sample value at 95% confidence limits. All spectra were depth corrected.

3.2   Hydrodynamic Measurements

The Marsh McBirney electromagnetic current meters were generally deployed at 30 cm above the seabed sampling at 4Hz. It has been estimated that measurements are made with an error of about ±10% and mean flow uncertainty of about 4 cm/sec (Huntley, 1979). Pre-deployment tow tank calibrations were performed and still water offset checks made during the field data collection programme.

The four Paroscientific pressure sensors were all mounted at 30 cm above the bed and were sampled at 4Hz. Two recording modes were used. The first entailed sampling continuously throughout the entire data run. The second procedure followed was for the pressure sensor to be switched every 10 to 15 minutes through the course of the data run. The overall precision of the instrument is estimated to be 0.01% (3mm) on a full scale range of 30m (Gillie, 1984 b).

Lagrangian current measurements were carried out by INRS - Oceanographic, University of Quebec (Drapeau and Pade, 1984). A number of experimental drifter types were used including Shear-stress type, Norwegian Floats, One-Metre (diameter) Drifters, Synthetic Horse hair Disc Drogues and WHOI Seabed Drifters. The last of these are saucer-like seabed drifters and have been used extensively for more than thirty years. They are composed of an 18 cm diameter perforated plastic saucer maintained vertically by a 55 cm long weighted plastic stem. It was these that were the most successful of all types and they were released in groups of 100 at preselected locations. It has been noted that whilst these were not designed for the nearshore zone they appear to be sensitive to rapidly varying hydrodynamic conditions.

3.3   Suspended Sediment Measurements

The Optical Backscatter Sensor was developed at the University of Washington, Seattle and has been in existence for a number of years (Downing, 1983). It consists of a five element array of optical sensors which consist of a high intensity infra-red emitting diode and a low capacitance solar cell. The effective sample volume is less than 3 cc and the lightpath is less than 2 cm (Gillie, 1984 b). The performance of the instrument was good with fouling by seaweed being an unavoidable problem.

The two transmissometers had 5 cm path lengths and were mounted at 30 cm above the seabed. The instruments measure the attentuation of a collimated beam of light from an LED source. The main problem encountered with this instrument was that high intensity background turbidity caused by mud often reduced the transmission to negligible levels rendering it unfit for its intended purpose of measuring suspended sand.

Two other systems were used in the experiment; a nephelometer and a simple pump sampler. The details of these are reported by Gillie, 1984 b and 1985.

3.4   Beach Volume Measurements

Beach profiles were measured from the survey baseline to the maximum depth of wading, which naturally depended on weather, tide and wave conditions. Conventional tape, rod and level methods were used taking levels at 3 m spacing or less if special features were present. The standard accuracy estimate for such measurements is 0.1 m for distances and 0.01 m for elevations. Bathymetric surveys were carried out using a Raytheon DE-719C echo sounder and an electronic distance measuring instrument in a Zodiac survey boat. The survey lines are shown in Figure 2 and distance fixes were made at 5 to 10 m intervals. The level of accuracy of this data is less than that of the beach survey data.

Radioactive sand tracer experiments were carried out by INRS-Oceanologie, University of Quebec. Detailed results are reported by Gillie (1984 b) and Long (1984). The experiment was conducted over 9 days and covered one of two storm events that occurred during the experiment. The dispersion of artificial sand, tagged with a short-lived radioisotope was monitored using a towed sled equipped with a radiation counter, thus providing information for mapping dispersion. The depth of burial of the tracer was also measured. By using the profile survey information to give changes together with the centre of gravity established from the dispersion pattern, estimates of the rate and direction of sand transport were made (Gillie, 1985).

4.   PREDICTIVE TECHNIQUES

4.1   Wind Hindcasting

A parametric wave hindcasting model was used which included options to use four different hindcasting models; SMB deep water, SMB shallow water, Derbyshire-Draper and JONSWAP. The wind data required is a time series of wind speed and direction, adjusted if necessary to a standard elevation above mean sea level. For each forward step of the hindcast an effective wave, defined by height, period and direction, is the resultant hindcast wave composed of a generated and a decayed wave train. The larger of the two components was treated as dominant in defining the wave period and direction.

Fetch lengths were defined as a function of azimuth from the hindcast site. For this study "straight fetches" (rather than effective) were selected on the basis of experience rather than by calculation. These are shown in Figure 4. The fetch lengths shown were selected on the basis that the width of the wave generating fetch should be considered in relation to the divergence of the wind field. This is why none of the fetches shown penetrate the full extent of the northeast sector of the Gulf of St Lawrence.

The effective wave for each timestep was computed by examining the wind data and dominant wave conditions for a preceding duration time equal to the selected maximum wave generation sequence. Thus, by backstepping through the data from the current record every combination of wind speed and duration is examined up to a limiting condition, in this case 2 days, or to a point where the wind direction differs from the average of the preceding sequence by an amount exceeding a defined wave divergence angle eg 22.5 or 45 degrees. The wave decay sequence was initiated either when a calm was encountered or when the wave divergence angle was exceedable. A number of wave decay functions were considered. (See Fleming et al, 1984).

4.2   Wave Transformations

The two principal components of the computational procedure are a wave refraction model and a post-processor that calculates the spectral transfer functions which in turn provide wave height coefficients and direction shifts from deep water to a shallow water point of interest.

The wave refraction model used was based on the highly efficient 'circular arc' technique allowing the high ray density backtracking for several frequencies from each point of interest to be carried out. The seabed was defined by ten digitized depth grids which varied in both overall dimensions and mesh size according to the water depth and proximity to the field investigation site. These are shown in Figure 5.

The process of refracting a wave spectrum is based on the assumption that the wave energy flux in a frequency band will remain in that band as each component of the wave spectrum is transferred inshore. It is thus possible to discretise a deepwater directional wave spectrum and independently transfer it inshore applying appropriate shoaling coefficients.

The principal output from the wave refraction analysis is the deep water direction of all wave rays for all wave periods considered. By computing the deep water energy flux present for each deep water direction/frequency combination it is possible to compute the energy flux that would be present in the inshore elements. The deepwater energy flux that is present at any deep water direction at a given frequency is dependent on the deep water directional spectrum. In this case the JONSWAP model was used. Its applicability to this site was verified by matching computed spectra with offshore spectra measured in 1982 (Fleming et al, 1984). The form of direction spreading function used was a power cosine function. Having constructed the inshore wave spectrum for a given deep water mean direction and peak wave period integration across the direction range and then the various movement across the frequency range provides the inshore wave height coefficients, direction shifts and peak shifts.

The direct application of this procedure to the offshore wave measurements led to a substantial over-estimation of inshore wave heights during one of the storms. This led to an investigation into energy dissipation mechanisms and as a result wave spectrum saturation for shallow water were incorporated into the wave spectrum transformation procedure. This implicitly allows for wave energy dissipation due to complex interactions including partial wave breaking. The shallow water equilibrium spectrum theory used is due to Kitaigorodskii et al, 1975. Comparative results are presented in Section 5.

4.3   Nearshore Sediment Transport

The inshore wave climate defined by the above procedure allowed calculation
of alongshore sediment transport rates for the same breaker line, surf zone
geometry and hence, alongshore current distribution for up to twelve sediment
transport models. Results from nine of these are presented here.

The magnitude and distribution of the wave-driven alongshore current in the
breaker zone depends on the momentum balance, which in turn depends on the
underwater profile, the incident wave characteristics and the wave breaking
mechanism. The alongshore current theory used for computations was developed
as an extension to the Longuest-Higgins (1970) formulation (Fleming and Swart
1982). Both bulk energy sediment transport predictors as well as detailed
predictors were applied for comparative purposes. These were as given in
Table 1.

The computational procedures were as follows:-

(i)   The beach profile was divided into a number of representative sections of
        variable size.

(ii)   The inshore wave conditions were used to find the breaker position and cor-
        responding properties. These were used in the bulk predictors.

(iii)   The mean wave height was calculated according to a Rayleigh probability
        function. If wave breaking was found to occur within or offshore of a profile
        section the mean wave height was found by integrating the Rayleigh distrib-
        ution truncated by the maximum possible wave.

(iv)   Wave and sediment characteristics were used to calculate ripple dimensions
        according to Swart (1976). This in turn provided data for the hydraulic
        bed roughness to be evaluated.

(v)   The wave and current parameters used to calculate the alongshore current
        distribution across the profile.

(vi)   Wave properties, currents and the various coefficients calculated were used
        to determine bed shear stresses which in turn are used to drive some of the
        detailed prediction models.

(vii)   Each wave condition was taken in turn and running totals of gross and net
        transport and relative distribution kept as a running total.

	MODEL	GENERAL CLASSIFICATION
1	CERC Formula (1977)	Bulk energy model
2	Swart modified CERC (1976)	Bulk energy model, includes grain size as a parameter
3	Engelund and Hansen adaptation by Swart (1976)	Based on bed and suspended loan concentrations with a background current, no incipient motion criterion
4	Willis (1978)	Adaptation of Ackers and White (1973), includes an incipient motion criterion
5	van der Graaf and van Overeem (1979)	Adaptation of Ackers and White (1973), includes an incipient motion criterion
6	Nielsen (1979) (for breaking waves)	Based on bed and suspended load concentrations with a background current, includes an incipient motion criterion
7	Nielsen (1979) (for non-breaking waves)	Ditto
8	Fleming (1977)	Ditto
9	Swart and Lenhoff (1980)	Adaptation of Ackers and White (1973), includes an incipient motion criterion derived from a wide range of wave and current conditions.

Table 1: General Description of Alongshore Sediment Transport Predictors

5   COMPARISONS OF MEASUREMENTS WITH PREDICTIONS

5.1   Offshore Waves

A number of wind data sets were available for possible use in hindcasting.
These included a data set from the Magdalen Islands, which lie in the middle
of the Gulf of St Lawrence, the wind speeds and directions at the offshore
CCIW wave buoy and a similar data set at the land based data acquisition
station.   The first of these was used for a number of interesting comparisons.
Firstly, the four parametric hindcasting formulations described in Section 4.1
were tested using different wind divergence limits and wave decay functions.
The difference between the various models was not large in respect of wave
heights, but showed some variation in wave period.   The selection of the best
overall predictor for this site was not clear cut.

Figure 6 shows three sets of hindcast and measured wave heights and periods
between the 21st and 31st October 1983 for one of the storms that occurred
during the experimental period.   Set (A) used the SMB deep water parametric
hindcast model.   It is seen that the hindcast wave heights were in reasonably
good agreement with the measurement with the exception of the peak of the
storm when the predictions grossly exceeded measurements.   The hindcast wave
periods were generally in good agreement with measurements throughout the period shown.   Set (B) used the SMB shallow water parametric hindcast model.
However, instead of using water depths characteristic of the entire fetch the
water depth at the wave recording site (16 m) has been used.   It may be circumstantial, but the hindcast wave heights show much better agreement with
measurements during the peak of the storm, but are little changed at other
times.   In contrast the hindcast wave periods became rather less than the
measured particularly through the storm.   At this point it might be noted that
combination of SMB shallow for wave heights and SMB deep for wave periods
would give a reasonably respectable result.   The final set (C) was obtained by
applying the SMB deep water hindcast model modified for spectral saturation
which is a function of depth (Kitaigorodskii et al, 1975).   Here both the
hindcast wave height and wave period show excellent agreement with measured
values.

5.2   Nearshore Waves

The nearshore wave measurements were made in a little over 6 m mean water
depth.   Both deep water measured and hindcast wave sets were transferred
to this shallow water site by applying the wave height coefficients and directions deduced from the spectral wave refraction model described in section
4.2.

Figure 7 shows three sets of predicted and measured wave heights and periods
between 21st and 31st October 1983 and can thus be compared directly with the
previous figure.   Set (A) represents the offshore measured waves transferred
to shallow water by direct application of the spectral transfer results.   It
can be seen that the predicted wave heights were largely excessive by up to
a factor of two for most of the record length, whilst the wave periods were in
reasonably good agreement.

Set (B) represents the offshore measured waves transferred to shallow water by
application of spectral refraction results modified to include shallow water
spectral saturation limits in each frequency bank.   Here it is seen that the
predicted wave heights show excellent agreement with those measured.   The wave
periods remain largely unchanged and in good agreement with measured values.

Set (C) represents the hindcast wave climate offshore which included saturation limits (ie Set (C) in Figure 6) transferred to the shallow water recording
site by application of spectral refraction results modified to include shallow
water spectral saturation limits as above.   The agreement on wave height
remains good but the wave period predictions appear to show a slight
deterioration.

5.3   Alongshore Sand Transport

The volume of sand transported during the two storms that occurred during the experimental period have been estimated by a number of investigators (Kooistra and Kamphuis, 1984, Morse, 1984, and Gillie, 1984) using both the profile surveys of the sand trap and adjacent beach as well as the radioactive sand tracer experiments. The estimate of net volume of sand transported during a particular event is dependent on whether only the sand trap itself is considered as the control volume or whether some part of the beach upstream of the sand trap should also be included. Considering the period of 22nd and 27th October estimated volumes vary from about 5000 cu m to nearly 11000 cu m. These figures do not account for efficiency of the sand trap. Table 2 summarises some of the most relevant results that have been achieved to date. The various model descriptions have already been given in Table 1.

Model	1	2	3	4	5
CERC	59,750	59,750	28,720	19,950	28,550
Modified CERC	58,950	58,950	28,330	19,950	28,460
Engelund & Hansen	351,500	125,700	61,640	10,390	9,550
Willis	284,700	118,200	60,720	12,970	14,000
Van de Graff	340,200	123,800	61,479	9,948	8,800
Nielsen (breaking)	36,280	22,700	11,820	6,965	8,150
Nielsen (non-breaking)	29,170	15,560	7,870	2,974	4,050
Fleming	104,600	35,720	16,560	2,555	2,470
Swart & Lenhoff	210,500	70,490	34,260	5,337	4,370

Key: 1 - Hindcast offshore wave data, friction factor variable
       2 - Hindcast offshore wave data, friction factor constant (0.01)
       3 - Measured offshore wave data, without saturation, friction factor
           variable
       4 - Measured offshore wave data, with saturation, friction factor
           variable
       5 - Measured inshore wave data, friction factor variable

Table 2 - Summary of Alongshore Sand Transport Predictions

Runs 1 to 2 used a hindcast deep water wave climate refracted inshore using standard spectral transfer techniques. Computed values are extremely high in relation to estimated volumes of sand transported during the storm. The large difference between 1 and 2 indicates the effect of friction factor on the alongshore velocity profile. Run 3 was identical to run 1 except that the measured deep water wave climate refracted inshore, was used corresponding to Set (A) in Figure 6. This demonstrates the very large effect that using the overestimated deep water wave climate had on runs 1 and 2. Run 4 used offshore measured wave data together with spectral transformation to shallow water but including spectral saturation considerations corresponding to Set (B) of Figure 6.

Here it is shown that the detailed predictors are within the range of measured volumes. Finally run 5 used measured inshore wave data. Here it is seen that the bulk predictors increased and some detailed predictors decreased slightly indicating some directional and wave height differences. However, the differences for the detailed predictors may be considered to be insignificant and merely indicate the close correspondence between predicted and measured nearshore wave data. The bulk predictors appear to be more sensitive when comparing runs 4 and 5 but somewhat less so over all tests.

The above computations were performed by dividing the nearshore profile into sediment transport zones within which all parameters such as wave height, wave direction, current velocity and sediment size related to variables were evaluated. Figure 8 shows a typical beach profile together with the estimated net distribution of sediment transport across the surf zone for run 5 plotted as the mean of all predictors considered.

6. CONCLUSIONS

Some aspects of instrumentation, measurement and analysis of data for the first stage of the Canadian Coastal Sediment Study have been discussed. A very large and comprehensive data set was successfully collected and this has been most useful in evaluation of a number of predictive models of wave hindcasting, wave transformation to shallow water, generation of alongshore currents and prediction of alongshore sand transport rates.

The offshore and onshore wave data sets were largely complete with few gaps in the data. Similarly the currents measurements gave a good spatial coverage of the site. That data indicated water currents at the site were dominated by semi-diurnal tidal components of low magnitude outside of the surf zone. Under storm conditions wave induced currents were typically between 0.5 - 1m/sec and the wind shear contribution to those currents is thought to have been significant due to their persistence beyond significant wave activity. However, the comparison of measured and predicted near-shore currents is the subject of ongoing studies.

Measurement of sediment transport rates in the field is a notoriously difficult task. There was some difficulty with field implementation of all of the techniques and the probable efficiency of each technique, whilst realisable qualitatively, is difficult to assess quantitively. Techniques such as profile surveys and radioactive tracer studies which effectively provide an integrated result have some advantage over instantaneous measurements particularly in terms of predicting net transport rates over specific periods. However, the instantaneous measurements, which show considerable temporal and spatial variability will in the long term provide a better understanding of the complex physical processes involved. Of the techniques used for suspended load measurements the OBS and pumped water sampler provided the most useful data. The response of other optical instruments was limited by high levels of background turbidity.

Wave hindcasts for the offshore measurement site showed that shallow water effects in 16 m of water were relatively large. The accuracy of the prediction could apparently be improved significantly by introducing depth dependent energy saturation principles into the hindcast models. Computation of nearshore wave heights and directions on the basis of two dimensional spectral refraction techniques did not provide good estimates for the larger of two storm sequences measured. Again the inclusion of depth-dependent energy saturation modifications to the transformation process improved the predictions considerably.

Predictions of alongshore transport rates have been made using both bulk energy and detailed predictors which rely on computed alongshore current distributions. A number of comparative estimates show (i) the sensitivity of the detailed predictors to the friction factors used which are indirectly dependent upon the ripple roughness model, (ii) the magnitude of error that would have occurred from using an unverified deep water wave hindcast model, (iii) the magnitude of error that would have resulted in using an offshore wave climate transformed by consideration of refraction alone and (iv) a close comparison in results between measured and predicted rates when the improved method of estimating the nearshore wave climate is used. It should also be noted that the best predictions of net volume transported were largely within the range of volumes deduced from various field measurements. Of all the detailed predictors the Nielsen models (breaking and non-breaking waves) were the most stable over the range of conditions tested and appropriately the breaking wave version gave results close to measured volumes. Clearly and not surprisingly quality of predictions of nearshore wave climate plays a dominant role in the level of success in predicting sediment transport processes. There is also considerable scope for improvement of the sediment transport predictors. Further field investigations have been carried out at Stanhope Beach and Prince Edward Island. These will be subject to similar treatment.

7. ACKNOWLEDGEMENTS

We gratefully acknowledge the support of Bert Pade and Peter Daniel, C2S2 study managers; Dave Willis, National Research Council; Chris Glodowski, Public Works Canada and other members of the ACROSES committee. The information gathered by the numerous investigators during the Study has been invaluable in providing the required data. It has been a privilege to have taken part in the C2S2 research programme.

8. REFERENCES

1. AUBREY D G, and SPENCER W D, 1984, "Inner Shelf Sand Transport Wave Measurements Pointe-Sapin, New Brunswick, Canada" NRC Canada, C2S2 Report C292-5.

2. CERC,1977, "Shore Protection Manual", Coastal Engineering Research Center, US Army Corps of Engineers.

3. DANIAL P E, 1975 "Data Summary Index 1983 Pointe Sapin, New Brunswick" NRC Canada, C2S2 Report C2S2-16.

4. DOWNING J P, 1983 "An Optical Instrument for Monitoring Suspended Particulates in Ocean and Laboratory" OCEANS '83, IEEE.

5. DRAPEAU G, and PADE B H G, 1984 "Lagrangian Currents and Seabed Drift Measurements at Pointe Sapin, New Brunswick" NRC Canada, C2S2 Report C2S2-6.

6. FLEMING C A, 1977 "The Development and Application of a Mathematical Sediment Transport Model", PhD Thesis, University of Reading.

7. FLEMING C A and SWART D H, 1982 "New Framework for Prediction of Longshore Currents" Proc 18th ICCE.

8. FLEMING C A, PHILPOTT KL and PINCHIN B M. "Evaluation of Coastal Sediment Transport Estimation Techniques Phase I", NRC Canada C2S2 Report C2S2-10.

9. GILLIE R D 1984a "Canadian Coastal Sediment Study, Site Maintenance Contract" NRC Canada, C2S2 Report C2S2-8.

10. GILLIE R D 1984b "Evaluation of Measurement Techniques", NRC Canada, C2S2 Report C2S2-9.

11. GILLIE R D 1985 "Evaluation of Field Techniques for Measurement of Longshore Transport" Proc Canadian Coastal Conference.

12. HUNTLEY D A, 1979, "Electromagnetic Flow Meters in Nearshore Field Studies" Proc Workshop on Instrumentation for Currents and Sedimentation in the Nearshore Zone, NRC.

13. KITAIGORODSKII S A, KRASITSKII VP and ZASLAVSKII M M, 1975 "On Phillip's Theory of Equilibrium Range in the Spectra of Wind Generated Gravity Waves" J Phys Geophys, 5.

14. KOOISTRA J and KAMPHUIS J W, 1984, "Scale Effects in Alongshore Sediments Transport Rates", NRC Canada, C2S2 Report C2S2-13.

15. MORSE B, 1984, "Volumetric Transport Rates at Pointe Sapin, New Brunswick, 1983" NRC Canada, C2S2 Report C2S2-11.

16. LONG B F, 1984, "Sediment Transport in the Breaker Zone. Results of the Tracer Experiment using Gold 198, Pointe Sapin", NRC Canada.

17. NIELSEN P, 1978, "Some Basics of Wave Sediment Transport" Institute of Hydrodynamics and Hydraulic Engineering, Tech Univ Denmark, Series Paper 20.

18. SKAFEL M G, 1984, "Offshore Wind and Wave Data 1983. Pointe Sapin, New Brunswick, CCIW Wave Direction Buoy", NRC Canada, C2S2 Report C2S2 -

19. SWART D H, 1976, "Coastal Sediment Transport: Computation of Longshore Transport". Delft Hydraulics Laboratory Report R968.

20. SWART D H AND LENHOFF L, 1980, "Wave Induced Incipient Motion of Bed Material" CSIR Research Report, NR10, Stellenbosch, S Africa.

21. VAN DE GRAFF J and OVEREEM T, 1979, "Evaluation of Sediment Transport Formulae in Coastal Engineering Practice", Coastal Engineering 3 (1).

22. WILLIS D H, 1978, "Sediment Load under Waves and Currents" Proc 16th ICCE.

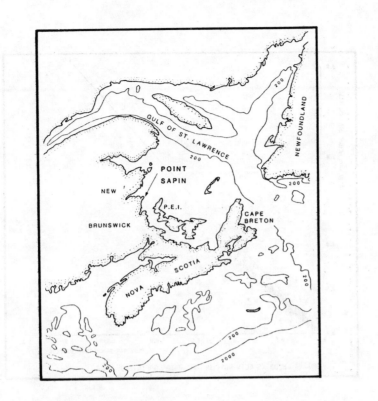

Figure 1: Location of Pointe Sapin

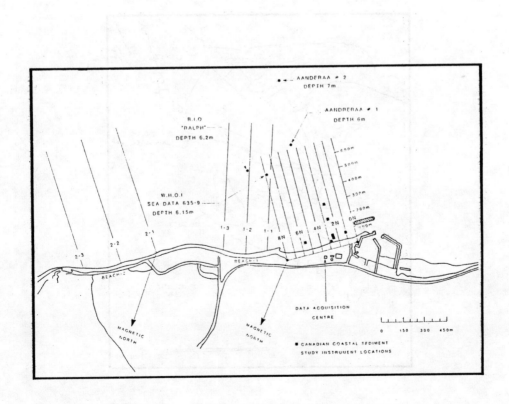

Figure 2: Survey Lines and Instrument Position, Pointe Sapin

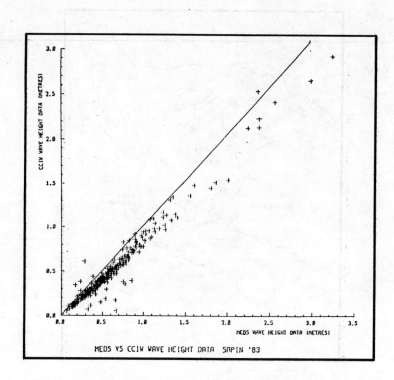

Figure 3:　Comparison of CCIW and MEDS Offshore Wave Data

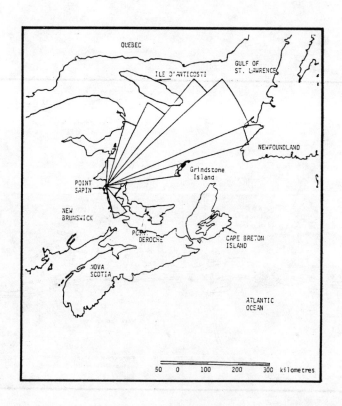

Figure 4:　Fetch Sectors for Wave Hindcasts

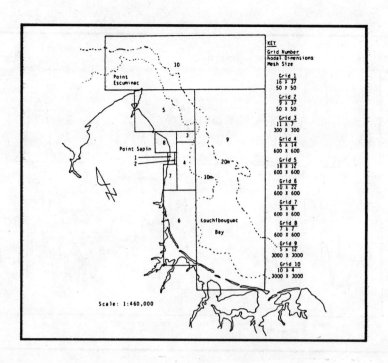

Figure 5:   Configuration of Wave Refraction Grids

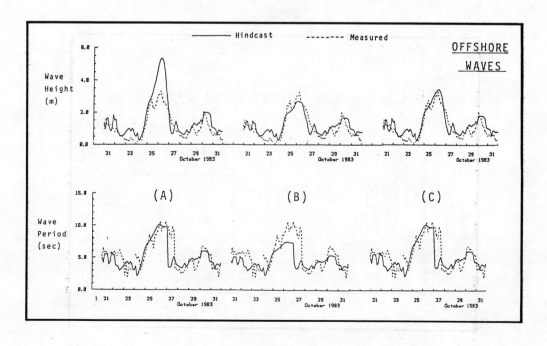

Figure 6:   Hindcast and Measured Offshore Wave Data

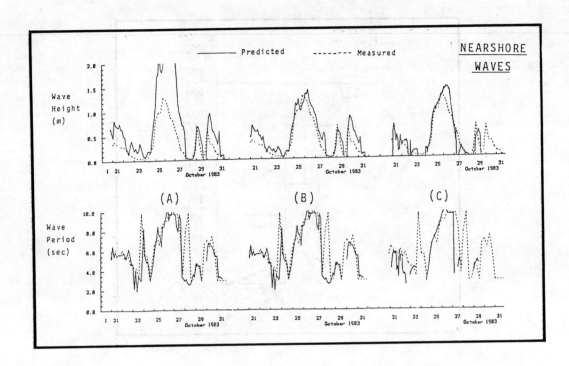

Figure 7:   Hindcast and Measured Nearshore Wave Data

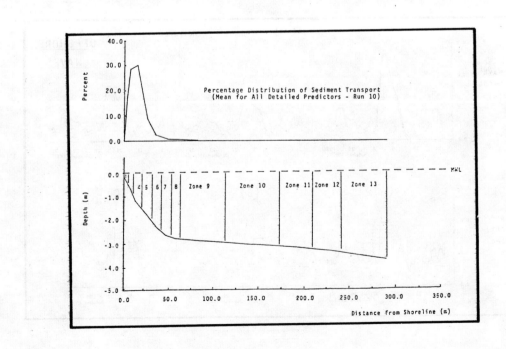

Figure 8:   Beach Profile and Distribution of Sediment Transport

International Conference on

Measuring Techniques

of Hydraulics Phenomena in offshore, Coastal & Inland Waters

London, England: 9-11 April, 1986

Full-Scale Measurements in the Surf Zone

Authors

Dr P C Barber

Managing Director, Ceemaid Ltd

Mr J G L Guthrie

Senior Engineer, Ceemaid Ltd

Mr K Binfield

Senior Engineer, British Maritime Technology Ltd

Held at Imperial College of Science and Technology, London. Organised and sponsored by BHRA, The Fluid Engineering Centre. Co-sponsored by the American Society of Civil Engineers and the International Association for Hydraulic Research.

Introduction

In the development and conservation of coasts and estuaries it is essential to understand the 'energy budget' for any particular location of interest. Energy is primarily input to a coastal site from winds, waves and tides. Tidal energy manifests itself by way of cyclical changes of water level and velocity over relatively long periods of time (>12 hrs) whereas wave energy manifests itself by way of short period oscillatory motion of the water (typically 3-12 second period). The addition of wind-induced currents and restricted water depth i.e. shallow water above usually mobile beds of granular or cohesive sediment makes the nearshore zone one of the most complex natural environments.

In an attempt to understand the nearshore area there has been considerable investment of time and finance into research of the major phenomena. However, understanding is still limited and significant resources continue to be wasted as nature reminds designers of their lack of knowledge. With the rapid progress of 'information technology' over the last decade, combined with the 'commercial momentum' provided by the offshore industry, it has been possible to make significant advances in techniques of measurement applicable along shorelines and within estuaries. This development has allowed a new and more confident approach to be taken in the determination of design criteria for coastal works.

The collection of high-level data at full-scale allows the accuracies of numerical and physical model prediction to be greatly improved by the use of such data to define model boundary conditions and, most importantly, to provide validation data for confidence testing of model results. As data sets of this quality accumulate it is possible to establish new theoretical relationships for engineering application, which allow the competent assessment of a wider range of solution options for any particular problem.

The development of a high-level system for the collection of data at full-scale is described below and covers a ten year period from 1975 to date during which the system has evolved considerably and now represents one of the most sophisticated capabilities available for measurements in shallow water (less than 50.0m).

Objectives: The primary objectives for the measurement system were set in 1975 and have remained unchanged throughout the development period:

1. To obtain measurements of sufficient quality and quantity for use in defining boundary conditions and validating numerical and physical models.

2. To ensure data integrity and undertake data analysis - both 'on-line' so that data capture is assured 'on-station'.

3. To provide a range of transducers to measure the major effects relevant to coastal works and, wherever, possible to provide, at least two types of transducers to measure the same effect.

4. To provide a mobile system capable of transport to any part of the world.

Development towards these objectives can be identified in three distinct stages.

1. 1975 - 1980   System A

2. 1980 - 1984   System B

3. 1984 -   System C

Each of these respective systems is subsequently described to illustrate the evolution towards system C.

<u>Outline Specification</u>

The major parameters for measurement were determined as:

Tide   -   Elevation

                 Velocity

                 Forces

Wave   -   Period, Length

                 Direction

                 Induced current

                 Run-up, set-up

                 Forces

Sediment -   Transport, Bed/Suspended

The range of instrumentation used to achieve the measurement of the above
parameters is considerable, especially when each parameter is measured by at
least two different devices. The water motion transducers were driven by
electrical energy and produced analogue signal outputs. The system
descriptions below are confined to the consideration of these types of
output.

A system may therefore be defined, for the purpose here, as the assembly of
equipment which receives the analogue output of a measurement transducer and
manages that data through conditioning, transmission, storage, integrity
check, analysis and results presentation.

Conditioning: Analogue signals are difficult to transmit
                       competently over large distances due to attenuation
                       from cable resistance, etc. Where long distance
                       transmission is envisaged (>100.0m) it is specified
                       that signals will be digitised.

Transmission: The most reliable form of signal transmission was by cable. Therefore cable transmission of digitised data is preferred for distances up to 2.0km and for ranges above, up to 20km, a radio-telemetry system is used.

Storage: Data is received by data logger or computer and is stored on magnetic tape in either spool or cartridge format. Raw data is always stored immediately after collection.

Integrity check: Following completion of data collection the data set is returned and checked for integrity before analysis is commenced.

Analysis: For computer-based systems the previous data set is analysed during subsequent data collection and results output in graphical and tabular format.

Results Presentation: Results are presented in graphical and tabular format with basic descriptions of the specific measurement task, diary of events, etc, before moving 'off-station'.

Calibration: Transducers are able to be checked in the field for range response and polarity (rigorous calibration is carried out in laboratory conditions).

In order to obtain satisfactory measurements it is necessary to provide 'rigid fixings' for the transducers but of suitable design to avoid 'regime interference'.

It is important to provide a 'clean power supply' and avoid 'fluctuations and spikes'.

The achievement of this outline specification for each of the three systems is detailed below.

System Evolution

System A

System A was developed in response to a specific need for the measurements of parameters as detailed above at a site on the North Wirral coastline. A typical engineering problem of erosion of a foreshore in front of a reflective sea wall was under investigation and in order to obtain a cost effective solution it was recognised that the natural processes needed definition.

The configuration of the system was constrained by the limited knowledge and availability of suitable instruments for measurements in the surf zone. Considerable effort was therefore put into identifying and appraising instruments and recording equipment. The type of analysis needed, together with the accuracy of measurements, was recognised and this then dictated the form in which the system developed.

Three types of instruments were selected:

The electromagnetic current meter (EM); a two axis instrument with very good sensitivity suitable for measurement of velocities near the bed.

Particle velocity meters (PVM); in higher levels in the water column 3 dimensional flow would occur and this precluded the use of the EM because of flow separation around the head. The PVM was particularly suitable and if arranged in pairs 3 orthogonal axis flow could be measured. The sensitivity was less than the EM but was adequate at the higher levels above the bed. 3 pairs of these instruments were deployed.

Wave Staff.   This was a flexible high resistance wire type gauge suitable

for measuring surface elevation both over short events for waves and in the

longer term for tides.

The analogue signals from these instruments were digitized and stored using

a microdata datalogger.

Low signal levels, micro volts in the case of the EM meant that the

instruments had to be clustered around a tower erected in the beach which

was then able to house the necessary instrument electronics to boost the

signal for transmission by cable to the shore where the data was recorded.

The tower also gave support to the Wave Staff and housed the battery pack

supplying power to the instruments.   Figure 1 shows the general arrangement

of system A.

The data retrieved from this system was of good quality and supplied the

information required for the engineering solution to the coastal problem.

Figure 2 gives an example of mean output from system A with Fig 3 showing

the way in which Field data was incorporated into the investigation.

However the system was of site specific use.   As the system was used

modifications were made, often being forced on us by problems that

developed.   Cable integrity was not good and led to housing the recording

equipment on the tower.   This led in turn to the automation of data

collection using a clockwork timer triggering in built triggering capability

in the data logger.   Analysis of the early data showed vibration noise on

the velocity measurements generated by the movement in the tower and

brackets. To overcome this detached frames were developed for the

instruments.   These developments coupled with the invaluable experience of

developing and maintaining a system encouraged us in the development of a

far more flexible system (system B) which was then used for the monitoring

of the influence of a detached breakwater, also on the Wirral coastline.

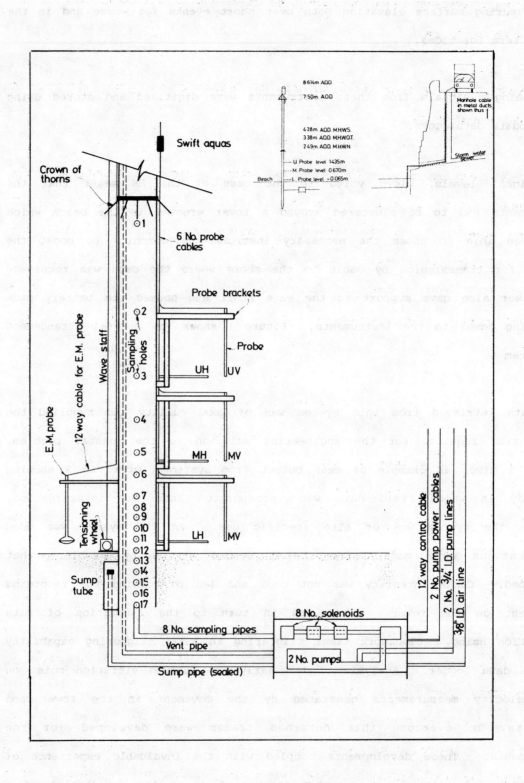

FIGURE 1

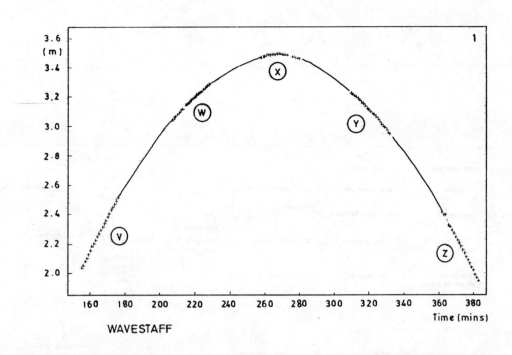

WAVESTAFF

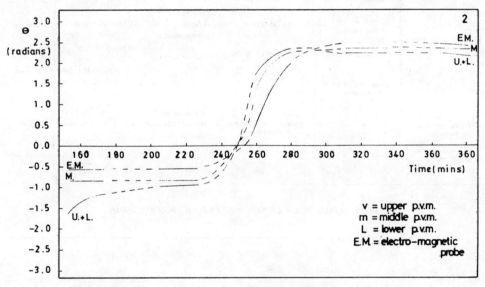

v = upper p.v.m.
m = middle p.v.m.
L = lower p.v.m.
E.M. = electro-magnetic
probe

VELOCITY RESULTANTS' ANGLE, THETA (for E.M. axes)

Run Number and Wind Data (kn)	Date	Tide Amplitude (m) A.C.D.	Time Relative to High Water (hrs)	Instrument
L3/ —	7 5 81	9.5	—	WAVE/TIDE STAFF

FIGURE 2

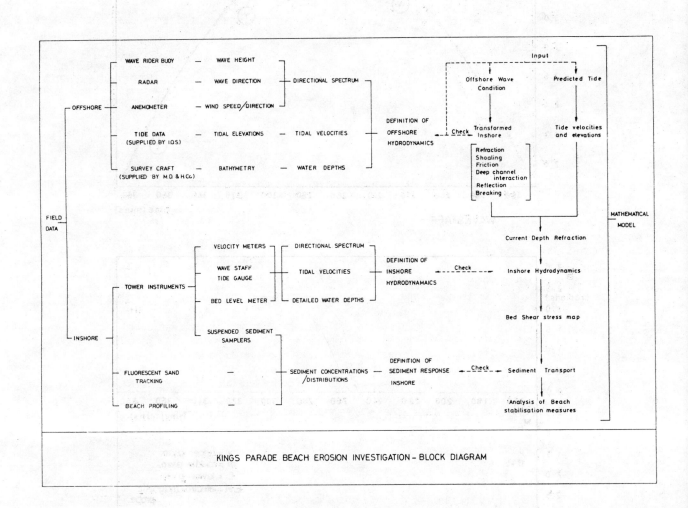

KINGS PARADE BEACH EROSION INVESTIGATION – BLOCK DIAGRAM

FIGURE 3

System B

At this time aims were limited by need and cost to principally the first of the objectives defined previously, that of collecting relevant, quality data, the second objective that of on-station data integrity checking and data analysis was recognised as desirable but beyond our capability however, to go some way towards it, data was examined but not analysed within a day of collection. Instrument behaviour was also monitored before and after each measurement run.

As yet the concept of a completely mobile system capable of using a wide range of instruments was outside the needs of the monitoring work.

It was obvious that to monitor any large structure a variety of instruments must be deployed around the area rather than have a single point measurement. It was decided to continue using the same types of instruments but to modify them to fit the system. This was done by housing all electronics and signal amplification within the instrument rather than just deploying the sensing elements. In this manner instruments were stand alone units that could be deployed 50m to 150m , depending on the instrument, away from the recording centre. Figs 4 shows the modified EM, PVM and surface elevation monitor. The recording centre was still tower mounted but the design of the tower and the recording/control housing were specificly designed for flexibility of numbers of instruments used and, because of the experience gained on system A, for ease and simplicity of operation and maintenance.

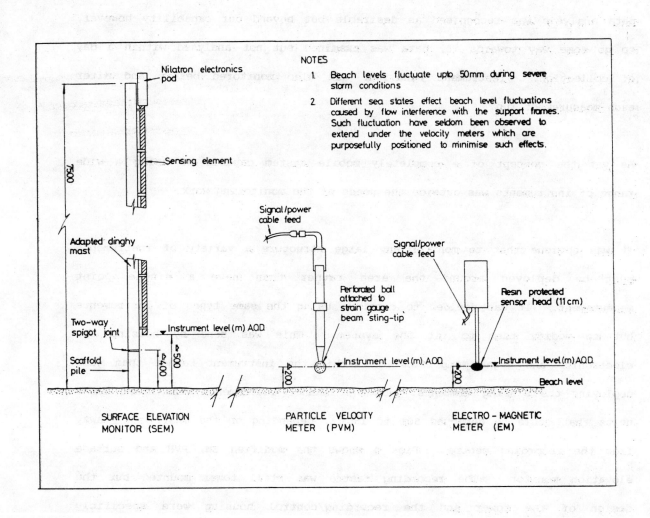

NOTES

1. Beach levels fluctuate upto 50mm during severe storm conditions

2. Different sea states effect beach level fluctuations caused by flow interference with the support frames. Such fluctuation have seldom been observed to extend under the velocity meters which are purposefully positioned to minimise such effects.

Nilatron electronics pod

Sensing element

Adapted dinghy mast

Two-way spigot joint

Scaffold pile

Instrument level (m) A.O.D.

Signal/power cable feed

Perforated ball attached to strain gauge beam 'sting-tip'

Instrument level (m) A.O.D.

Signal/power cable feed

Resin protected sensor head (11cm)

Instrument level (m) A.O.D.

Beach level

SURFACE ELEVATION MONITOR (SEM)

PARTICLE VELOCITY METER (PVM)

ELECTRO – MAGNETIC METER (EM)

FIGURE 4

Cables were obviously a very important link in the system, various types were examined before a selection was made. The effort expended in this area was rewarded by the results. For one part of the project a total of 1.5km of cable was laid on the beach in lengths of 50m to 100m for a period of 6 months with only two cables being damaged during that time. Connections were made with special underwater connectors which enabled the whole system to be dismantled in easily handled sections. The cabling was generally left deployed to allow a fast response to storm warnings.

Data recording was still carried out using a microdata data logger with the data being recorded on data cartridges. The data logger was housed in a weather proof steel box positioned approximately 4m above maximum still water level.

Because of the limitaions of battery power supply and data tape capacity it was necessary to service the system every tide, day and night during typical two week measurement periods. This did have the advantage of monitoring instrument condition every tide but played havoc with the social life of all those involved. The recording box also controlled the system using a panel of switches. Switch positions were carefully logged each run as was other data essential for analysis.

Each recorder box was designed to have connected into it up to 5 EMs, 12 PVMs and 15 SEMs or any combination of these, the total number being limited to 36 channels of data , by the data logger.

Two recording centres were used at the breakwater site to monitor offshore of the breakwater and behind it. A typical layout of instruments is shown in Fig 5. Data was normally collected for four twenty minute periods collecting at a regular time interval from 2 hours before high water to 2 hours after. For the layout shown in Fig 5 a total of 30 channels of data were collected each channel being digitized at a sampling rate of 3Hz.

As previously mentioned the data cartridge was retrieved after each run and the raw data were examined for instrument performance. The data were transferred to magnetic tape and selected data sets were analysed at Liverpool University.

In conjunction with designing the system configuration new arrangements had to be made for the rigid mounting of instruments. It was decided to develop frames using standard scaffold for the current meters, Fig 6. Frames of this type were found to be robust and rigid as well as being easily deployed. The surface elevation monitors were mounted on dinghy masts and the masts were scaffold compatible and could be easily rigged on driven scaffold tubes using a further three scaffold tubes as anchors for stay wires. The scaffold tube has been found to create very little scour problem and also because of its slenderness created very little disturbance to water flow and hence no significant regime interference.

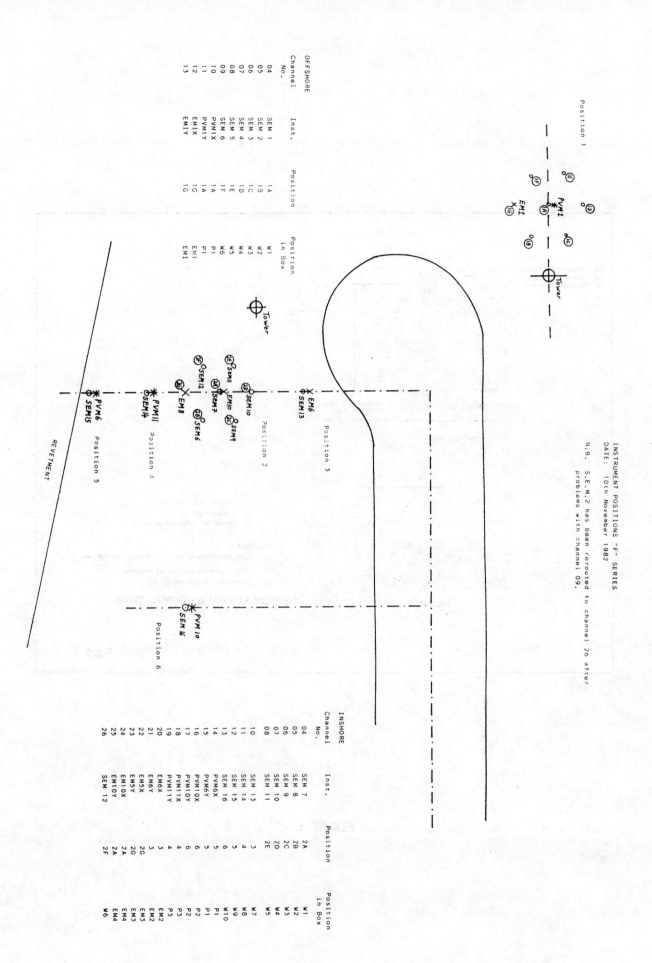

INSTRUMENT POSITIONS "F" SERIES
DATE: 10th November 1982

N.B. S.E.M.2 has been rerouted to channel 26 after
problems with channel 09.

FIGURE 5

217

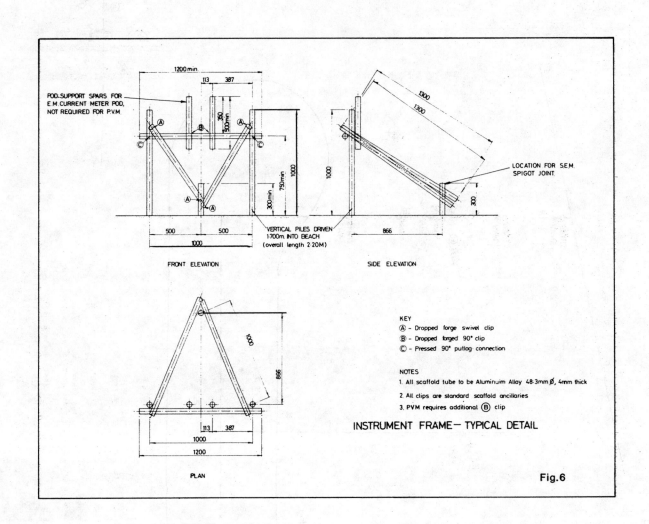

INSTRUMENT FRAME— TYPICAL DETAIL

Fig.6

FIGURE 6

As the need for full scale measurement became recognised so the need for the system to be adapted to other sites became essential. The system had up until now been used on a coast with a high tidal range enabling acccess to instruments and recording centres at low water. Other sites did not necessarily conform to this arrangement and it became obvious that the system should yet again evolve. The first steps towards this used the basic system B approach but linked back to a shore based recording centre where a computer was used to store data and control the system. This transition between systems B and C resulted in a cumbersome hybrid which although fulfilling the same objective as system B was not using the introduction of computers to the full.

System C

Having established that the three types of instruments were satisfactory the main direction of development had to be in the data collection record, control and analysis. BMT had some 2 years previously designed a data logging system for major gas dispersion trials and this was used as the basis of system C. The combined experience of BMT and the staff involved with the previous two systems was used to overcome the major problems of developing a highly sophisticated system for use in an environment which can only be described as an electronic engineers nightmare.

The system is shown schematically in figure 7 and the specification of the individual components are described below.

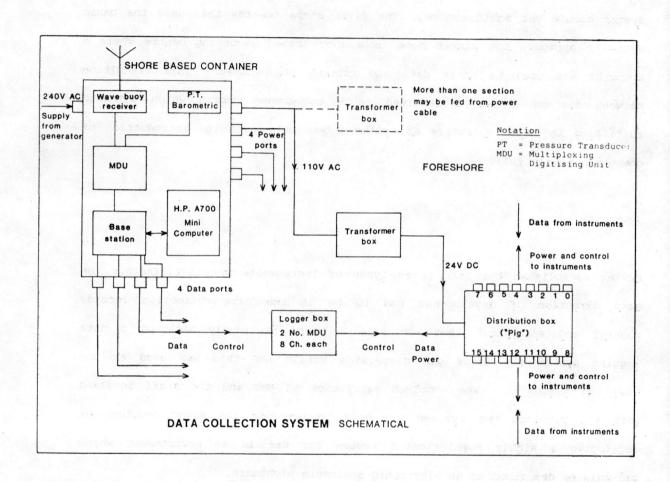

DATA COLLECTION SYSTEM SCHEMATICAL

FIGURE 7

Data loggers

Data loggers / Multiplexers (MDU) are housed in a water proof case capable of being deployed below the high water mark using specificly designed connectors (valeport dart connectors). Analogue signals are allowed into the logging unit and the resultant multiple serial data stream is buffered and transmitted at 9600 band to the base station and data integrity unit at the shore side. Cable lengths up to 4km have been utilised with no degredation of the transmitted data. The analogue cables from deployed sensors to the data logger were limited to 100 metres to reduce the possibility of noise pick-up in the system . Each data logger accepts up to 8 analogue inputs in the range ± 10 volts these are multiplexed to the 12 bit analogue to digital converter which samples each channel 20 times a second. The shorebased microprocessor controls the operation of the system which is controlled overall by a synchronised pulse transmission to all data loggers from the base station; data is only transmitted on receipt of a sync pulse.

The control function of the data loggers is operated via a duplex engineering link. This allows control command from the Base station to be carried out and the results of such commands to be sent back to base over separate cable pairs. Thus it is possible to remotely close relays, check status of parts of the system, and obtain single readings from single data channels.

Since all signals are passed through the data loggers their accuracy is fundamental to the acccuracy of the whole system. As well as stringent tests under laboratory conditions the system has been designed for rapid calibration on site. A calibration box was devised that could be plugged into any of the transducer ports and accurate voltages passed through the data loggers to the shore. This box also incorporates a continuty checker to ensure that no damage has occured to the logger or distribution boxes during transportation to site.

Base System

The data cables from up to 4 datalogger boxes are terminated in a primary multiplexer at the Base station.

Its major functions are

i)   Synchronize the data samples

ii) Clean up the electronic form of the digital signals

iii) Check the validity of the data received

iv) Route the individual data streams to the correct ports of the main computer

The station is capable of handling up to 128 channels but the system as a whole is designed for a maximum of 64 channels.

Computer Equipment

At the heart of the system is a H.P. A700 Minicomputer. This computer features a Multi-User disc based Real-Time Executive operating system and is The majority of the software is written in FORTRAN, the exceptions being the special device drivers which are written in Assembler code.

System Software

Three main programs are available for data collection

i)   ODANY. Any or all transducers can be sampled for short periods at any time and a printout obtained . This programme is used for general monitoring of system performance.

ii)   ODAUTO. Data can be automatically collected at a preset time for a period, normally 20 minutes. These data are stored and a printout given in terms of Max, Min, Mean for each channel. The system then lies dormant for a preset interval when another period of data is collected. This pattern is continued for as many records as are required.

iii) ODCONT. This is basically the same as ODAUTO except that as soon as the data from each period is stored another period starts. In this way a nearly continuous data set may be obtained.

The organizing of data uses a housekeeping file which contains transducer information such as location, height, station type, calibration and offsets as well as bearing of velocity axis and level of instruments to a selected datum. This file is created on site as instruments are deployed and can be updated when necessary during the measurement period. (Table 1.)

Analysis Software

A suite of analysis software has also been constructed for data reduction and is mounted in the computer on site – these programmes are designed to check the data integrity, divide the data into its respective channels and using the housekeeping file to convert the data from volts into engineering units together with execution of analysis for each channel. The various processed data files are then stored and graphical output of the analysis produced. All this can be done on site concurrent with data gathering on the computer system used. At the end of each set of measurements mean process analysis can be output in numerical and or graphical form.

223

TABLE 1 : Housekeeping File

```
*  HSKP1  CHANNEL,T'DUCER & CAL INFO UPDATED (830402.0952)
*
* ALL DISTANCES MEASURED IN METRES & ANGLES IN DEGREES
**                    AINSDALE DATA ( 4/3/85 - 10/3/85 )
*
*DT  CH      LOCN      HT   ST  TYPE   CAL    OFFSET  S.NO  LEV    AXIS    PR
*I2  I2   FFF.F FFF.F  F.FF AA  AAAA  FFF.FFF FFF.FFF  AA  FF.FFF AFF.FFF I2
*
*
 01  00  915.0 689.0  0.57 PC   EMX  -2.000    0.000  05  0.616 x 34.00 99
 01  01  915.0 689.0  0.57 PC   EMY  -2.000    0.000  05  0.616 v124.00 99
 01  02  915.0 689.0  0.00 PC  PRES  -3.411    0.126  05  0.036 A000.00 99
 01  03  000.0 000.0  0.00 AA   SP    1.000  000.000  00  00.000 A000.00 99
 01  04  914.0 740.0  0.50 PD   EMX   2.000    0.000  07  0.456 x301.00 99
 01  05  914.0 740.0  0.50 PD   EMY   2.000    0.000  07  0.456 v 31.00 99
 01  06  914.0 740.0  0.00 PD  PRES  -3.406   -0.400  07 -0.124 A000.00 99
 01  07  000.0 000.0  0.00 AA   SP    1.000  000.000  00  00.000 A000.00 99
*
 00  00  915.0 649.0  0.52 PA   EMX   2.000    0.000  01  0.381 x311.00 90
 00  01  915.0 649.0  0.52 PA   EMY   2.000    0.000  01  0.381 v 41.00 90
 00  02  915.0 649.0  0.00 PA  PRES  -3.403    0.240  01 -0.199 A000.00 90
 00  03  915.0 688.0  0.45 SC   SEM  -1.800   -8.300  01  0.480 A000.00 99
 00  04  915.0 659.0  0.53 PB   EMX   2.000    0.000  04  0.458 x117.00 99
 00  05  915.0 659.0  0.53 PB   EMY   2.000    0.000  04  0.458 v207.00 99
 00  06  915.0 659.0  0.00 PB  PRES  -3.450    0.600  04 -0.122 A000.00 99
 00  07  914.0 808.0  0.36 SE   SEM  -1.800   -7.600  02 -0.416 A000.00 99
*
 09  00  000.0 000.0  0.00 BA  BARO   1.000  000.000  01  00.000 A000.00 98
 09  01  000.0 000.0  0.00 BU  BUOY   1.000  000.000  01  00.000 A000.00 98
 09  02  000.0 000.0  0.00 AA   SP    1.000  000.000  00  00.000 A000.00 99
 09  03  000.0 000.0  0.00 AA   SP    1.000  000.000  00  00.000 A000.00 99
 09  04  000.0 000.0  0.00 AA   SP    1.000  000.000  00  00.000 A000.00 99
 09  05  000.0 000.0  0.00 AA   SP    1.000  000.000  00  00.000 A000.00 99
 09  06  000.0 000.0  0.00 AA   SP    1.000  000.000  00  00.000 A000.00 99
 09  07  000.0 000.0  0.00 AA   SP    1.000  000.000  00  00.000 A000.00 99
*
```

System Hardware

The shore based equipment is all housed in a '10ft' industrial container which can also house a wave sensing radar system which may be controlled by the computer. The power for the system is supplied either from normal mains or from a specially designed generator. Within the container the 240v AC supply is transformed down to 110v AC which is fed down to the beach where it is further transformed and rectified to 24v DC. This is then the standard voltage used throughout the beach measurement configuration. Separate cables are used for power and data and these have been specifically designed for the system and include integral staining wires to minimise damage due to movement under the flow conditions.

All equipment on the beach is housed in substantial steel boxes with cathodic protection and these boxes are secured on the beach by clamping to 2m long scaffold tubes driven into the bed.

The whole system is easily transportable and can be deployed typically within one to four days depending on the size of any experiment and number of transducers.

Instrument Development

The PVM and SEM used on previous systems remain basically unchanged apart from adapting them to 24v. The EM however has now been combined in a beach pod with a pressure transducer so that in the one housing velocities and surface elevation may be measured at a point. The system has been designed to allow a wide variety of instruments to be used and with simple modifications it is intended that other parameters may be measured such as salinity, temperature, suspended sediment, dissolved oxygen, wave run-up and loads.

In this way although system C is still developing it now matches well to the specification for measurement formulated some ten years previously.

Data Integrity

Throughout the system development the integrity of the data (output from the transducer through to storage on magnetic tape) has been constantly and rigorously checked and preserved.   It is still necessary to examine the recorded data, however, to detect 'spikes' or similar distortions as a result of 'relay flutter' or similar phenomenon of electronic circuitry.

The data is examined to procedure:

(i)   Accumulate data valves; determine mean and standard deviation of data valves and slopes.

(ii)   Re-accumulate data valves; identify locations and extent of data exceeding specified thresholds for magnitude and/or slope.

(iii)   Appraise 'bad data'; determine extent relative to whole record , identify sections for substitution.

(iv)   Fit substitute data; using hermite polynomial or similar piecewise cubic interpolation routines.

(v)   Re-accumulate data valves; determine new mean and standard deviation and repeat procedure till no further 'bad-points' identified.

For work in shallow water the selection of thresholds for this check on data integrity is critical and has involved extensive analysis of data sets from different transducers measuring the same phenomenon.   Standard thresholds obtained from Gaussion distributions and used in deep water work have been found to be inaccurate for work in shallow water resulting in the loss of good data by 'clipping' of the record.   Table 2 shows the presentation of the data integrity process.   This has been designed for rapid assessment.

Data integrity is finally examined between channels to identify if any

system behaviour is evident upon all channels recorded – this is undertaken

by straight time history comparisons, Fig 8 and from cross-phase

relationships between data channels.   In the latter operation use is made

of known relationships to be expected between particular sensors – this

work also provides a cross-check on transducer calibrations.

Data Analysis;

With the data integrity estabilshed, various analyses are executed:

(i)   To determine the statistical moments of each measurement and to

       compare the data with standard distributions for value, crests,

       troughs, ranges etc. Fig 9.

TABLE 2 : "Wild Point Editor" output of SEM channel (example of output)

```
==================================================================
CHANNEL  13 ( 8 of 28)  SEM
DATA VALUES:: Mean = -994.23 Upper limit = -870.56 Lower limit = -1117.9
SLEW VALUES:: Mean =    .002 Upper limit =  24.713 Lower limit = -24.709

                           SLEW GRAPH
6xSD :     *  *                        *        *        *
     :    * * *     *       * ***       **      * **    *  * *
     :  ***  *** ** **   ***** ** * * ****** ** * **** **** ***
     :********************* ************************************
     :************************************************************
1xSD :************************************************************
                           DATA GRAPH
6xSD :                                              *
     :      *        * **               **     *   *
     :   * * ***  ** **  * ***       *  **** ** ****  * ***
     :*************** ************ ************************ *****
     :************************************************************
1xSD :************************************************************
                           REPLACEMENTS
      :::::::R::::::::::::::::R::::::::::::::::::::::R::::::R:::::::
   ---+-------------------------------------------------------+
    1                                                         12000
CHANNEL 13 COMPLETE:     46 Records replaced
 12000 Records    4 Bad Blocks    20 Slew fails    20 Data fails
==================================================================
```

Notes

i) Data Values: These are the basic statistics of the data ordinates
                  derived from the raw time history and describe the
                  mean value and accepted limits of the measured process.

ii) Slew Values: These describe the mean value and the upper and
                  lower limits of the accepted rate of change of the
                  measured process over the digitising interval.

iii) Asterisk   The plots represent the 12000 data points from
     Plots: the 20 minute time history as 59 blocks of asterisks
                  against a vertical scale representing the range
                  of standard deviation SD form 1 x SD to 6 x SD.
                  The two graphs shown represent the maximum 'Slew rate'
                  and maximum 'Data value' time histories for each block
                  of the data.

iv) Replacements: If the recorded data fails against pre-set criteria
                  the analysis routine contains a facility whereby
                  sets of data points (or single points) can be replaced
                  through fitting a polynomial to points on either
                  side which are within limits specified. The routine
                  indicates replacements as 'R' at the relevant block
                  in the time history and states in tabular form;
                  the number of 'records replaced', the number of
                  'bad blocks' encountered and the number of 'Slew
                  fails' and 'Data fails' in the sample.

File:- XX15 (SD= .227 Mean=-.000)

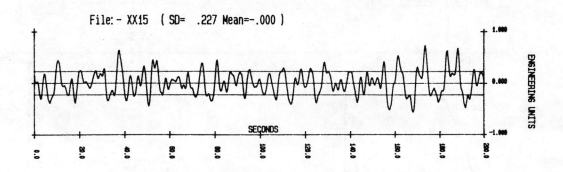

File:- XX16 (SD= .150 Mean=-.000)

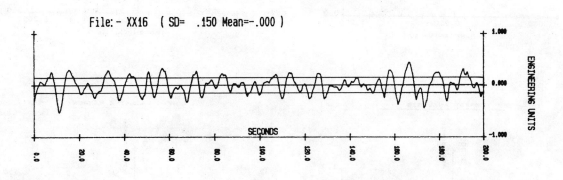

File:- XX17 (SD= .147 Mean=-.001)

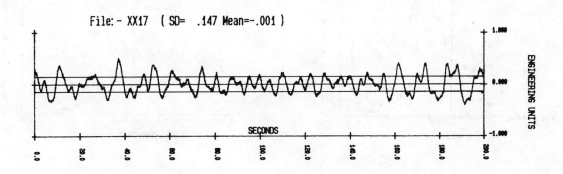

TIME HISTORY PLOTS

SEA PALLING DATA (Run No.78)									
ANALYSIS OUTPUT DATE :- Thu Aug 15, 1985 11:50:24 am									
File	Description	Type ID	EAST	NORTH	LEVEL	Calib	O/Set Hght	Bearng	S/N
XX15	Zero Mean Data	EMY P8	470.0	87.5	-.62	-.13	0.000 .82	0.007	05
XX16	Zero Mean Data	EMX P8	470.0	87.5	-.62	-.13	0.000 .82	0.007	05
XX17	Zero Mean Data	PRES P8	470.0	87.5	-1.21	3.41	-.001 .22	0.007	05

FIGURE 8

229

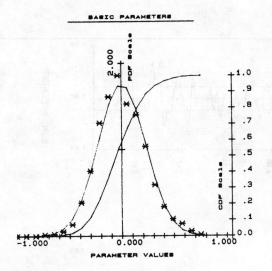

BASIC PARAMETERS

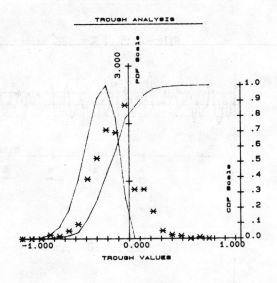

TROUGH ANALYSIS

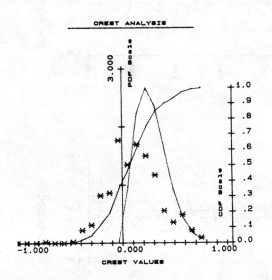

CREST ANALYSIS

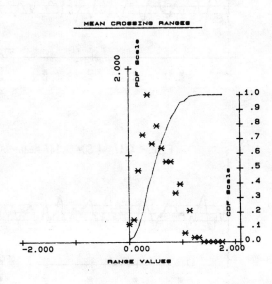

MEAN CROSSING RANGES

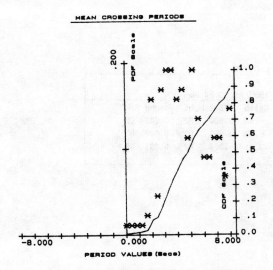

MEAN CROSSING PERIODS

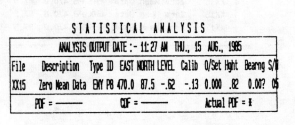

STATISTICAL ANALYSIS

ANALYSIS OUTPUT DATE :- 11:27 AM THU., 15 AUG., 1985									
File	Description	Type ID	EAST	NORTH	LEVEL	Calib	O/Set	Hght	Bearng S/N
XX15	Zero Mean Data	EMY P8	470.0	87.5	-.62	-.13	0.000	.82	0.00? 0S

PDF = ——	CDF = ——	Actual PDF = X

FIGURE 9

230

(ii) To examine the frequency composition of the data by the calculation of auto and cross-spectra, coherence etc. Fig 10.

More specific analyses to check theoretical simulations to be used in models are also possible on this high-level data. These specific analyses might typically involve checks on the validity of linear wave theory in shallow water for the prediction of wave refraction, water motion within wave groups. The data can then be assembled into suitable format to serve as boundary input or for validation of numerical and physical models.

Further Development;

With the system C developed attention has now returned to the transducer end of the measurement operation. The experience of development and operation of the high-level system is allowing the competent development of simpler lower cost systems by their validation against the high-level system. The development of better high-level transducers is now needed however espcially for the determination of sediment motions.

Work is well advanced upon the development of a mobile measurement platform to be used with system C - this assembly will permit movement of the measurement station during a measurement period - it will operate both through and beyond the surf zone but will be able to maintain a rigid posture when required for accurate data capture.

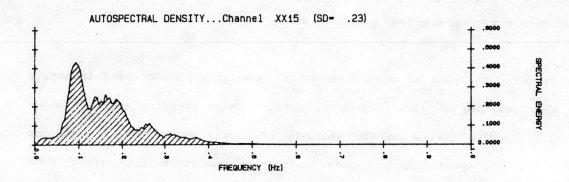

AUTOSPECTRAL DENSITY...Channel XX15 (SD= .23)

FREQUENCY (Hz)

SPECTRAL ENERGY

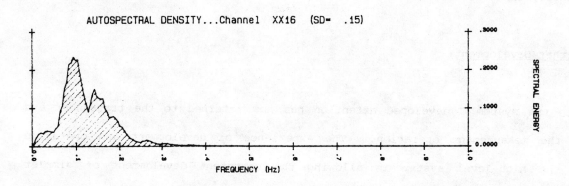

AUTOSPECTRAL DENSITY...Channel XX16 (SD= .15)

FREQUENCY (Hz)

SPECTRAL ENERGY

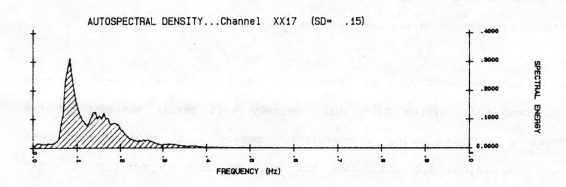

AUTOSPECTRAL DENSITY...Channel XX17 (SD= .15)

FREQUENCY (Hz)

SPECTRAL ENERGY

FREQUENCY ANALYSIS

File	Description	Type ID	EAST	NORTH	LEVEL	Calib	O/Set	Hght	Bearng	S/N
	SEA PALLING DATA (Run No.78)									
	ANALYSIS OUTPUT DATE :- 11:46 AM THU., 15 AUG., 19									
XX15	Zero Mean Data	EMY P8	470.0	87.5	-.62	-.13	0.000	.82	0.00?	05
XX16	Zero Mean Data	EMX P8	470.0	87.5	-.62	-.13	0.000	.82	0.00?	05
XX17	Zero Mean Data	PRES P8	470.0	87.5	-1.21	3.41	-.001	.22	0.00?	05

FIGURE 10

Conclusions

The evolution of a measurement system has been described to record data at full-scale in the surf zone and beyond along coastlines and within estuaries. The difficulties of establishing and maintaining such a system have been identified together with the need to constantly check whether you are measuring what you think you are measuring and whether what you have measured is reaching tape storage in uncorruped format.

The use of proven data has been examined for checking theoretical simulations and for its subsequent use in modelling work. By obtaining quality measurements at full-scale it is possible to assemble numerical models which, from their validation against field data, may be used with greater confidence to predict effects from coastal works.

In this way a procedure has been formed by which developers and conservationalists can obtain much more information upon which to base their decisions. By collecting data at full-scale immediate plans can be more competently formulated to the community benefit and as the data base expands the understanding of the natural phenomena will improve and lead onto both further economies for developers and improved capabilities for conversationalists in the future.

Acknowledgements

Between 1975 and 1980 the system was developed by British Maritime Technology Ltd (then NMI) for the Wirral local Authority on Merseyside and was therefore developed by consortium funding from the Science and Engineering Research Council through Liverpool University, the Wirral Authority and BMT Ltd. Since 1984, to data the system has been developed by BMT Ltd in association with its subsiduary CEEMAID Services.

DEVELOPMENT OF THERMAL FIELDS IN WATERS NEAR POWER STATION OUTFALLS

G.C.C. Parker and I.J. Todd

Central Electricity Generating Board
Generation Development & Construction Division
Barnett Way, Barnwood, Gloucester GL4 7RS

Summary

Typical cooling water requirements to condense the exhaust steam of turbines at a 1200MWe power station can amount to $50m^3 s^{-1}$ and can result in a temperature rise of $10°C$ in the discharged cooling water. In order to achieve a satisfactory station thermal efficiency, it is necessary to maximise the dispersion of this discharged water to minimise recirculation back to the station intake where direct cooled systems are employed. These direct systems use a large river, estuary or coastal water for the supply and return of the cooling water. It is also necessary in certain cases to consider the effect of the warm discharge water on the local marine ecology. Work has been done to gain an improved understanding of the mechanisms of dispersion, the formation and behaviour patterns of artificial thermal fields, and the site related effects of natural thermal fields.

This paper outlines some of the findings which have resulted from hydrographic surveys, and how the results have influenced the location of offshore cooling water intake and outfall disposition at coastal sites. The paper shows the importance of survey equipment choice and disposition and refers to future developments.

Held at Imperial College of Science and Technology, London. Organised and sponsored by BHRA, The Fluid Engineering Centre. Co-sponsored by the American Society of Civil Engineers and the International Association for Hydraulic Research.

1.   INTRODUCTION

In the most common steam cycle used for power stations, large volumes of water, in the order of 50m^3/s, are needed for condensing the steam at the exhaust of a turbine, so that the condensate, which is the water derived from the steam cooling process, can be re-cycled through the boiler. A condenser consisting of a large number of parallel tubes, with water flowing inside the tubes and steam on the outside, carries out the condensing process.

The efficiency of the steam cycle is very important in the costs of station operation and small changes in efficiency have very significant effects on the station fuel consumption and operating costs. The cooling water temperature is a major factor in achieving a good efficiency. For example, a 1°C decrease in cooling water temperature, achieved by correct juxtapositioning of intake and outfall structures, can result in an operating cost benefit of several million pounds over the life of a nuclear station. This is the result of the cooler water improving vacuum conditions at the back end of the turbine and thereby increasing its efficiency.

As the cooling water flows through the condenser tubes, its temperature is raised and this could vary between 8 to 14°C. This warmed water must then be discharged to the cooling source and be dispersed in such a manner as to minimise its recirculation back into the cooling water intake and attendant loss of steam cycle efficiency. Because it is necessary to use so much cooling water to obtain adequate dispersion in the cooling source, the preferred location for a power station from the cooling water viewpoint, is near a large river, estuary or coastal site.

It will be understood from this description that one of the key problems facing the cooling water system designer is to provide the optimum separation between the cooling water intake point and the outfall. Another important requirement is to design a system which has the minimum effect on marine ecology. In this connection it is necessary to ensure that warm water is adequately dispersed to avoid harmful effect on marine life.

The acquisition of information on currents and water temperatures over a large area is vital for these cooling water studies, and this paper describes the behaviour of warm water at two specific locations, associated survey techniques, and the consequent analysis of results.

2.   DISPERSION OF OUTFALL WATER

Eventual loss of heat to the atmosphere is a lengthy process and in the intervening period the dispersion of warm water discharged from a station outfall can be identified in a number of separate stages. The first, or near field stage, is represented by the immediate mixing of newly discharged warmed water into the ambient sea.

After a brief transitory period, a second or midfield stage is represented by a buoyant plume of warm water lifting away from the sea bed and spreading outwards at a rate determined by gravity currents, momentum effects and the action of the tidal stream. A midfield plume can eventually reach several hundred metres in width and can extend in length for 1 to 2 kilometres in the direction of the tidal stream (Figure 1). The plume depth below the surface can decrease from about 3m to 1.5m in moving alongshore. The dominant mixing process appears to be bed generated turbulence which erodes the underside of the plume and warms the lower waters[1]. Eventual result of this process is complete vertical mixing of temperature in the water column leading to a small rise in ambient water. Approaching slack water, the plume broadens considerably and at tidal reversal the trailing plume may partly detach from the shore and hinge around the outfall position thus providing considerable advection of warm water offshore. The normal practice is to minimise recirculation by physical separation of the intake and outfall structures. Civil tunnelling costs may limit the degree of protection that can be afforded by this practice, but additional protection can be sought by designing the intake structure to minimise drawdown from an overhead plume, to ensure maximum possible depth of water over the period of coverage, and to minimise the period of coverage.

CEGB surveys have identified a third stage in the heat dispersion process at a number of sites. During periods of calm weather conditions it has been observed that sequential flood and ebb movements of the midfield plume alongshore can develop a far larger pool of warm water and Figure 2 shows this condition at Sizewell on the Suffolk coast. Equilibrium conditions of spatial spread and thermal development was attained within 8 to 10 days with temperatures up to 1.5°C above natural ambient conditions recorded from seabed to surface. It will be seen that the spatial spread extended a full tidal excursion alongshore and several kilometres offshore. This far field plume is also moved alongshore by the reversing tidal stream and an amount of secondary recirculation cannot be avoided in the example shown as the cost of separating the intake and outfall structures exceeds the recirculation penalty. It has also been found that a far field plume is quickly dissipated when wind strengths rise above 5m/s.

Since the discovery of this third stage in the dispersion process, subsequent design studies have had to identify the possible coincident periods of maximum contribution from both the mid field and far field to the maximum intake temperature, in order to provide the appropriate design criteria for station plant.

3. NATURAL TEMPERATURE WARM WATER FIELDS

In addition to the separate dispersion stages of the power station warm water discharge, it is necessary to consider the natural temperature rises in ambient waters due to solar radiation which is particularly significant at inshore positions. The number and relative strengths of these solar fields can vary considerably with changes in topographical features and season of the year, and can have significant effect upon the intake temperature.

Deep sea ambient temperature normally follows an annual rise to a peak about two months after maximum possible solar radiation in June, the time lag being due to the thermal inertia of the deep sea to respond to insolation. The difference in deep sea maximum temperatures achieved in a hot summer compared with a cold summer can be up to 3°C. During periods of increased solar radiation and light winds the shallow inshore waters warm more quickly than the deep sea. Survey readings have shown that the shallow inshore waters can approach an equilibrium value of 3°C above deep sea ambient after about 9 days of calm, sunny conditions.

Estuarine temperature gradients, local warming over mud flats and sand banks inside bays and harbours, can all contribute to the thermal structure of intake temperature and each requires an individual assessment of its maximum temperature contribution and relative changes during the tidal cycle.

4. SURVEY AND ANALYTICAL METHODS

It is important that survey operations should be conducted over a long period to ensure that the eventual design of the cooling water offshore works is founded upon a data base that sufficiently represents the variable meteorological and tidal current conditions local to the site. It is equally important that the survey period should include the calmer and warmer conditions of the summer months when the natural and artificial thermal fields are most likely to reach a combined maximum temperature.

Some hydrographical information will be available for proposed sites near to an existing power station. However, the increase in size of new developments, the need to place the new offshore works in correct juxtaposition to existing structures, and the need to ensure that eventual combined discharges will not adversely affect local ecology, will still require additional survey operations.

The initial step in planning a new survey is a preliminary desk study of all available data. A valuable information source on an existing site is the historical record of condenser intake temperatures. Changes in these temperatures may correspond with variations in meteorological conditions, tidal state and station load, and might provide a significant contribution to understanding the complex variation of the separate natural and artificial temperature fields which together compose the thermal structure of the intake temperature. A preliminary action to any recent main survey has to supplement any existing operating station equipment with a measuring system capable of recording intake temperatures at ten minute intervals to an accuracy of $0.2^{\circ}C$.

The second step is to decide upon the actual survey operations. A survey will comprise an array of moored instrumentation to record continuous data of flow patterns and water temperature changes throughout the survey period and a number of individual operations generally limited in time to a single tidal excursion. The moored array can include current meters, tide gauges and thermistor stringers which, together with an onshore automatic meteorological station, provide an overall record of data to improve understanding of the results from individual survey operations. These individual operations can include float tracking, infra red photography, dye release, thermal plume profiling, and temperature/current/salinity profiling. Alongside these activities, which are mainly designed to assist in evaluating thermal plume behaviour, the survey will contain the necessary echo sounding, side scan sonar, seismic work, sea bed sediment sampling and wave recording to supply the information required by civil engineers for designing the station structures.

Improvements in survey operations and measuring equipments have been paralleled by development of mathematical modelling techniques. Experience however, has clearly demonstrated the complex problems involved in both modelling the separate temperature fields that make up the thermal structure of a body of water and of estimating the relative contributions of these temperature fields at different sites and under differing tidal and meteorological conditions. Consequently, despite some good correlation between survey results and model predictions, the CEGB continues to regard the hydrographic survey as the primary tool in present investigations. In particular it provides a validation source for the flow predictions of a mathematical model and the only satisfactory means of identifying the natural temperature fields.

The CEGB approach to measuring techniques in offshore waters and the subsequent analysis of data is best illustrated by examples drawn from actual surveys. The two sites which are described in this paper are Sizewell, in Suffolk and Hinkley Point, in Somerset. Nuclear stations were built on these sites during the mid 1960's and mid 1970's, and the surveys were required to provide design information for further power station development at both locations. The 1975 Sizewell survey has been chosen as it represents the first major study of flow patterns over a large area of open sea and resulted in the first clear identification of a far field plume. The 1983 Hinkley survey, in contrast, investigates cooling water discharge into the far larger spring tidal range of the Severn Estuary and illustrates the first use of the thermistor stringer as a major survey tool.

4.1   <u>Sizewell Survey</u>

In the locality of Sizewell station the Suffolk coastline runs
generally north and south. Offshore from the site the sea bed shelves
gently downwards and then rises again to form the channel between the
shore and Sizewell Bank. This shoal feature roughly parallels the
coast for about 3km. Further north the channel is continued by Dunwich
Bank and widens out into Dunwich Bay. These two banks are located
about 1.5 to 2km offshore and are totally submerged at all times.

This composite structure of channel and banks rests upon a relatively
shallow inshore area reaching about 3 to 4km out to sea and stretching
about 20km along the shore. Depth of water is generally less than 9m
below chart datum, the exception being a deeper trough formed inshore
of the banks.

4.1.1   Pre 1975 Survey Information

Collation of information on sea bed movements dating back to
1824 indicated areas of accretion and erosion along the channel
feature and some evidence of a wave ripple pattern transverse
to the shore with wave crests about 4km apart. This
investigation concluded there were no progressive long term
changes in the area and that the present 'A' station offshore
works had not had any significant effect on the sea bed
topography.

A Ministry of Agriculture, Food and Fisheries investigation[2] of
bottom residual drift in 1968 established that there was a
general southerly drift of water within a coastal band south of
Yarmouth and suggested that this band width could extend 16km
seaward of Sizewell in the summer time, and sweep on southwards
to form part of the circulatory movements in the Thames
Estuary.

During 1970 to 1972, data was obtained from current meters
moored at four positions along the Sizewell Coast. Owing to
poor availability it was only possible on two occasions to
obtain coincident readings from a maximum of two current
meters. Drift measurements showed a persistent southerly drift
at all positions within the channel with values varying
between 0.061 and 0.094m/s (5.25 - 8.1km/day) which, when
compared with peak tidal currents of 0.73m/s and 0.36m/s for
spring and neap conditions, represents an appreciable rate of
water loss from the area.

A considerable amount of information existed regarding the
spatial spread and thermal contours of the Sizewell 'A' station
mid field plume. This information was gained in 1967 by
means of a number of thermistors attached to a pole mounted
vertically to the side of a small boat. In this manner
simultaneous readings of water temperature at a range of depths
were obtained as the boat traversed the plume. This technique
normally requires 30 to 40 minutes to complete sufficient
traverses to obtain full detail of the plume and is obviously
far too slow to correctly identify the sequential changes in
spatial plume spread around tidal stream reversal.

In 1973 another 'thermistor pole' survey of the mid field plume was supplemented by infra red scans of the surface spread of the plume. These relatively instantaneous images of the plume were later analysed to provide thermal contours. Figure 1 illustrates detail of thermal contours within a plume about 30 minutes before low water slack. The plume has spread further offshore due to lessening of the tidal stream and the surface plume boundary has just begun to float over the intake position.

The 1973 intake survey also included temperature depth profiles obtained by lowering a measuring probe at incremental depths to the sea bed, near the position of the intake. These depth profiles showed the arrival of the buoyant plume over the intake structure towards the end of the ebb tide and the subsequent period of coverage, until the plume began a pivotal movement and swung southwards with the flood tide. This depth profiling operation also identified a regular temperature change extending from the surface to the sea bed. This change reached a maximum of about $1^{\circ}C$ during each tide.

4.1.2   Survey Equipment and Field Work

With sufficient detailed information on the mid field plume behaviour, the field work in the 1975 survey was designed to improve understanding of large scale water currents and associated drift patterns. Mathematical modelling work was proceeding in parallel so that this could be used as a design aid in future work, and it was recognised that much of the survey data would be invaluable in providing modelling criteria and validation.   It was also essential to identify the full spread and thermal contours of the fully mixed down temperature pulse detected by the depth profiling runs and to establish whether this temperature rise represented variations in natural conditions or a third stage in the eventual dispersion of artificial warmth into the atmosphere.

This latter case, if proved, would have important bearing on design considerations and, in order to define the spatial limits of this possible far field effect and to determine any formative changes due to variance of meteorological conditions, an array of current meters, each fitted with a thermistor, were moored over about $150km^2$ of sea in the vicinity of the station. A meteorological station was erected onshore. Readings were recorded at ten minute intervals and, with this interrogation frequency, meter batteries and magnetic tapes were renewed at monthly intervals during the six month survey period.

The meters were fitted with flume calibrated impellers with a velocity range of 0.05m/s to 2.5m/s, accurate to +2%, and a starting velocity of typically 0.03m/s. Silt degraded bearings, fouled impeller, marine encrustation, are all risks to be faced during immersion and, as a safeguard, the contractor replaced the impeller unit at each tape exchange during the summer. Occasional mechanical damage from collision or adverse weather was reported and one case of encrustation was observed, but the major concern was the considerable external interference with mooring buoys and disappearance of a number of meters from the extensive area of sea involved. In view of these difficulties it is now CEGB policy to change tapes at shorter intervals and to maintain a regular patrol of moored equipment.

As a supplement to the current meter thermistors recording temperature changes at a number of single depth positions, hourly measurements of sea water temperatures at varying depths from sea bed to surface were attempted at two locations. In order to achieve these measurements a surface buoy supporting a weighted vertical wire fitted with thermistors at 1 metre intervals was moored at each location. These temperature sensors were connected by a seabed armoured cable to a power supply/bridge circuit/recording unit mounted on the existing 'A' station cooling water intake structure. At hourly intervals this unit sequentially scanned all the sensors and printed out for each a sensor related voltage. These values were subsequently related to the sea water temperature by a calibration curve for each thermistor determined prior to installation. These two prototype equipments represented the first attempt by the CEGB to obtain continuous recordings of temperature with depth over a long continuous period. In the event, no readings were obtained from one location and the second unit failed after five days because of damage to the electrical cables.

Extensive tracking operations were carried out with groups of floats released at varying depths simultaneously along lines transverse to the coast and extending up to 5km offshore. These releases were followed over four consecutive tidal excursions.

4.1.3   Results and Analysis

During the 1975 Sizewell survey the float tracking observations clearly established that under light wind conditions, the flow of water tends to be parallel to the general north/south run of the coastal features for at least 8km offshore.

It was established that the Lagrangian value of southerly residual drift obtained from these float tracking exercises showed close agreement with current meter Eulerian predictions of drift obtained by summation of current vector readings over the same tidal period. This measure of agreement was considered to be due to canalisation of flow alongshore owing to the presence of offshore shoals paralleling the shoreline. It was therefore reasonable to assume that the current meter predictions would continue to remain reliable indicators of residual drift throughout the survey period.

The large amount of information received from the current meters was processed by CEGB computers into various analogue displays. Figure 3 represents information obtained from a current meter moored about 50m offshore of the present intake structure in July 1975. Residual drift traces indicate a predominantly southerly loss of water of up to 6km over a tidal excursion. Vertical lines have been added to assist in comparing separate traces and also represent the periods of high water slack. This particular meter was moored at a depth of 5m below Ordnance Datum and it is not considered that the regular fluctuations of sea temperature recorded by this instrument would have been unduly influenced by the occasional approach of the buoyant mid field plume around low water slack.

Despite the relatively short period of recording, the prototype thermistor stringer, moored further offshore, was able to confirm that these regular fluctuations of temperature occurred from sea bed to surface on the flood and ebb tidal excursion and also indicated changes in the field strength with variations in solar radiation, wind speed and station generation over subsequent tides.

These regular temperature changes were repeated at all other current meter positions, including several many kilometres distant from the area of sea swept by the mid field plume. The amplitude of these regular temperature fluctuations reduced in magnitude in recordings taken from meters further offshore. This confirmed the temperature field was contained within a localised inshore area and was not part of some natural temperature gradient in the main sea.

Based upon the close agreement between the measured float runs and current meter predicted float tracks, it was assumed that current meter thermistor temperatures plotted upon individual current meter predicted float tracks, could be used to generate a surface thermal profile of this far field plume.

The accuracy of this profile would be affected by any actual divergence from the predicted tracks and by any subsequent surface heat exchange during the tidal period and, to obtain the actual far field temperature, it would also be necessary to subtract the value of deep sea ambient temperature, and natural rise in sea temperatures close inshore. This analytical process was carried out on a number of occasions and a typical far field plume at Sizewell is represented by the temperature profile in Figure 2. This estimate of the far field at the end of the flood tide, forms a bulge of warmed water attenuating about 6km seaward of the station site and about 20km alongshore. The centre of this warm water field, with temperature rising to about 1°C above local natural ambient conditions, moved about 8km northward during the subsequent ebb tide. The approximate spatial spread of the mid field plume has been added for comparison.

From Figure 3 it is evident that there was a sudden reduction in the southerly residual drift value on the 8th July. This change was recorded at other current meter locations within the channel and offshore the shoals. This change remained for the rest of the summer. It is unlikely that the local wind strengths involved could have produced these dramatic changes in drift patterns and evidence from the total array of current meters seems to point to a progressive interruption of a general southerly drift by a developing resisting current from the south. Confirmation of a change in the residual drift pattern is provided by a statistical analysis of Sizewell data by MacQueen[3].

Coincident with the change in residual drift patterns there occurred a fundamental change in the relative position of the tidal excursion of the far field plume alongshore. Prior to the change, tidal excursion of the far field thermal contours resulted in maximum far field plume contribution to the cooling water intake temperature around low water slack. This coincided with maximum contribution from the mid field plume as it passed over the intake position. Figure 3 shows that within 3 days the peak contribution of the far field to intake temperatures shifted to the HW slack period. During the same period, the effective overall tidal excursion of the far field was displaced approximately 8km northwards alongshore. This change in pattern of contributions to intake temperature also remained constant for the remainder of the summer. There is some evidence of reversal of this process during late September but temperature fields were increasingly disturbed due to adverse meteorological conditions during this period and it was not until calmer weather conditions later in the autumn that the far field plume was able to reform, and it was possible to confirm that the tidal temperature changes due to this temperature field had reverted to the previous early summer pattern at all meter positions.

The 1975 survey data represented the only detailed evidence of seasonal change in the pattern of movement of the far field plume and in view of the important effect this could have on the thermal structure of the intake temperature it was decided to arrange for further measurements to be recorded in the summer/autumn period of 1980.

The characteristic temperature variations recorded in this summer were similar to those measured in the same period in 1975 and eventual southerly displacement of the far field took place in 24-28 September 1980. The 1980 phase change provides clear evidence of similarity in seasonal changes in the relative position of the far field plume during the unusually warm conditions in 1975 and the rather poor weather conditions of 1980.

Study of the survey data had shown that at a near approach to equilibrium stage in its thermal development, the far field plume had provided an 80% contribution to the total artificial recirculation during a tidal excursion and could add 1.5°C to the maximum temperature recorded during a tide. As stated in Section 2, the civil engineering costs ruled out any attempt to avoid recirculation by physical separation of new offshore works.

4.1.4   Mathematical Modelling

The extensive area of sea involved, the great difficulty in representing macrocirculatory current flows and residual drifts, and the problems of matching hydraulic and thermal modelling scales, ruled out any physical modelling process. The CEGB therefore commissioned the development of a depth integrated numerical modelling technique utilising a large 2km computing grid to generate boundary conditions around a smaller 500m grid closer to the station. This model was able to provide good agreement with survey findings[4] and has since been used to forecast far field developments for increased heat rejection at Sizewell and to estimate similar temperature fields at other sites of potential interest. This numerical model of far field temperature was also run with an overall heating term equivalent to typical solar radiation and again showed good agreement with the survey measurements of the inshore solar field.

4.2   Hinkley Survey

Survey operations were carried out in coastal waters up to 20km alongshore of the Hinkley Point Power Stations which are sited about 15km from Bridgwater in the County of Somerset.   The site is about 9km south west of Burnham-on-Sea in the Severn Estury and about 6km east of Lilstock. Beyond the cliff/sea wall line, at low tide, is a wave cut platform. This area is bedrock in front of the power stations, but moving westwards, some superficial sediment overlies the rock locally. Eastwards are extensive areas of sands and mud flats.

4.2.1   Pre 1983 Survey Information

Hinkley is another site of prime importance for future possible development and surveys have been carried out on a number of separate occasions over a number of years. These have included extensive float tracking operations, use of current meters and dye releases which together have provided a good understanding of flow patterns in the area.

Preliminary investigations at Hinkley had also recorded detailed changes in the spatial form of surface temperatures of the midfield plume. This was achieved by using an infra red camera from flying heights of approximately 1000m (Fig.4). The progressively darkening areas of the plume represent increasing temperatures in the surface water. As a general guide to scale, the intake structure is approximately 750m westwards of the point of channel discharge into the estuary. During the flight period the A and B stations were generating 420MW and 550MW respectively with an associated heat rejection totalling 2240MWt. This amounts to about 60% of the combined generation figure during most of the 1983 survey period.

These infra red scans show that around 2 hours before LW the channel discharge jets about 180m offshore before moving downstream on the ebbing tide. About 200m westwards of this discharge position the warm water is directed further offshore by an anti-clockwise gyre which results in the majority of the primary plume passing offshore of the intake structure at this state of tide.

An hour later the gyre has been dissipated due to the diminished tidal stream. This reduction in tidal velocity has also allowed the plume to spread about 330m offshore and to move westwards in such fashion as to spread over and beyond the intake position. Therefore at first sight, this plume would appear to represent the major influence on fluctuations in the intake temperatures.

Station recordings had occasionally indicated considerable changes in intake temperatures during a tidal excursion and prior to commencement of the 1983 survey and throughout the period of survey operations, additional temperature recording equipment was installed and operated by the CEGB to provide more detailed information on the varying patterns of temperature changes.

4.2.2   Survey Equipment and Field Work

The 1983 survey was designed to supplement previous work and to provide an understanding of the highly variable changes in the intake temperature of the present station. On this occasion a spatial array of seven thermistor stringers was utilised, operating continuously over the summer months and so arranged as to provide a three dimensional understanding of changes in plume form under varying conditions of tide, weather and station generation. An additional thermistor stringer was moored just upstream of the station site and two more about 8km west of Hinkley Point.

The thermistor stringers used on this project were specially designed to support 14 temperature sensors and 4 depth sensors. The interrogation period was set to ten minutes, giving a useable tape capacity of approximately 25 days data. The spring tidal range at Hinkley can lead to tidal currents in excess of 1m/s. Because of these difficult conditions, a mooring design was specially devised to relieve the sensor string and instrumentation of any mooring load during either ebb or flood currents. In addition, the string was made nearly neutrally buoyant and of high drag coefficient, so that even at times of quite low current it would tend to lift clear of the sea bed and take up a catenary shape covering the full water depth. In practice, the mooring arrangement worked well. The sensors, spaced approximately 1 metre apart along the string, covered the full depth fairly uniformly, except for a brief period of slack water at low tide, when much of the string was suspended near to the sea bed. The surface buoys suffered some physical damage from persistent wave action and one thermistor stringer was lost during the survey.

Installation of these moorings in the strong currents experienced was a difficult task.   The stringers were deployed with the vessel heading upstream to obtain more vessel control. Care had to be taken to ensure that the thermistor string did not foul up. There were persistent and recurring instrumental faults, principally with the sensor strings, but because the instruments and mooring arrangement allowed in-situ checking of the sensors and recorders, most of these faults were detected and rectified within five or ten days of their occurrence.

The thermistor stringers remained in position throughout the summer months of an exceptionally warm year and provided the main survey tool. Two current meters were moored near the station site and a number of short term and long term float tracking operations were carried out at spring and neap tides to study changes in tidal stream flow and to measure tidal excursion and residual drift. A meteorological station was set up on the station site.

4.2.3   Results and Analysis

Figure 5 examples a graphical plot of intake temperature recorded early in July and beginning a few days before commencement of main survey operations on the 7th July. The temperature peaks and troughs approximate respectively to the periods of LW and HW slack of successive tides and these temperature fluctuations illustrate the response of estuarial water to meteorological conditions and discharge of cooling water from the A and B stations.

From this graphical plot it can be seen that the 3rd July afternoon tide established a peak temperature rise of about 1.7°C above the equivalent morning period. On subsequent tides there is a diminishing difference in peak temperatures, with the morning peak rising and the afternoon peak falling, until both tides attain similar peaks around the 9th July period. This suggested a solar influence and it was noticeable that the variation and magnitude of temperature fluctuation between LW and HW periods did not appear to be affected by a considerable fall in generation during this period. This pattern of change was observed at other periods through the summer, and, in each case during the survey period, the maximum differences between morning and afternoon peaks occurred about 2 days before the neap tide, with a progressive fall in these maximum difference values which is consistent with reducing solar radiation as the summer progressed.

Previous study of satellite thermal imagery of the estuary around LW showed a warm area immediately eastwards of the station site. This area corresponded with the shallow water zones stretching between Stert Flats and Sand Bay and it has been estimated that peak solar radiation onto this area can approach 20GW on a summer's day. The efficiency of energy absorption and the attainment of maximum water temperature from summer sunshine, must be influenced by the relative stage of the tide with respect to the sunshine period and to the position of the tide in the spring/neap cycle.

The overall length of the Stert Flat/Sand Bay shallow water area roughly equals the average tidal excursion past the site and it is considered that warming of shallow waters over these local flats and sands contributes towards these differences between morning and afternoon LW slack peak temperatures.

An additional temperature gradient can also develop along the estuary as the shallower water upstream responds more quickly to solar radiation. This estuarial gradient could be clearly observed on satellite infra red scans of the Severn Estuary during the survey period and study of these scans and knowledge of tidal excursion allowed an estimate to be made of the estuarine gradient on station intake temperatures. A subsequent study of the contribution made by cooling water discharged into the area revealed that these two natural shallow water solar fields could provide up to 40% of maximum diurnal changes in the intake temperature.

Temperatures recorded at ten minute intervals at ten stringer
positions provided a formidable mass of data and to facilitate
study the information was processed into graphical time plots
on a tidal basis. An example of this format is shown in Fig.6
which provides detail of temperature/tide changes at a position
50m directly offshore the present intake. The occasion was a
neap tide on the 20th July. Station heat rejection on this day
averaged 3571MWth. Each ordinate set of values represents
detail of temperature changes with depth at the position at a
particular time interval. Isotherms have been constructed
across these separate time intervals but these only serve to
indicate the sequential changes in plume structure with time
across the intake position and by no means represent a plume in
real form.

On this basis it can be observed that the buoyant primary plume
penetrates to a depth of about 2m below the surface at this
inshore position, with surface temperatures rising to nearly
$3^{o}C$ above sea bed values.   Below the surface plume the water
temperature remains fairly uniform down to the sea bed but
temperatures at these lower levels do show a tendency to rise
to higher values around mid tide.   To provide an estimate of
direct recirculation from the mid field plume it was assumed
that this stringer provided a close estimate of seabed
temperatures near the intake.   The seabed fluctuations would
have comprised the total effect of the fully mixed down solar
fields and any contribution from a far field plume. These
seabed fluctuations were then matched with comparative intake
temperatures and the difference value provided an estimate of
mid field recirculation.

The 20th July tide temperature changes occurred at the end of a
progressive rise in local sea ambient temperatures during a
calm period that began on the 7th July. From previous
experience at other sites this should have provided ample time
for development of a far field plume. An overall reduction in
sunshine prior to the 20th July effectively reduced the natural
temperature gradient up the estuary to a negligible value and,
it was considered unlikely that either the estuarine gradient
or solar warming of mud flats and sands would have produced any
significant variation in sea bed temperatures during a morning
tide. Under these conditions any variation in sea bed
temperatures due to a far field plume was more likely to be
detected.

Fig.7 details the changes in sea bed temperature recorded at
the thermistor stringer near the intake position and at
another position a further 750m directly offshore on the
morning tide of the 20th July. Temperature changes at both
positions show a distinct rise above local sea ambient on both
ebb and flood movements, peaking about 2 hours either side of
LW slack. Apart from surface waters immediately affected by
primary plume temperatures, it was observed that these tidal
fluctuations in temperature were unchanged from sea bed to
surface.   These temperature fluctuations on the 20th July tide
indicated tidal movement of a fully mixed down large warm water
field across the station frontage and the slightly longer
period between peak temperature and LW slack on the flood tide

suggests that the overall cycle movement has an easterly bias, probably due to residual drift in that direction. Subsequent studies revealed that variations in residual drift could, on occasion, bring the maximum contribution from this far field plume into phase with the mid field plume contribution at LW, but there was no repeat of the distinctive seasonal shift in far field traverse as observed at Sizewell.

Peak far field plume temperatures above deep sea ambient at these two positions were around 1.0°C. Temperatures remained at this level for about an hour and the overall tidal passage of the far field plume across the offshore position lasted about 2 hours. It has been estimated that this warm water field extended about 10km along shore and evidently showed little reduction in field strength up to 1400m offshore.

This was the first occasion that the CEGB had employed a massed array of thermistor stringers and, despite the adverse conditions provided by the large tidal range and high current velocity, these thermistor stringers operated successfully over the summer months and the creditable 80% data return helped to provide a valuable understanding of the response of mid field and far field temperatures to changes in station generation and meteorological conditions. It was possible to show that the two artificial heat fields each provided a contribution of about 30% to the maximum variation of intake temperatures during a tidal excursion.

4.2.4   Mathematical Modelling

Once again the depth integrated numerical modelling technique developed for the CEGB provided a good comparison of temperature range and alongshore spread of this far field plume. The model also predicted that field temperatures would approach an equilibrium value within 7 to 8 days.

An additional mathematical model study[5] commissioned by the CEGB, was able to confirm that solar heating of the extensive shallow water areas to the east of the station could provide considerable diurnal change in the temperature of waters adjacent to the station site.

5.   FUTURE DEVELOPMENTS

Past trends in CEGB design of CW offshore works, illustrated at both Sizewell and Hinkley Point is the positioning of the outfalls inshore while the intakes draw water from the cooler deeper levels further seawards. This arrangement is the most economic, but there is a possibility that the relative positions of the intake and outfall will have to be reversed at some locations, if future environmental stipulations necessitate considerable dilution to further limit the temperature rise of water in the immediate vicinity of the outfall. Mathematical modelling work is proceeding with the intention of providing many comprehensive predictive tools. It is hoped that one model will be capable of simulating the near, mid and far field configuration, including the effects of wind and solar heating.

Future survey work at a new 'greenfield' site may include the long term monitoring of deep sea ambient temperatures in the years preceding station operation. This might involve remote acquisition of temperature data which is telemetered to shore and then transmitted via telephone to a CEGB office. Satellite imagery gathered in thermal infra red wavelength using the NOAA and LANDSAT series of satellites also allows the collection of water surface temperatures synoptically over a large area and CEGB Research Scientists are now collaborating [6] with the National Remote Sensing Centre at Farnborough to exploit this new source of information.

Survey operations are now being complemented by use of a CEGB helicopter providing aerial infra red photography and work is proceeding to develop a vertical 'look down' facility [7].

It is CEGB policy to continue to seek further development of measuring techniques and modelling processes and to support improvement in the design of survey equipment to enhance understanding of the formation and behaviour of warm water fields to maximise the efficiency of planned station generation and ensure minimum ecological disturbance from future warm water discharges.

In carrying out this policy, the CEGB will no doubt continue to receive the present high level of support provided by Consultants, Contractors and Research Organisations who, in their separate ways, have assisted in planning survey operations, have carried out the necessary fieldwork, working on occasion under very difficult operating conditions, and have supplied supportive evidence of survey findings and have considerably advanced modelling techniques within the last decade.

6.   CONCLUSIONS

The CEGB investment in large scale hydrographic surveys and mathematical modelling studies has provided considerable understanding of the highly variable behaviour of natural and artificial warm water fields. It has shown that simple temperature measurements at one or two locations can give a totally misleading indication of the causes of increased water warming.

Survey results from a number of sites have shown how the effects of shoreline features, sea bed topography, varying tidal streams and changing meteorological conditions can bring about various combinations of these separate thermal fields at different times of the tidal cycle.

Experience has shown that each survey should be carried out over a prolonged period to ensure that the final selection of cooling water intake and outfall positions is based upon information that is sufficiently representative of local meteorological and tidal current conditions. This policy has led to a number of findings significant to power station design.

During the recent decade the CEGB have provided considerable assistance in the development of mathematical modelling studies, but, at this stage, the hydrographic survey remains the primary investigative tool in any new site development.

7.   ACKNOWLEDGEMENT

This work was done at the Generation Development and Construction Division, Barnwood and is published by permission of the Central Electricity Generating Board.

References

1. RODGERS I.R. The vertical turbulent transport of heat within a power station cooling water discharge. CEGB Report TPRD/L/AP/107/M83.

2. RILEY J.D., RAMSTER J.W. The pattern of bottom currents along the Coast of East Anglia. Fisheries Laboratory, Lowestoft.

3. MACQUEEN J.F. and PARKER G.C.C. Tidal currents measured near Sizewell Power Station. Nucler Energy. Volume 20. Number 3. June 1981.

4. MILES G.V. A numerical model for background temperature fields. Hydraulic Research Station Report EX806. February 1978.

5. MILES G.V. Inshore solar heat field with and without the Severn Barrage. Hydraulic Research Limited. Report EX1298. March 1985.

6. Private communication from Dr. A.G. Newlands. Central Electricity Research Laboratories.

7. Private communication from Dr. I.R. Funnell, Central Electricity Research Laboratories.

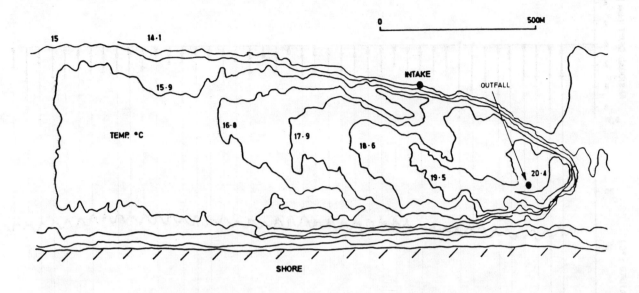

Fig 1. Typical midfield plume surface contours about
30 minutes before low water slack at Sizewell.

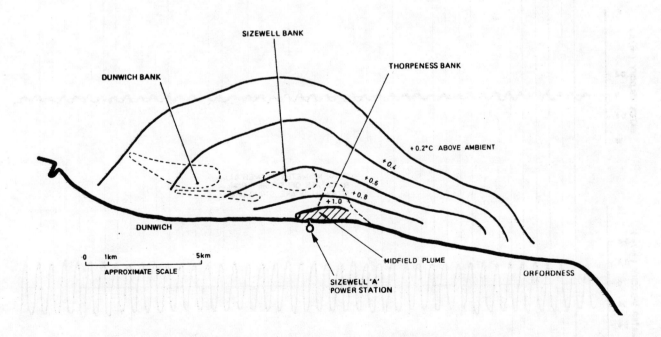

Fig 2. Typical summer far field at end of flood tide.

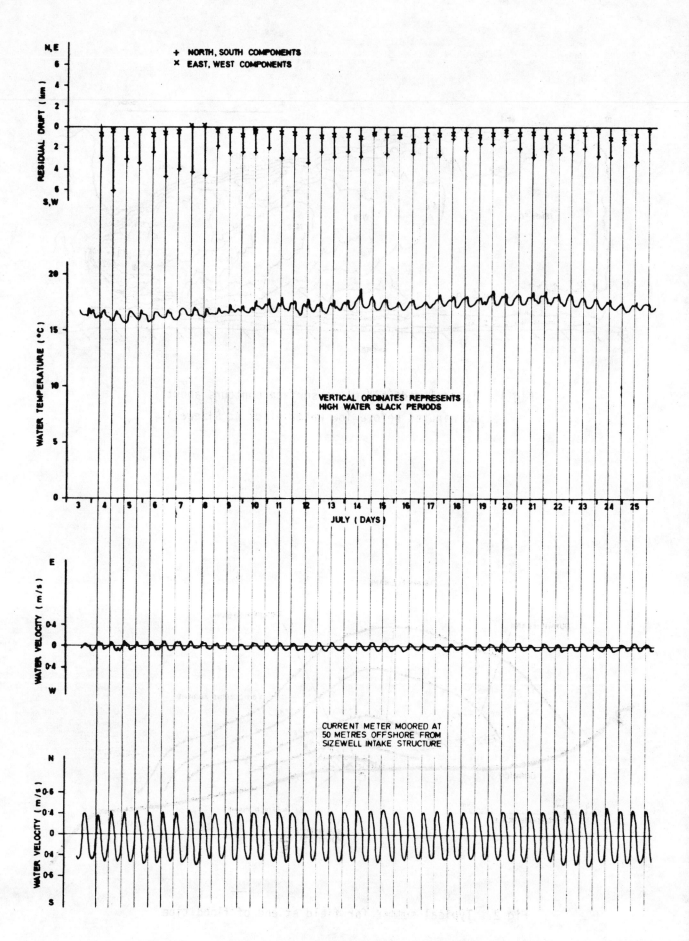

Fig 3. Extract from data presentation of current
meter readings.

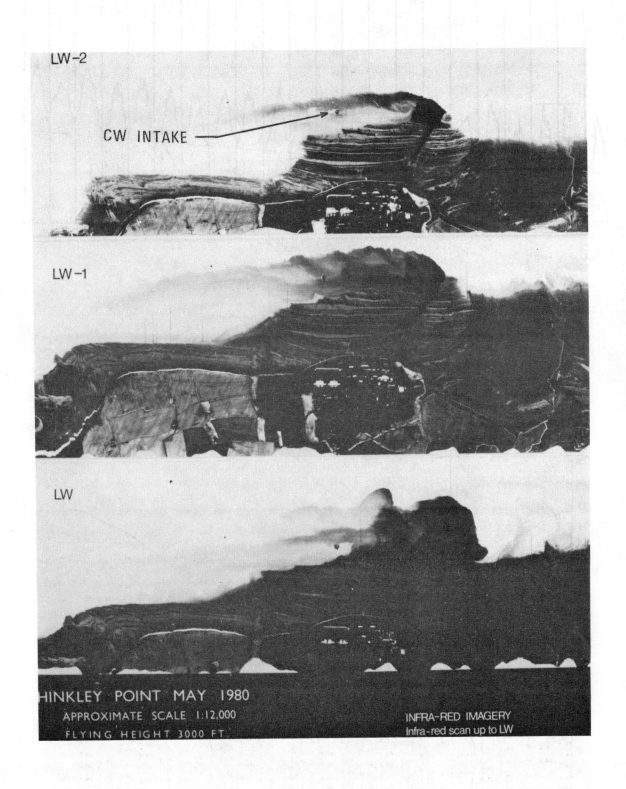

Fig 4. Infra red scan of primary plume before low water
at Hinkley Point.

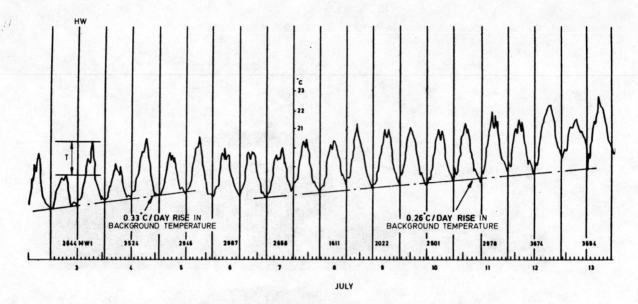

T IS MAXIMUM DIURNAL CHANGE IN LOW WATER TEMPERATURES DUE
TO SOLAR HEATING OF FLATS AND SANDS IN BRIDGWATER BAY.

VERTICAL LINES REPRESENT TIMES OF HIGH WATER

A AND B STATION COMBINED HEAT REJECTION GIVEN IN MW THERMAL.

Fig 5. Graphical plot of condenser intake temperature.

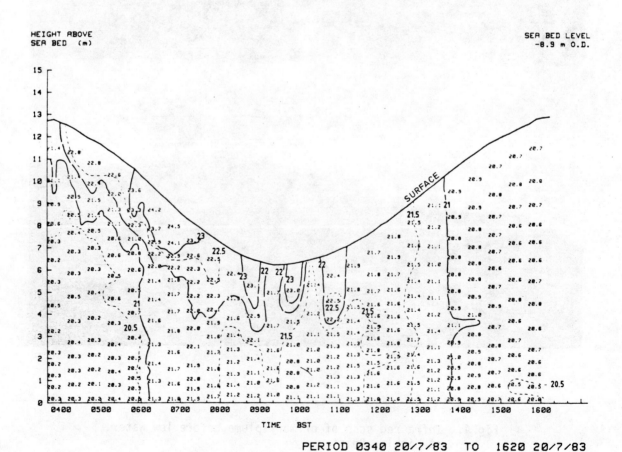

PERIOD 0340 20/7/83   TO   1620 20/7/83

Fig 6.   Graphical temperature/time plot at thermistor position
near present intake at Hinkley Point.

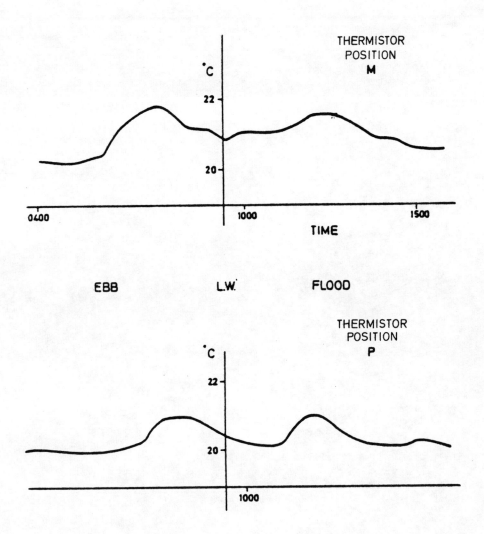

POSITION M IS 50 METRES DIRECTLY OFFSHORE FROM INTAKE
POSITION P IS 800 METRES DIRECTLY OFFSHORE FROM INTAKE
INTAKE IS 570 METRES DIRECTLY OFFSHORE COASTLINE AT HIGH WATER MARK

Fig 7. Far field plume temperature changes at positions directly
offshore intake on 20th July at Hinkley Point.

International Conference on
Measuring Techniques
of Hydraulics Phenomena in offshore, Coastal & Inland Waters
London, England: 9-11 April, 1986

International Conference on
MEASURING TECHNIQUES
OF HYDRAULIC PHENOMENA IN
OFFSHORE, COASTAL & INLAND WATERS
London 1986

HIGH RESOLUTION METHODS FOR INSPECTION OF
UNBURIED AND PARTLY BURIED PIPELINES

AUTHORS:

H. Meister, Head, Survey Division
I. Nording, Electronic Engineer, Survey Division

Danish Hydraulic Institute
Agern Allé 5
DK-2970 Hørsholm
Denmark

Held at Imperial College of Science and Technology, London. Organised and sponsored by BHRA, The
Fluid Engineering Centre. Co-sponsored by the American Society of Civil Engineers and the
International Association for Hydraulic Research.

1. INTRODUCTION

Submarine hydrocarbon transmission pipelines are essential for the oilgas exploitation in the North Sea and in other major offshore regions.

The integrety of these lines is closely matched by means of on-line monitoring systems and regular submarine inspections.

The protection level is influenced by the occurrence of free spanning pipe sections and the degree of pipeline trench backfilling.

Over the last decade much emphasis has therefore been put on developing accurate and economic-in-use sensors for determining these two parameters in particular.

The present paper discusses the principles of two sensors recently developed at DHI for use in submarine pipeline inspection:

- The Triple Head Sonar System (THSS)
- The High Resolution Trench Profiler (HRTP)

Both systems have been used extensively in practice and proceed their value in several inspection projects in the North Sea and in the Mediterranean.

The THSS

The THSS is an ROV mounted free span detection device for use in a longitudinal pipeline survey. The system generates a continuous record showing the pipe seabed longitudinal profile highlighting possible free span sections.

The HRTP

The HRTP is an ROV mounted trench profiling system designed for pipeline cross-profiling yielding accurate information on trench condition, pipe seabed interface and interface between unconsolidated backfill material and the original trench bottom.

Photos 1 to 4

Front of Scorpio with Triple
Head Transducers (1)

Aft of Scorpio with High Resolution
Trench Profiler (3)

Transducer Assembly (2)

Transducer Assembly (4)

2. BASIC ACOUSTIC PRINCIPLES

2.1 General

To enhance the understanding of the techniques behind the THSS and the HRTP, some background information relevant to distance measuring using acoustical principles and the interaction of acoustic signals with the seabed will briefly be presented. The equipment used is sometimes referred to as SONAR (Sound Navigation and Ranging) devices or systems. Typical acoustic equipment currently in use is:

- Echo sounders
- Sub-bottom profilers
- Side- and sector scanning sonars
- Subsea position fixing and location systems

2.2 Sonar Technique

This paper deals with a high accuracy echo sounder and a shallow, high resolution sub-bottom profiler also designated a penetration echo sounder. The sonar technique discussed in the following is specifically related to these instruments. Both systems are based on the same principle of operation which is: measuring the time interval between the transmission of an acoustic pulse from a transducer and its reception at the transducer after reflection at the seabed or from shallow layers in the seabed.

Three sub-systems and their characteristics shall be discussed:

- the transducer
- the signal processing
- the data interpretation

2.3 The Transducer

Commonly applied acoustic sources produce a burst of narrow band acoustic energy which is directed into a limited well-defined angle or beam towards the seabed. The sudden change in acoustic properties at the water/sediment interface and/or the shallow sediment/"hard" bottom interface cause a proportion of the incident energy to be reflected. Usually, but not always, the receiver is the same transducer as the source and in such cases, parameters which are important in determining the received signal level are the normal incidence reflection coefficients and the back scattering characteristics. The acoustic energy returned to the receiver for the seabed thus depends on the acoustic impedance of the sediment, the interface roughness and the geometry of the situation.

The transducer specifications for the THSS and the HRTP differ significantly and will be treated separately in the following.

The transducer layout for the THSS is based on piezoelectric elements (change in the dimension of the crystal or ceramic materials occur as a result of an applied electrical potential or field and produce pressure waves and vice versa). The shape of the beam is a property of the transmission frequency and the geometry of the transducer. The beam width of a transducer indicates the degree of directional concentration of the pressure waves produced and is defined as the angle between the points at which the sound intensity has fallen to half that along the centre axis of the beam. The area of the seabed from which reflections are received is obviously directly proportional to the beam width. As the THSS is designed for high accuracy (1 cm), short range (15 m) and to be used from an ROV, the side transducers (see later) were selected with a beam width of 3 degr. and a resonant frequency of 710 KHz. The centre transducer (see later) was designed as two transducers in one housing with two different beam angles 60 degr. and 20 degr. both with the resonant frequency of 300 KHz. Further details in Chapter 3.

The scope of the development of the HRTP did not allow for any refinement beyond known methods and principles regarding the transducer. The improved resolution over the first 0.5 to 1 m penetration for a trenched pipe partly covered was obtained in the signal treatment and the data presentation.

The transducer layout for the HRTP was based on a magnetostrictive transducer (ferromagnetic material - changes in the dimensions occur as result of magnetization and produce pressure waves and vice versa). The depth resolution of the transducer depends primarily on the frequency, duration of the acoustic pulse and on the beam angle. As a compromise a frequency of 30 KHz and pulse not exceeding 0.15 msec. were chosen. Again an ROV was the sensor carrier and considering roll and pitch the transducer beam angle was chosen to be 7x13 degrees with the narrow beam in the pipe crossing alignment.

The term "depth resolution" has in the present context two interpretations:

(1)   The minimum depth below the seabed at which a soil interface can be detected.

(2)   The minimum vertical discrimination between two interfaces at any depth greater than in (1).

In case (1), the beam angle is critical, but in case (2) it is much less important. This will be apparent from the well known diagram in Fig. 1, which is worth repeating because it is critical to the data interpretation.

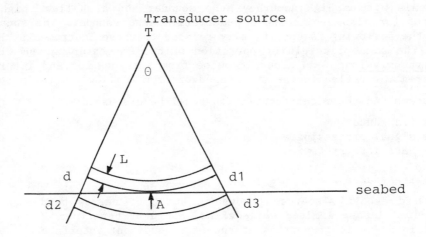

Fig. 1 "Diagram"

A spherical wave of pulse length L radiates from source T within beam angle θ, and its leading edge d-d$_1$ encounters the seabed at A. Seabed reflections are received continuously until the wave reaches position d2-d3 and will "mask" all shallow subbottom returns. Other factors such as seabed roughness and receiver characteristics will have some influence, but the relative effect of beam angle is easily assessed. The theoretical "masking effect" or echo length from the HRTP system used on an ROV will be approximately 0.20 m compared with approximately 1 m from commonly used subbottom profilers. The practical results obtained in nature appear to be somewhat better than the theoretical considerations (Fig. 2 and Fig. 3). The diagrams below show the envelope of a double rectified single seabed return with two distinctive reflectors, as obtained during an actual survey.

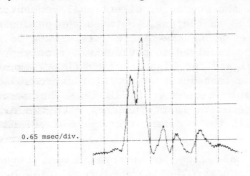

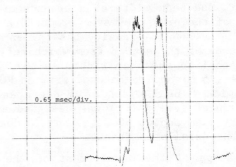

Fig. 2 "Seabed Soft Silt/Mud"   Fig. 3 "Trench Crossing with
                                          Natural Backfilling"

In Fig. 2 the seabed consists of soft silt/mud on clay and in Fig. 3 a trench crossing with natural backfilling covering a 36 inch pipe is presented.

An other important factor concerning beam width and sounding rate should be looked upon which mostly have to be taken into account in the data interpretation, it must be the influence of the differences on the direct reading of an echogram in a longitudinal direction of the sounding and the true object size in nature. In general written as:

$$l = Vt(N-1)-Rs\theta$$

where

              l = length/width of object
              V = velocity of transducer over the bottom
              t = sounding interval
              N = number of soundings
              Rs = depth according to the recorder
              θ = transducer beam angle in radians

It has to be remembered that this expression only is valid if there is at least 100% bottom coverage.

2.4   Signal Treatment

Both systems have a common philosophy – the signal treatment has to be optimized and carried out to as large an extent as possible by microprocessor or micro computer techniques. Stored data for post-processing has to be raw data.

The THSS has a general, but highly sensitive receiver with a time variable gain (TVG) amplification, variable gain slope, step selected transmitting pulse width, variable receiver gate time, variable receiver enable time delay and a step selected "filter" for acceptable differences between two readings, all controlled by the sub unit microprocessor. As only the first incoming echo from each sounding is of interest, the microprocessor measures and logs the acoustic pulse travelling time, compares 3 incoming echoes with previous echoes from the same transducer and the first (timewise) incoming echo having a travelling time parallel in the previous sounding sequence is accepted and transferred to the surface. For further details on the special signal treatment and data processing in the THSS, see Chapter 3.

The HRTP's acoustic signal treatment is carried out by a modified standard sediment sonar receiver combined with a microprocessor. The sonar receiver is provided with TVG. Slope and sensitivity adjustments and the sounding rate are upgraded to 20 per second, with a digitizer of 8 bit resolution and a sampling rate of 12.75 micro seconds. As all the above options are standard options for sediment receivers except the specifications and the introduction of the microprocessor, the special signal or data processing is done in the surface desk computer and will be discussed in further details in Chapter 4.

2.5   Data Interpretation

Data interpretation for the THSS is a straightforward processing technique where the operator can specify the requirements for the actual survey and get the presentation on tailormade drawings. Software treatment of data and examples of presentation are shown in Chapter 3.

As regards the HRTP data interpretation we will revert to the equation $l = (Vt(N-1)-Rs\theta$. If we use this equation on the recorded example of raw data shown in Fig. 4, we will find that the recorded pipe diameter is not 1.55 m as measured on the record, but only 1.15 m which corresponds to the actual pipe diameter incl. concrete within 10 cm.

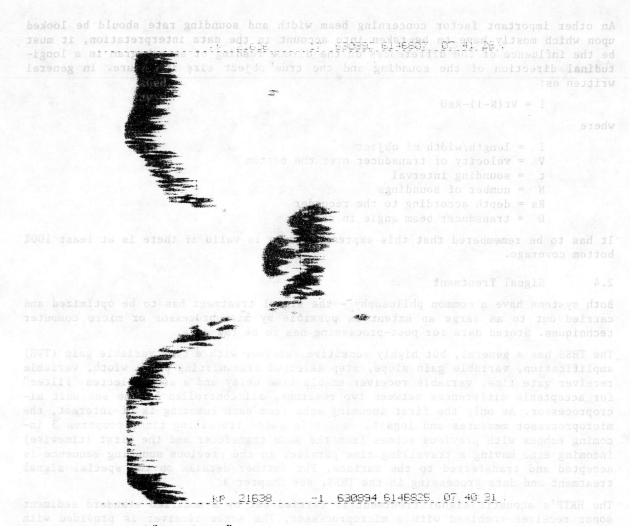

Fig. 4 (reduced size) "Raw Data"
V = 0.217 m/sec.   t = 1/20   N = 135
Rs = 2.5 m   θ = 7 degr.

The reason for the 10 cm inaccuracy can be found by looking at the transducer angle
and its side loops. Because the ROV was flying too close to the seabed inducing
transducer side-loop effects the total beam width was 8.5 degr. and not the 7 degr.
as the main beam angle is. The same effect must be taken into account when we look at
the trench width and the trench slopes. These corrections could easily be done in the
computer post processing.

At this state of the HRTP system there is not yet developed a final computer post
processing system, but a preliminary computer interpretation program has been used
successfully during the survey as an on-site check. This computer interpretation program will be covered in Chapter 4.

3. TRIPLE HEAD SONAR SYSTEM

3.1 General

The DHI Triple Head Sonar System measures the distance from a remotely operated ve-
hicle (R.O.V.) to the top of an unburied pipeline as well as the distance from the
R.O.V. to the surrounding seabed.

The THSS works in conjunction with the on-line navigation computer on the survey vessel and transfers data to the surface via the R.O.V. umbilical, from the light weight
microprocessor equipped subsea unit mounted on the R.O.V.
The surface computer (HP 9836 or equivalent) sends a telegram down the R.O.V. umbilical to a subsea microprocessor in order to control the distance samplings, filter
setting and other housekeeping information. This telegram also commands the micropro-

cessor to start interrogating the three high frequency echo sounders.

To avoid interference from one transducer to the other, the echo sounders are inter-rogated in sequence: Center, port, center, starboard.

The received echoes are then processed in a controlled filter and transmitted up the R.O.V. umbilical as a digital signal to the on-line computer. Furthermore the subsea microprocessor outputs a signal via a digital to analog (D/A) converter to the R.O.V. auto-altitude control.

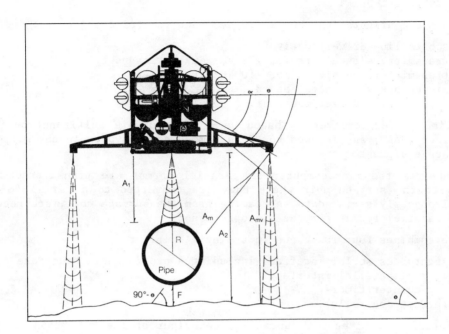

Fig. 5 R.O.V. Cross Section.

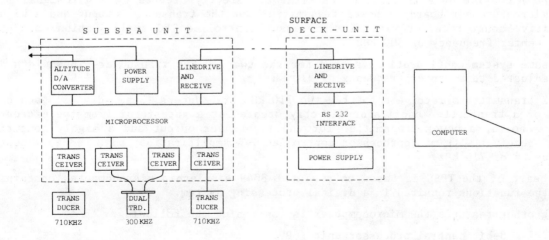

Fig. 6 Block Diagram for System Set-Up.

In the on-line computer the data is further processed using: Data history, compensa-tion signal from the R.O.V.'s roll and pitch sensor, sound velocity corrected distance measurement and the dual signals from the center dual beamwidth transducer.

Guidance information to the R.O.V. operator may be given by a graphic video picture generated on a remote colour video monitor indicating the R.O.V. location relative to the pipe and the seabed. All raw THSS data from the on-line computer will be logged together with all navigation data on the on-line data tape recorder for later processing.

The onboard off-line processing system has the capability to define the exact location of the pipe relative to the R.O.V. and to display any vertical and horizontal information of free spans of the exposed pipeline.

3.2   Hardware Set-Up

The THSS hardware configuration can be divided into the following units:

- 1 subsea line driver/receiver
- 2 transmitters/receivers for 300 kHz dual transducer
- 2 transmitters/receivers for 710 kHz transducers
- 1 microprocessor control board
- 1 surface line driver/receiver.

The subsea line driver/receiver is based on a programmable multifunction SAB 8256A serial interface (UART) and is used for data communications between the microcomputer and the surface equipment.

Data received from the microcomputer in parallel format are transmitted as serial data streams to the surface unit where they are converted to a serial RS 232 data format. Simultaneously serial data, streaming from the surface unit, are received and assembled to parallel format for the microprocessor.

The interface combines four functions in a single chip:

- Asynchronous serial interface including a programmable baud rate generator.
- Universal parallel interface.
- Event counter/timers.
- Interrupt controller.

Like other interface devices the functional behaviour of the SAB 8256A is programmable by the system software. Thus it exhibits the high degree of flexibility required by the THSS.

The half duplex asynchronous serial signal is transmitted in a current loop configuration with a baud rate of 4800.

The 2x300 KHz transmitters/receivers for the dual 300 kHz transducer (beamwidth 20° and 60°) consist of a transmitter oscillator, directly steered by a gatesignal from the microprocessor board, a power transmitter for the transducer output and a highly sensitive sonar receiver with microprocessor controlled TVG amplification, adjusted to a center frequency of 300 kHz.

The same system configuration is used for the two 710 KHz transducers (beamwidth 3°) but adjusted to a center frequency of 710 KHz.

The 2 transmitters/receivers for the two 710 kHz transducers (beamwidth 3°) each consist of a transmitter oscillator directly steered by a gatesignal from the microprocessor board, a power transmitter for the transducer output and a highly sensitive sonar receiver with microprocessor controlled TVG amplification, adjusted to a center frequency of 710 kHz.

The heart of the THSS is based on an Intel 8049 singlechip microcomputer containing all the functions required in a digital processing system.

Among other features the microcomputer is containing the following:

- 8-bit central processor unit (CPU)
- 2 K x 8-bit ROM programme memory
- 128 x 8-bit RAM data memory
- 27 input/output (I/O) ports
- 8-bit timer counter function

- oscillator and clock circuit
- reset circuit
- interrupt structures

With an internal cycle time of 1.36 micro second and a repertoire of over 90 instructions each consisting of either one or two cycles and a single 5V supply requirement, the Intel 8049 is one of the most advanced microprocessors presently available.

To be able to transmit and receive the digital signals through an umbilical or similar cable the communication is made as a 25 milli amp. current loop serial data transmission. This communication is driven from the surface line driver and is galvanically separated by means of an opto coupler unit in the subsea line driver/receiver.

The surface unit also contains a current loop transmitter and receiver converting the signals from the current loop to RS 232 level for communication with the navigation computer.

3.3 On-Line Data Processing

After power-up the THSS will start in its command mode and be ready to receive the system's parameters generated from the software in the on-line computer. When all system data have been received and controlled, the echo sounders are sequentially interrogated for distance measurements in following steps:

- Step 1, Center distance 20° beamwidth
- Step 2, Port   "   3° "
- Step 3, Center   "   60° "
- Step 4, Stbd   "   3° "
- Step 5, Processing and transmission of data to survey vessel as well as outputting altitude information to R.O.V. pilot.
- Step 1, Center distance 20° beamwidth.

An echo cycle will typically be as follows:

The microprocessor operates a 4.88 micro seconds counter and delivers a gatepulse to the transmitter oscillator, after which the microprocessor logs the travelling time for the first 3 incoming echoes. These are now compared with the previous echoes from the same transducer. The first (timewise) incoming echo having a parallel in the previous sounding sequence is accepted and transferred to the line drive circuit for transmission via R.O.V. umbilical to surface line drive receiver and further to the on-line navigation computer.
On-line processing by the navigation computer is divided into 3 main software routines. First the data is received and checked for errors such as missing data or missing contact with the pipe.

Secondly the data is corrected for roll and pitch variations sensed by the R.O.V. roll and pitch gauges. Conversion of echo travelling time, created in the subsea microprocessor as 4.88 micro sec. intervals and the correction for sound velocity will also be done in this routine.

Third the data is filtered through a prediction filter taking the previous data history into account and assuming that the pipe within one or two joints is a straight line. This routine also contains a certain procedure for comparing the measured distances from the center transducer's narrow and wide beams to improve the determination of the R.O.V.'s sidewards position above the top of the pipe. The resulting on-line data is used to create a colour video picture for guiding assistance for the R.O.V. operator. To flag the probability of a free span on the pipeline, the side transducer's data and the above mentioned filtered data are used in a non-storing software routine.

All raw data are logged before any filtering on the on-line data tape for later off-line processing.

3.4 Off-Line Data Processing

In the off-line processing system the data from the center transducer is treated by special software routines locating the points on which the R.O.V. has been exactly above the top of the pipe. These points are then used together with the positioning data and the knowledge of pipe parameters such as maximum curvature, etc. to define the exact location of the pipe. These data are then further processed together with the data from the two side transducers and the R.O.V. roll and pitch correction data, either for plotting and/or hardcopy print-out. Any obtained information on free span of the pipeline will after operator selected horizontal and vertical free span criterion be transferred to the final QA drawing which is plotted on a high speed multi-colour plotter. A library with run-line histogram plot showing percentages of free span lengths is also a part of this final routine.

3.5   System Specification

Range .... 15 metres.

Resolution .... 1 cm.

Accuracy .... Standard deviation of measured distances equal to 1 cm or 0.5% of measured distance, whichever is greater.

Repetition rate .... Subsea unit 5/sec.
Surface computer 1/sec.

Transducers .... Sides 710 kHz, 3 degr.
Centre 300 kHz, 20 and 60 degr.

Transmit power .... 10 to 20 W.

Receiver sensitivity ..... 710 kHz input 10 micro V,
300 kHz input 20 micro V.

Digital output: Time counts.
Subsea unit .... Passive 25 milli amp. current loop.
Deck unit .... Active 25 milli amp. current loop. Including current loop to RS 232 converter.

Output D/A converter ..... Analog 0 to +5 volt, 8 bit resolution.

Power .... 115 volt ac, 50 or 60 cycles. 6 VA.

Max. pressure .... 50 bar (= 725 psi) = 1626 ft. = 495 m.

System weight in water ... 4.0 kg.

3.6 Example of Data Presentation

See Figs. 7 and 8 on the following pages.

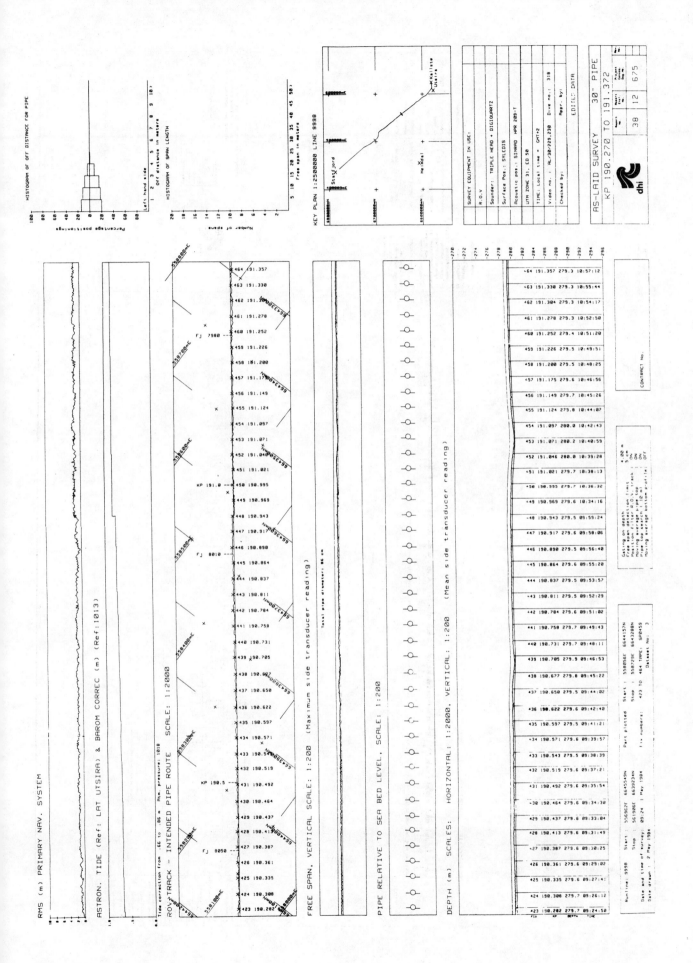

Fig. 7

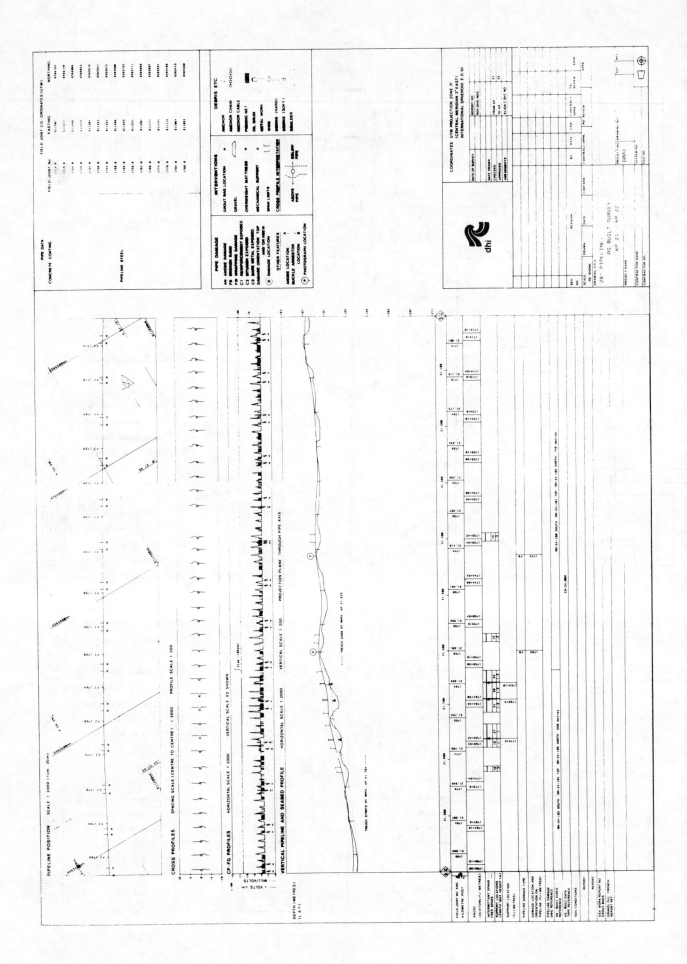

Fig. 8

4.   HIGH RESOLUTION TRENCH PROFILER SYSTEM

4.1   General

The sonar system is designed for ROV cross profiling of a pipeline trench in order to obtain a sonar record showing the open trench including approx. 5 m of natural seabed on each side. If the pipe is covered or partly covered by natural backfilling it is the aim to detect the original trench bottom and the surface of the backfill material if the nature of the natural backfill is loose or very soft sediment.

4.2   Hardware Set-Up

The system set-up is a 30 KHz transducer with a beam width of 7 x 13 degr. mounted on the ROV frame with the transducer cable attached up the ROV umbilical to the survey vessel. A modified ELAC sediment echo sounder combined with a micro processor is used as a transmitter/receiver module with data display on a video monitor for on-line monitoring.

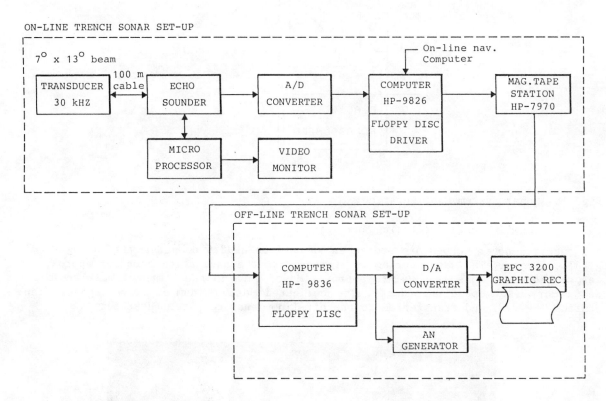

Fig. 9 On/off-Line Hardware Set-Up.

The sonar processor consists of the following:

- Microprocessor controlling echo-key pulse and display of the first echo.
- Opto-couplers from micro to echo-sounder and HP-GPIO.
- Amplifier and active double-rectifier for echo signals.
- 8 bit resolution, analog to digital converter for interfacing of echoes to HP-computers.

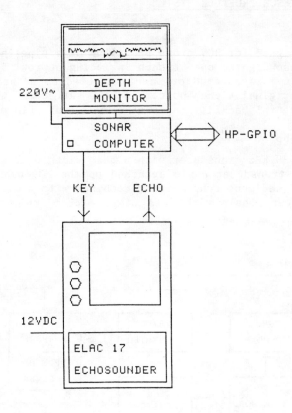

Fig. 10 Sonar System Configuration.

4.3   On-Line Data Monitoring and Storing

All received sonar echoes are stored in an on-line HP9826 computer via analog to digital converter. The data storing is started from the navigation computer approx. 5 m
before crossing. This is equal to approx. 1MB of data in the computer with an extended memory (ROV speed dependent). The collected data is dumped on a magnetic tape
station (HP7970) and transferred to the off-line sonar computer (HP9836).

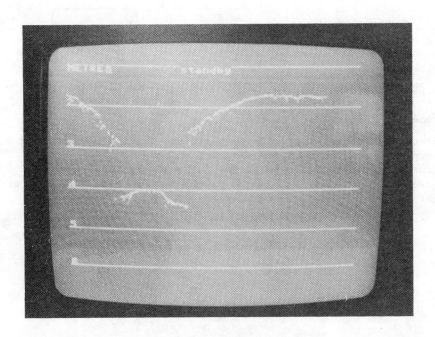

Fig. 11 Video Monitor Example.

4.4   Off-Line Data Processing

The off-line sonar computer dumps the sonar records on both channels on a graphic EPC3200 recorder. Channel one displays raw echo data and channel two will include preliminary computer interpretations of the obtained echo signals. The graphically recorded trench crossing is annotated with UTM coordinates at the start and the end of the crossing.

The UTM coordinate corresponds with the ROV position when starting the crossing and is stored and recorded together with the raw sonar signals.

The preliminary computer interpretation and data presentation of the obtained echo signals is made in the off-line HP9836 computer by a software program written in Pascal.

The software interpretation of reflected sonar data from the trench crossing sonar system is in principle made in the following way:

- Each sounding consists of 512 points. The first 175 are omitted from further calculations.

- Each sounding is filtered using a moving average to obtain smoothing of data.

- Each sounding is searched to find significant points i.e. maximums of the curve.

- From these points single points without realistic relation to the surrounding points are taken out (noise suppression). This leaves one or more curve tops as significant echoes from one shot.

- By plotting these tops on the EPC-recorder, one shot after another, a realistic picture of the reflection from the horizontal layers is obtained. Or to be more specific sonar reflection obtained by differences in bottom layer density.

- The vertical position of a specific data top within the total number of possible points (176-512) in one shot is the relative depth of the reflection.

For each scan plotted on the EPC-recorder it is possible to expand data in horizontal as well as vertical direction.

4.5   System Specification

ECHO-SOUNDER:

Model	:	Elac 17
		Sediment transceiver LVG 12 R 01
Transmit Power	:	Approx. 400 Watts
Output Pulse	:	0.15 msec.
Transducer	:	LSE 53, 30 kHz.
Beamwidth	:	7*13 Degrees at -6dB.
Transducer Cable	:	150 Metres
Modifications	:	Paper feed on/off
		External keying. (opto)
		External fixmark/annot. (opto)
		5 V positive keypulse output
		(opto)
Power. Supply	:	12 V dc.

SONAR CONTROLLER/DISPLAY:

Micro-processor	:	CBM 6510, 64 kB ram.
Program Language	:	COMAL 80, Danish version.
Program Loading	:	Autostart, prommed storage.
Sounding Rate	:	20 per seconds.
Range Vertical	:	5 Metres at soundvel. 1485 M/sec.
Resolution	:	2.5 cm.
Offset	:	1 to 13 Metres.

Range Horizontal	:	1600 soundings.
Resolution	:	5 soundings.
Screen Update	:	Each 1600 soundings or by external pulse.
Echo Input	:	+/- 5 V analog.
Keypulse Output	:	+5 V, 1.8 to 2.0 m/s.
Digitizer	:	8 bit resolution. 0 V => 000, +/- 5 V => 255. 12.75 micro-sec. sample rate.
Interface	:	HP GPIO compatible.
Power Supply	:	220 V ac, 50 Hz.
Options	:	External fixmark with 16 character annotation. Auto offset shift with scale annotation.

4.6   Example of Data Presentation

In the below examples, the on-line computer interpretation program has been used to present preliminary interpretated records in different program modes (see Figs. 12, 13, 14 and 15 on the following pages).

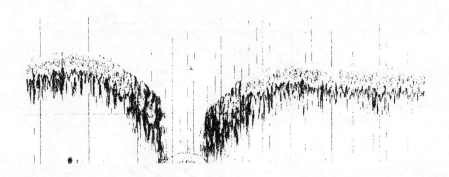

Fig. 12. "Data Presentation Example"

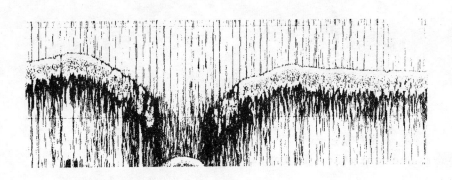

Fig. 13   "Data Presentation Example"

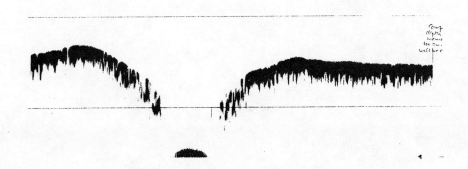

Fig. 14   "Data Presentation Example"

The examples on Figs. 12, 13, and 14 are expanded vertically with a factor 1.5.

Fig. 15. "Data Presentation Example"

In the example on Fig. 15 a vertical expansion factor of 3 is used.

5.   ACKNOWLEDGEMENT

The authors are very grateful to our colleague Mr. P.I. Hinstrup, Head, Offshore Department for his assistance and helpful criticism during the preparation of this paper.

Measuring Techniques

of Hydraulics Phenomena in offshore, Coastal & Inland Waters

London, England: 9-11 April, 1986

DEVELOPMENTS IN HYDROGRAPHIC DATA COLLECTION FROM COASTAL

POWER STATION SITES

A.G. Newlands

Central Electricity Research Laboratories of the CEGB,

Leatherhead, Surrey, KT22 7SE

SUMMARY

Part of the cooling water studies programme at the Central Electricity
Research Laboratories has been to monitor water temperatures and tidal currents in the
vicinity of coastal power stations. This data is useful in understanding the physical
processes leading to mixing of the heated water discharged from the station outfall,
and in proving mathematical models of the subsequent heat dispersion, both needed for
site investigations and station design purposes. Also of interest are naturally
occurring temperature variations, i.e. heat fields caused by solar heating and
sandbanks. Any artificial temperature rise will be superimposed on these naturally
occurring variations, and may affect both power station efficiency and local ecology.

This paper describes developments in data collection since a small
hydrographic survey off the East Anglian coast in 1982. Recording current meters and
thermistor strings were used, but data was lost due to damage to the instruments. In
1983 a new design of commercial current meter was evaluated at the Dungeness Power
Station site. This instrument incorporated several novel features for protecting the
data, including a UHF telemetry link to shore. The results of this trial are
described, together with some local hydrographic features of Dungeness.

With regard to the importance of data security, a system was then developed
at CERL for monitoring remotely collected data directly from base. Software was
written to enable a data logger to be controlled by a minicomputer. The data logger
could be situated at a remote site, but accessed from CERL via a modem and ordinary
telephone lines. The system incorporated monitor and playback facilities, so that
incoming data could be inspected and recorded at CERL. Visits to the site were thus
necessary only in the event of equipment failure, or other malfunction. In this case
the system was set up to monitor the cooling water intake temperature at Hinkley Point
Power Station on the Severn estuary, and results of this trial are described. A power
spectral analysis of a typical intake temperature record is discussed.

Taken together, the two systems described could be combined to provide a
means of ensuring the greatest possible security of data being collected, either from
shore or sea based instruments. This data may be monitored from a laboratory remote
from the site using commercially available equipment and ordinary BT telephone lines.

Held at Imperial College of Science and Technology, London. Organised and sponsored by BHRA, The
Fluid Engineering Centre. Co-sponsored by the American Society of Civil Engineers and the
International Association for Hydraulic Research.
©BHRA, The Fluid Engineering Centre, Cranfield, Bedford MK43 0AJ, England 1986.

1. INTRODUCTION

A problem of concern to the CEGB during the planning and design of direct cooled coastal power stations is the magnitude and duration of any change in the cooling water intake temperature. Typically, approximately 2600 MW of waste heat will be rejected from a 1300 MW(e) power station, so that a cooling water flow rate of 60 m^3s^{-1} would be required for a design temperature rise of 10°C across the condensers.

The temperature at the power station intake will be the result of contributions from the following:

(i) The naturally occurring background temperature, i.e. the temperature in the sea which would exist in the absence of the power station. This temperature is determined by the solar and atmospheric heat fluxes.

(ii) Direct recirculation, due to the impingement of the warm, buoyant discharge on to the intake structure.

(iii) The far field effect of the station, i.e. the overall background warming produced as the buoyant plume becomes vertically mixed over several tidal cycles.

Information about the movement and dispersal of these heat fields has been obtained by large hydrographic surveys done at various power station sites throughout the UK, e.g. Sizewell (Parker[1], Macqueen and Parker[2]), Morecambe Bay (CEGB[3]), Dungeness (CEGB[4]), and Hinkley Point (CEGB[5]). A problem common to all these surveys is that conventional moored equipment such as a current meter or thermistor string, is susceptible to loss or damage due to harsh weather conditions, collision with fishing or other vessels, and of course, theft. Furthermore, with data recorded internally on tape, the loss of an instrument will result in the additional loss of potentially valuable data. In 1982 a small hydrographic survey involving three recording current meters and two thermistor strings was designed to study naturally occurring temperature variations off the East Anglian coast (Newlands[6]). Although most instruments were eventually recovered, all moored equipment was found to have disappeared on a routine servicing visit one month after deployment.

Desirable features for future surveys were identified from these experiences. Firstly, telemetry of data from offshore equipment to an onshore receiving station would greatly increase the security of the data. Secondly, the ability to monitor this data from one's laboratory or company headquarters would enable immediate action to be taken in the event of equipment loss or failure. Major cost savings could be achieved, since manpower need be deployed on site only in the event of equipment malfunction.

The first part of this paper is concerned with developments which are taking place in current meter design. Following discussions between CERL and a representative from Valeport Developments (Dartmouth) in 1982, several modifications were made to a prototype current meter, the Series 1000, which was subsequently brought out in Summer 1983. Section 2 describes a trial of this instrument at Dungeness Power Station, and Section 3 gives some results regarding tidal currents and temperatures in the area. Although the system included a UHF link to a shore station, equipment for transmitting the data to CERL was not available during the trial. The second part of the paper describes subsequent developments at CERL on the collection of data from remote sites. Section 4 describes a system whereby data received at such a site may be monitored or recorded at CERL. Normal telephone lines are used for the transmission of data. The system was set up to monitor the condenser intake temperatures at Hinkley Point Power Station on the Severn estuary, and results of this trial are described. Section 5 discusses the physical processes leading to variation in the cooling water intake temperature with reference to a power spectrum plot. Overall conclusions are given in Section 6.

2. DUNGENESS TRIAL – DESCRIPTION OF THE EQUIPMENT

The vicinity of Dungeness was considered to be a good location for a trial of any moored equipment, since marine currents are high and the wave climate is rough. The Valeport Series 1000 current meter consisted of a moored subsea unit (the fish), a large toroidal surface buoy 1.7 m in diameter, and a shore based control unit. A choice of sensors was available, and for the purposes of the trial it was decided to measure current speed, current direction, water temperature, water conductivity, and the height of the water column above the meter. All data management was performed by a microprocessor module inside the fish. This allowed logging intervals to be set up, and also converted raw sensor data to physical units.

The fish was moored between the sea bed and the large surface buoy, which was designed to prevent accidental removal of the equipment. The buoy supported a small wind generator to ensure that all power supply batteries were kept charged, and a UHF transmitter/receiver to provide a telemetry link between the microprocessor and the shore based control unit. Data from the sensors was transmitted over this link, and read on line by the shore based unit. Although there was a facility for sending command signals from the shore station to the microprocessor, it was found that in this mode certain control codes interfered with the operation of the Columbia tape reader which recorded the data. Thus the microprocessor was programmed before deployment of the equipment. To ensure additional security of the data should the telemetry link fail, logged data was stored in a solid state memory module before being transmitted ashore.

The shore based station was installed in a building overlooking the sea belonging to HM Coastguards, and to which access was obtained. The equipment layout is shown in Fig. 1, (Rodgers[7]) and consisted of a UHF receiving unit, the current meter surface control unit, a Columbia tape reader, and a Hewlett Packard HP85 desk top computer. Data was output from the control unit via an RS232 interface to the Columbia reader, where it was recorded on to an ECMA 46 cartridge. The HP85 computer was used as a display and plotting device only. Each set of readings received at 10 minute intervals was displayed in full on the screen, and the current speed, water temperature, and battery voltage were plotted on the thermal printer. The output was designed so that any faulty data or instrument malfunction could be recognised immediately. This feature proved its worth during the trial when the current speed impellor became entangled. The zero current speed reading was noticed by the coastguard on duty who phoned the information to CERL within two hours of the event.

Fig. 2 shows a map of the area around the current meter, which was positioned in water of depth 8 m (c.d.) approximately 400 m offshore, and 70 m north west of the Dungeness Power Station intake. The instrument produced good data from 20 October–1 November 1983, when the current speed sensor became entangled. The other sensors continued to function until 15 November, when failure of two diodes in the wind generator resulted in a power failure, and termination of the trial. Again, this malfunction was immediately evident at the shore-based station.

Summarising the advantages of this system over conventional recording current meters are as follows:

(i) Malfunction, or other damage is speedily detected.

(ii) Data from the sensors is telemetered to a secure location on shore for storage, and also for display and monitoring.

(iii) Additional security of data is provided by recording data on to a solid state memory in the subsea unit.

(iv) All data management, including the setting up of logging intervals and the conversion of new sensor data to physical units, is handled by a microprocessor inside the subsea unit.

(v) The surface buoy is designed to prevent accidental removal of the moored system.

(vi) A wind generator on the buoy maintains the battery charge.

(vii) The RS 232 output on the surface control unit allows interfacing to a standard modem, and hence transmission of data over normal British Telecom lines (not used in our trial).

3. CURRENTS AND TEMPERATURES IN THE DUNGENESS AREA

Results from the Dungeness trial are illustrated by considering tidal currents and temperatures obtained from the current meter.

(i) Currents

Figs. 3, 4 and 5 show the current direction, current magnitude, and current east component for the spring tidal period during the trial (maximum tidal amplitude occured on 23 October 1983). The positions of high and low water are also shown. It is seen that the tidal height is asymmetrical, with a longer duration ($7^1/_2$ hours) of the ebb tide, than the flood tide (5 hours). Furthermore, high water is seen to occur at approximately mid east flowing tide, and low water at mid west flowing tide. These results are found throughout the whole tidal record, and agree with generally accepted tidal behaviour in the Dungeness area. By comparison, at many UK coastal sites of interest to the CEGB, the time of high water either coincides with the time of slack water, or leads it by intervals of up to 1 hour. The unusual behaviour of the tidal currents at Dungeness is believed due to the influence of the North Sea on tidal waves in the English Channel.

(ii) Temperatures

Fig. 6 shows temperature records taken over the spring tidal period. The considerable fluctuation is believed due to noise in either the sensor or associated electronics. The CEGB survey (CEGB[4]) would suggest that at the distance of the current meter from the Dungeness outfall, the excess temperature of the primary plume is 1-2°C. If the fluctuation on the record is due to eddies detaching from the plume, a temperature variation greater than the observed value of 0.5°C would be expected. Furthermore, the time scale of the fluctuations is approximately $^1/_2$ hour, so for a mean tidal velocity of 0.5 ms^{-1}, a typical length scale for the fluctuation is 900 m, greater than the extent of the primary plume.

The effects of plume behaviour on the temperature record are not immediately obvious. The peak at 14.00 h on 23 October occurs on an east flowing tide, and is almost certainly due to the plume. The peak at 21-30 h on the same day occurs at the changeover between west and east flowing tides, and may be due to ponding, i.e. the large increase in area of the primary plume which occurs at slack water.

Further inspection of the temperature records shows evidence for some tidal variation, i.e. temperatures appear to rise at the start of a west flowing tide, and reach their peak at slack water as the tide changes from west flowing to east flowing. This would be consistent with a heat field lying to the east of the instrument, e.g. within the shallow regions of Roar Bank. Confirmation of this was obtained from the cross correlation of temperature $T(t_i)$ with the current east component $u(t_i + \tau)$, where t_i is the time of measurement, and τ the delay time. Fig. 7 shows the evidence for a weak tidal variation in the temperature record. The first maximum occurs for a delay time of 3 hours ($^1/_4$ tidal period), i.e. the temperature maxima will occur on average $^1/_4$ tidal period before corresponding maxima in the current east vector. This is the time of slack water at which tidal currents are changing from west to east.

4. DATA COLLECTION AT REMOTE SITES

The previous sections have described advances taking place in marine current monitoring, and the trial of a new instrument at Dungeness. In general, however, a full hydrographic survey at a given site may involve at least 10 instruments, each producing around 80 bytes of data at 10 minute intervals. Apart from possible equipment failures, regular visits to the site are required for routine monitoring, and changing data tapes. Considerable saving in manpower and time would result if survey data could be monitored from a central laboratory or company headquarters. Furthermore, any equipment malfunction could be spotted immediately, and appropriate remedial action taken to minimise the loss of data. This section describes the trial of a data logger which was controlled remotely from CERL. The equipment was set up to monitor the cooling water intake temperature at Hinkley Point Power Station on the Severn estuary, and data recorded locally could be monitored from CERL. Further details are given below.

(i) Description of the Equipment

The data logger in question was a Microdata Prolog 1680L. Software was written at CERL to allow this device to be controlled by an HP85 desk top computer (Morgan and Church[8]). Details of the logging system are shown in Fig. 8. Command signals generated by the HP85 at CERL were transmitted to Hinkley Point via a modem and ordinary telephone lines. Data was returned to CERL via the RS232 interface on the logger and the modem at Hinkley Point. The following logger operations could be selected:

(a) Control Mode. The logger must be in this mode in order to respond to signals from the HP85 through the RS232 interface.

(b) Status Mode. This mode will cause the status of the logger to be displayed by the HP85. For example, if control mode is set, the HP85 will display

> Front Panel Enable
> Power OK Switch On
> Control Mode Set

(c) Real time clock. In this mode the logger clock may be set up or changed as required.

(d) Time base. This enables logging intervals for individual channels to be set up or changed.

(e) Record mode. In record mode data is recorded locally on to ECMA46 cartridge. This is the normal mode of operation for the logger on site.

(f) Tape mode. Normal tape operations, including rewinding, moving forward or back a block or file, and changing tracks, may be done in this mode.

(g) Replay mode. On rewinding the logger tape, this mode allows previously recorded data to be replayed through the RS232 link, and hence to be received at CERL. This mode is selected when downloading data from the logger tapes for recording at CERL.

(f) External mode. Selection of this mode enables data to be viewed as it is being either recorded or replayed. If the logger is recording, it may be necessary to wait an interval of time equal to the logging interval in order to view the data.

(g) Monitor mode. This facility enables the instantaneous reading on any given channel to be output by the logger, and hence transmitted to CERL for inspection. By contrast to external mode, the result is obtained immediately in real time.

(ii) Details of the Trial at Hinkley Point

The system was arranged to log the cooling water intake temperatures at set number 3, Hinkley Point Power Station. The telephone, modem, and data logger were installed in an air conditioned hut in the condenser hall basement, and the water inlet temperature was obtained from a platinum thermometer positioned in a pocket in the inlet pipe. Environmental conditions in the hut were monitored by logging the air temperature, and the logger performance was continuously checked by monitoring the

value of a standard (fixed) resistance. The system worked successfully from
25 May 1984 to 11 April 1985, when the last successful transfer of data took place.
Failure after that date occurred due to the tape jamming in the cartridges.

The Microdata 1680L logger possesses 150 available channels, and since two
channels are required for day number and time, up to 148 separate instruments may be
monitored sumultaneously. In practice, however, handling such large amounts of data
may cause problems. For example, approximately 80 bytes of data were recorded every
10 minutes during the Hinkley trial, and for a data transfer rate of 300 baud, the
time taken to transfer this to CERL is around 40 minutes per week. Although higher
baud rates can be used (600, 1200), it was found that buffer overflows tended to
occur. One disadvantage of the system is that the replay/record modes cannot be
engaged simultaneously, so that data logged over the period of transfer will be lost.
This could be a problem if a large number of channels was being monitored. Thus the
system is best suited to monitoring small amounts of data at remote sites.

5.   STUDY OF THE INTAKE TEMPERATURES

Sample intake temperatures at Hinkley Point for 27 July – 9 August 1983 are
shown in Fig. 9. This data was obtained from a previous study at Hinkley Point using
conventional logging equipment, and has been analysed by Proctor[9]. In Fig. 9 the time
of successive high tides are shown as vertical lines. Spring and neap tides occur on
25 July (day 206) and 3 August (day 215) respectively. By contrast to Dungeness, the
temperature record at Hinkley is seen to have a marked tidal behaviour and peaks up to
4°C above the background temperature may occur. There is little phase difference
between tidal heights and currents at Hinkley Point, and so Fig. 9 shows that the
temperature pulse occurs on the ebb (west flowing) tide, and reaches its peak around
low water slack. The outfall at Hinkley Point Power Station is a channel discharge to
the east of the station intake, so a temperature pulse of this magnitude on a west
flowing tide is almost certainly due, at least in part, to the primary plume. Fig. 10
shows the temperature record for the 28 week period from 6 July 1983 – 22 January
1984. A substantial tidal oscillation is seen to occur over the whole period, and
even in December this pulse may exceed 2°C. The isolated peaks in the record occurred
when the set was taken off load, so that only the ambient air temperature was measured
in the inlet pipe.

The envelope of the troughs shown in Fig. 10 will give the local ambient
temperature upon which all tidally varying temperatures will be superimposed. These
include the immediate effects of the power station plume, the station far field, and
possible solar contributions. The seasonal variation in this temperature, from 21°C
to 6°C, is clearly seen in Fig. 10. Further inspection of the temperature records
shows evidence for a fortnightly spring – neap temperature fluctuation.

To investigate the physical processes contributing to variation in the
intake water temperature, a power spectrum for temperature data over the above period
was obtained. Fig. 11 shows the plot of the power spectral density function

$$G(f) = \left| \int_{-\infty}^{\infty} T(t) \, e^{-2\pi i f t} \, dt \right|^2$$

where T(t) is the temperature at time t. The frequency f is measured in units of
(tidal periods)$^{-1}$, and Fig. 11 shows the spectrum in the range 0–2.5 tides^{-1}, or
periods of oscillation from ∞ to 0.4 tide. The (normalised) tidal contribution to the
record is clearly seen. The small blip at twice tidal frequencies may be due to the
effect of ponding of the power station plume at each slack tide. There is seen to be
almost no daily effect on the temperature record (corresponding to a frequency
0.52 tide^{-1}). Several authors (Edinger and Geyer[10]) have discussed the effect of
variations in solar input on bodies of water, and it is seen that daily changes occur
too rapidly to induce any significant response in water several metres deep. However,

a much larger temperature response can occur for lower frequencies, e.g. seasonal changes. Fig. 10 shows that a seasonal variation of 15°C in the background temperature may result due to changes in the surface heat exchange.

The low frequency end of the spectrum is shown in greater detail in Fig. 12 for a frequency range 0-0.25 tide^{-1}. It is likely that the peaks which occur between 0.005 and 0.015 tide^{-1} (104 and 35 days) are due to the seasonal trend in the temperature. Any effects due to the spring – neap cycle would be expected to occur at a frequency of 0.035 tide^{-1}, corresponding to a period of 14.786 days, or 28.571 tides. There is seen to be a small but significant peak at 0.037 tide^{-1}, the displacement being due possibly to the effect of harmonics introduced by the seasonal trend. One possible reason for a spring – neap effect may be the small but regular addition of stored solar heat into the covering water from the intertidal regions of Stert Flats, to the east of Hinkley Point. Since given states of the sun and tide repeat at intervals of the spring – neap cycle, effects due to mud flats are expected to vary with this period. Spring neap variations in other parameters, e.g. residual flow, (Uncles and Jordan[11]) may also contribute to variation in the temperature record, and analysis of this is an ongoing research project.

Finally, the contribution to the power spectrum around 0.06-0.07 tide^{-1} (period 7.4-8.6 days) remain unexplained. They may be caused by cycles in local weather conditions.

6. CONCLUSIONS

This paper has discussed advances in the collection of hydrographic data from remote sites, such as coastal power stations. Loss or damage to conventional recording current meters frequently resulted in the loss of valuable data. The trial of a new design of current meter was described, in which data was telemetered to a secure location on shore. Further aspects of the system included microprocessor control of the logging processes, back up recording facilities, a wind generator, and the facility to display the incoming data. Some results relating to tidal currents and temperatures in the Dungeness area were given.

Considerable savings in time and manpower would result if data collected at remote sites could be monitored and recorded at one's central laboratory or company headquarters. With this aim a system using standard "off the shelf" equipment was developed at CERL. Since suspect readings could be identified immediately from headquarters and the required corrective action taken, loss of data due to equipment failure was kept to a minimum. The system was set up to monitor the cooling water intake temperature at Hinkley Point Power Station, and results of the trial were described. The power spectrum of a typical intake temperature record was obtained, and the seasonal, tidal, and spring – neap contributions to the variation were discussed. The importance of site topography in cooling water studies is seen on comparing temperature records at Dungeness and Hinkley. By contrast to Dungeness, the effect of the primary plume gives rise to a marked tidal variation in the temperature record at Hinkley Point.

The remote data collection system has access to 150 channels, so that a large number of instruments can be controlled, in principle, from one central location. In practice, however, considerations like the time taken to transfer data from the remote site will limit this number. It is planned to extend the monitoring at Hinkley Point to include records of the intake temperature together with comprehensive meteorological data, e.g. solar and net radiation, wind speed and direction, and wet and dry bulb temperatures.

In conclusion, both trials discussed in this paper were designed to maximise the security of data collected in a remote or hostile environment.

7. ACKNOWLEDGEMENT

This work was done at the Central Electricity Research Laboratories, Leatherhead, and is published by permission of the Central Electricity Generating Board.

8. REFERENCES

1. Parker, G.C., "Study of factors affecting sea water temperatures at Sizewell Power Station - an appraisal of the 1975 Sizewell hydrographic survey", CEGB (GDCD) Report, 1977.

2. Macqueen, J.F. and Parker, G.C., "Tidal currents measured near a British coastal power station", Advances in Water Resources, 2, 1979, pp 113-122.

3. CEGB (Generation Design and Construction Division), "Heysham Nuclear Power Station state II hydrographic survey 1978", Commercial report, 1978.

4. CEGB (Generation Design and Construction Division), "Dungeness 'C' Power Station hydrographic survey, 1983", Commercial report, 1984.

5. CEGB (Generation Design and Construction Division), "Hinkley Point 'C' Power Station hydrographic 1983 survey", Commercial report, 1984.

6. Newlands, A.G., "An experiment to detect a temperature pulse off East Anglia", Internal CERL Memorandum, 1984.

7. Rodgers, I.R., Private Communication, 1984.

8. Morgan, C.J. and Church, P.R., "Instructions for remote operation of a Microdata 1680 data logger from HP85 computer", Internal CERL Memorandum, 1985.

9. Proctor, M., "Condenser inlet temperatures at Hinkley Point, April 1983 to January 1984", Internal CERL Memorandum, 1985.

10. Edinger, J.E. and Geyer, J.C., "Cooling water studies for Edison Electric Institute, Project No. RP49, Heat exchange in the environment", The John Hopkins University, Baltimore, 1965.

11. Uncles, R.J. and Jordan, M.B., "A one dimensional representation of residual currents in the Severn estuary, and associated observations", Estuarine and Coastal Marine Science, 10, 1980, pp 39-60.

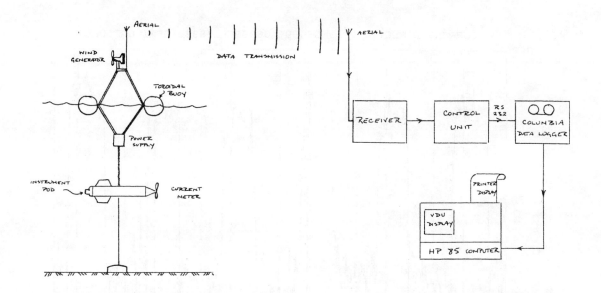

Fig. 1a   Mooring arrangement   Fig. 1b   Shore station details

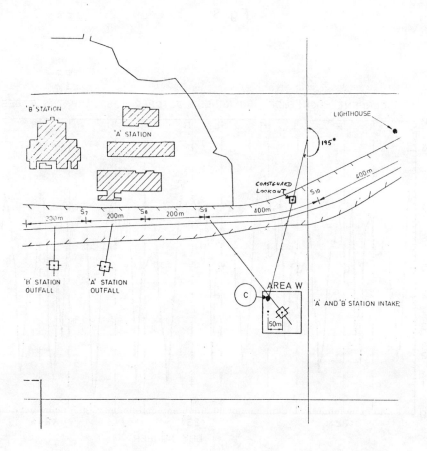

Fig. 2   Location of series 1000 meter at Dungeness

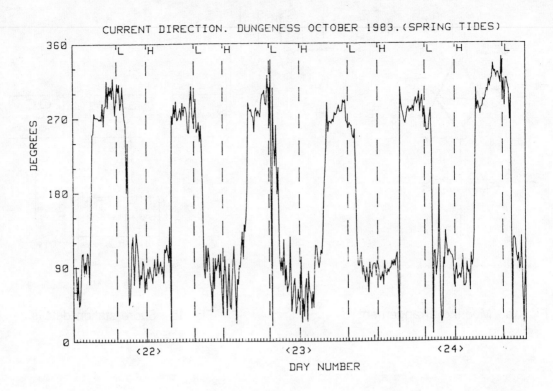

Fig. 3

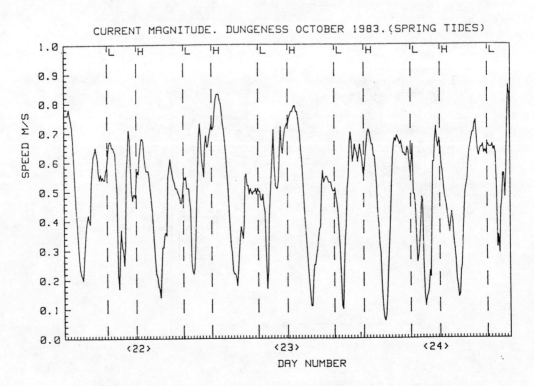

Fig. 4

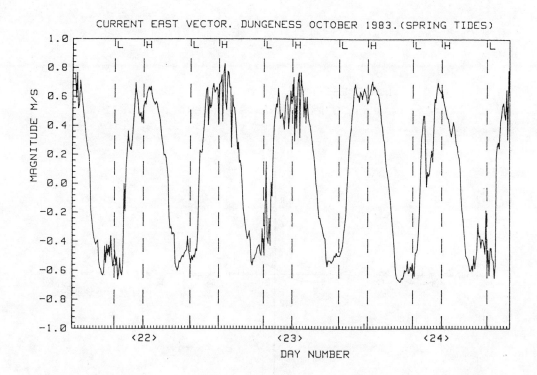

Fig. 5

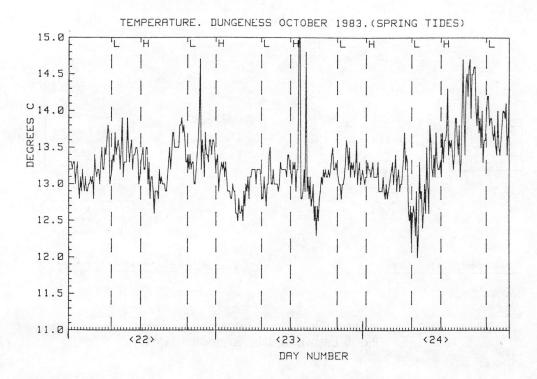

Fig. 6

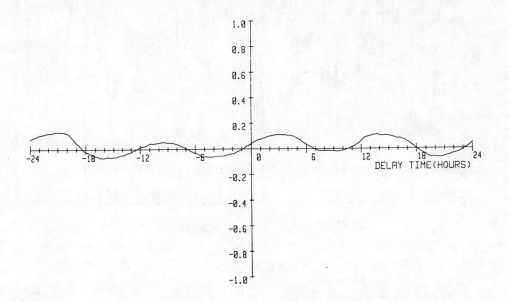

Fig. 7

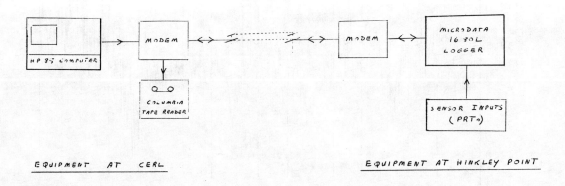

Fig. 8 Layout of data acquisition system

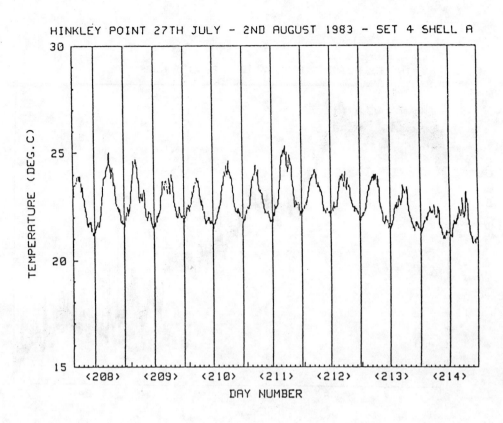

HINKLEY POINT 27TH JULY - 2ND AUGUST 1983 - SET 4 SHELL A

DAY NUMBER

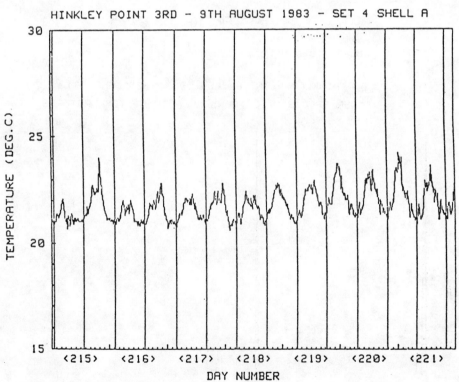

HINKLEY POINT 3RD - 9TH AUGUST 1983 - SET 4 SHELL A

DAY NUMBER

Fig. 9

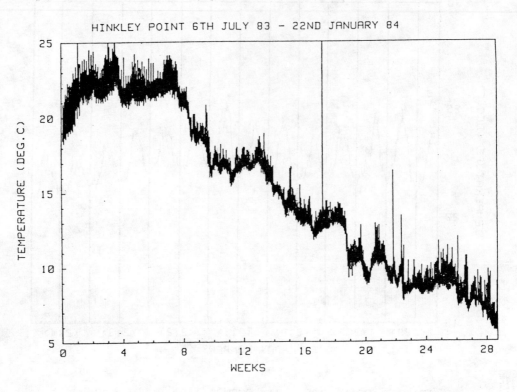

Fig. 10

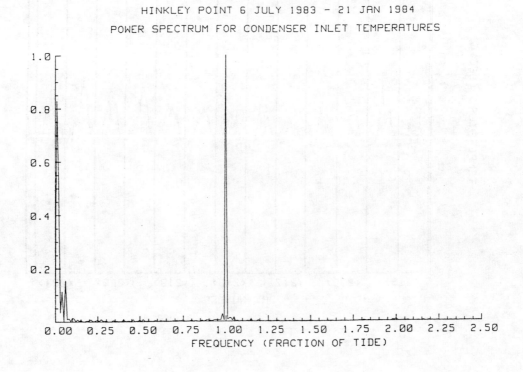

Fig. 11

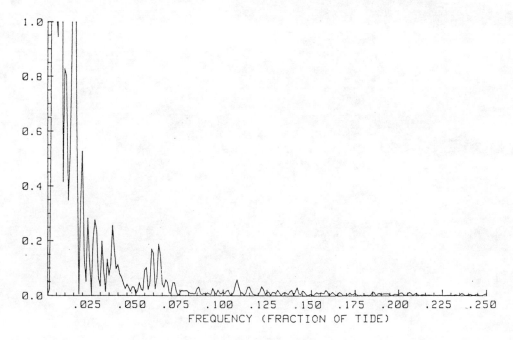

HINKLEY POINT 6 JULY 1983 - 21 JAN 1984
POWER SPECTRUM FOR CONDENSER INLET TEMPERATURES

Fig. 12

International Conference on

Measuring Techniques

of Hydraulics Phenomena in offshore, Coastal & Inland Waters

London, England: 9-11 April, 1986

PAPER G1

INSTRUMENTATION FOR HYDRAULIC MEASUREMENTS AT REMOTE INLAND SITES OVERSEAS –
AUTOMATIC SUSPENDED SEDIMENT SAMPLERS AND CAPACITANCE WATER-LEVEL RECORDERS

G R Pearce and C J Teal

Hydraulics Research
Wallingford, Oxon OX10 8BA

SUMMARY

Research studies in developing countries have highlighted the need for robust
and reliable field instrumentation that can be used to collect field data from remote
sites on water use, soil erosion, etc. This paper describes two such systems
developed by Hydraulics Research; the first for the measurement of suspended sediment
concentrations in ephemeral rivers and runoff from small catchments, and the second
for discharge and water-level monitoring in studies on irrigation schemes. Future
intentions are to combine the two techniques so that the data needed to calculate
total sediment flux can be collected by means of one combined system controlled by a
micro-processor.

Held at Imperial College of Science and Technology, London. Organised and sponsored by BHRA, The
Fluid Engineering Centre. Co-sponsored by the American Society of Civil Engineers and the
International Association for Hydraulic Research.

1   INTRODUCTION

Suitable and appropriate instrumentation for monitoring discharges and sediment concentrations in the irrigation canals of developing countries and the river systems that supply them is an area which has only in recent years received the attention it deserves. Considering the vast total area devoted to irrigated agriculture in the developing countries of the world, it is surprising how little monitoring of irrigation scheme performance is carried out, also on the problems of soil erosion and reservoir sedimentation. This can in part be attributed to the dearth of instrumentation that firstly can stand up to the hostile environmental conditions often encountered and secondly is sufficiently economically priced to allow deployment in the large numbers required.

The Overseas Development Unit (ODU) at Hydraulics Research (HR) has been promoting the development of this class of field instrumentation for use on research studies in developing countries. Such instruments are suitable for hostile conditions since they are robust and reliable, and are appropriate for remote sites since they are powered by rechargeable batteries and are micro-processor controlled wherever possible for automatic control or ease of data retrieval. A suite of various low-cost sensors is being developed – (i) stage/water level measurement by capacitance gauge-board, float-operated transducer and pressure-transducer; (ii) velocity/discharge measurement by ultrasonic current meter and vane-deflection flowmeter; (iii) sediment concentration by single-point and multi-point sediment sampler; (iv) borehole water-level recorder, and (v) crop cover meter. These sensors are mainly based on low-cost, solid-state data-loggers. At the moment they are intended as research tools for overseas development research projects, but they or similar sensors all have potential for widespread general use.

This paper describes two of the more developed systems, firstly the automated sediment sampler and secondly the capacitance gauge-board. It concludes by describing how the two instruments may in future be combined.

2   AUTOMATIC SEDIMENT SAMPLER

The automatic sediment sampler has been developed to provide sediment load data for either canal sedimentation studies on a regular (eg daily) basis or for investigations of short-duration transient floods on an intensive basis. It obviates the need for trained staff, apart from the collection and processing of the sample-bottles, and enables measurements to be made in very remote locations and at the unpredictable times at which flood events occur. For studies of soil erosion it is particularly important to take samples whenever intensive storms occur so that soil loss rates can be quantified.

Sediment samplers are not a new idea, several devices are commercially available, but the special developments included in this design are novel. Primarily these are : (a) suitability for use in hostile, remote sites and (b) capability for abstracting samples of suspended sediment with particle sizes into the sand range. The automatic sediment sampler makes reliable measurements possible in areas where previously they could not have been taken at all.

2.1 METHOD OF DEPLOYMENT

Because the vertical distribution of sediment sizes in a channel is non-uniform, single-point sampling can only be carried out at locations where good mixing occurs, such as a weir (see Fig. 1). Alternatively, if such a location does not already exist a drop structure may have to be installed to provide a mixing zone. This is a feature of soil erosion studies where small catchments are being monitored. ODU standard practice (Amphlett, Ref. 1) is to utilize the well-mixed zone in the drop-stilling basin of H-type flumes installed for discharge measurement. Plate 2 illustrates such an arrangement set up in Malawi. Total sediment load is calculated from the discharge rate and the sediment concentration measured at a single point. Fig. 2 illustrates a typical result. At any given measurement location it is important to check that complete mixing is taking place over a variety of discharge rates, to substantiate the use of this calculation procedure.

In larger streams or rivers where it has not been possible to find or create a mixing zone, multi-point sampling techniques are desirable. This approach is more appropriate to the steady flow conditions of large canals or rivers. The approach has been fully described by Sanmuganathan et al (Ref. 2) and Bolton (Ref. 3).

2.2 SAMPLER SYSTEM

The automatic sediment sampler developed at Hydraulics Research is capable of storing up to 24 half-litre samples working on either pre-programmed intervals or on a sampling schedule triggered by rising water-level. The system can lift sediment samples up to 8 m and deliver them into sample bottles without altering the particle size distribution. Cross-contamination between successive samples is prevented by automatic self-purging, also if sediment concentrations are high the system can be programmed to back-flush after each operation to prevent blockages occurring. Since batteries are used for power, the system can be left running at remote sites, and where appropriate solar panels can be utilised for recharge.

Fig. 1 is a diagrammatic representation of the major components and control system of the automatic sediment sampler. Those components are described as follows :

2.2.1 Sampler head and pump

The sampler can handle suspended sediment particles in the silt to fine sand range. For effective pump-sampling, velocities of not less than 1 ms^{-1} are maintained throughout the system to prevent heavier particles from dropping out of suspension before delivery to the bottle. This is based on the earlier development work of Crickmore and Teal (Ref. 4) on pump samplers carried out at Wallingford which as well as determining this minimum line velocity, also showed that matching the intake velocity to the instantaneous stream velocity seen at the intake was unnecessary since the concentration and particle size distributions were unaffected (for particles up to 200 μm diameter) by flow ratios of between 0.5 and 5.0.

A Jabsco submersible pump is used for good performance over vertical lifts of up to 8 m. It is self-priming even if air-bubbles are entrained, and wear-resistant to sand in the water. Also reversible drive is available to facilitate back-flushing if required. For lifts of less than 2 m a rigid coupling is used between the pump and the 24V DC motor to keep the motor clear of the water. For greater lifts it is replaced by a flexible coupling which also enables any attitude to be used, depending on site conditions. In circumstances when improved discharge/lift head stability is required a Mono pump is used with its internal voids filled to prevent sediment trapping.

2.2.2 Bottle rack

The pump delivers to a Rock and Taylor bottle rack modified to cope with high line-velocities and incorporating a suitable distribution arm and solenoid pinch-valve. Internal hoses are arranged to prevent flow constriction and sediment deposition. The rack holds 24 bottles in two cylindrical tiers and is mounted in a weatherproof GRP housing.

2.2.3 Programmable control unit

The system is controlled by a programmable controller comprising a 2K EPROM chip, a clock oscillator and trigger units. Programmed instructions actuate eight relays which in turn operate the various solenoid valves, pumps, sample distribution arm and, if required, an event marker connected to a water-level recorder.

The completion time for filling all 24 bottles can be varied from 12 minutes to 11 hours. The time interval between each sampling operation is independent of the others, progress through the program is monitored on LED displays.

Sampler start-up can be controlled by various means. A water level switch is most commonly used, and allows the controller to be synchronised with the onset of transient floods (for applications such as catchment run-offs or ephemeral streams). For applications requiring long periods between sampling operations (such as canal sedimentation studies) an external timer can be used to extend interval times to 24 hours or more.

2.2.4 Power

The system is battery powered so that it can be used at remote sites. It is designed to run from two 12v car batteries which are readily available throughout the world and are not usually difficult to recharge or replace. Solar panels can be used for continuous trickle recharge, thus reducing battery maintenance to only periodic replenishment of the electrolyte.

2.3 SYSTEM PERFORMANCE

The automatic sediment sampler is being used successfully in applications ranging from soil erosion studies in Malawi and the Philippines, canal sedimentation research in Nepal and Thailand to studies on sediment control in spate wadis in North Yemen. In the Nepal study, the sediment sampler was modified to incorporate 3 sampler heads at different depths in an unmixed reach of the canal. The main difficulty encountered has been the need to design for hot, humid conditions, others being unreliable maintenance of batteries and having to guard against theft and vandalism.

A typical measurement is shown in Fig. 2 of the instantaneous soil loss rate from a Malawian study catchment during a significant storm. Sediment fluxes are calculated from the H-flume's stage record and the laboratory analysis of sediment concentrations in the samples.

2.4 FUTURE DEVELOPMENT

Apart from developing the potential of the micro-processor control system, which is explored at the end of this paper, there are two main areas identified for future work both concerning the accuracy of taking representative samples particularly with regard to the sand fraction.

Firstly the accuracy can be improved by taking larger samples. This can already be achieved by using, say, 4 bottles instead of 1; however it is intended to do this more acceptably by using one litre bottles instead of half-litre, also by developing a cascade system where a controller can be linked to more than one bottle rack.

Secondly work has already been carried out by Kiff (Ref. 5) on the use of a hydrocyclone to concentrate the sand fraction in the sediment sampler developed by HR for transport in tidal waters. It is intended to incorporate a suitably dimensioned hydrocyclone in the automatic sediment sampler's bottle rack.

3   LOW COST CAPACITANCE GAUGE-BOARD

In irrigation water management studies there is often a requirement to monitor a large number of discharge rates and water levels at many locations throughout a complex irrigation scheme. This would be prohibitively expensive if conventional water level recorders were utilized, particularly when the cost of digitising chart records is also considered. There is also a need for instrumentation that can function at remote sites in harsh climates. The system developed by Hydraulics Research for measuring and recording water levels is essentially a custom-built research tool which fulfils these special requirements.

Certainly the potential for such a system is believed to be very large; irrigation schemes and drainage schemes throughout the world would benefit from increased operational information which at present cannot be justified on cost grounds.

3.1 METHOD OF DEPLOYMENT

The simplest way of measuring water-level is to use a gauge-board. However it is usually impossible to arrange for observers to take frequent and regular measurements over both a large area and a long time-span, so for the continuous monitoring required in research studies some form of recorder is required. The capacitance gauge-board and data-logger system that has been developed at HR was designed with this purpose in mind.

The main principle of use is that a gauge-board can only be used in conjunction with some form of hydraulic structure, such as a weir or a flume, for which the Q/h relationship is known. Plate 3 shows a typical location on the small-scale irrigation study in Zimbabwe. In an irrigation scheme this requirement can entail building structures where data is needed. However the approach used on ODU projects is to minimise disruption to the established flow regime on the irrigation scheme by utilising existing structures such as control gates. Here instead of one stage reading, the head difference across the gate is required as well as the area of gate opening. This means using two gauge-boards upstream and downstream of the control gate, and monitoring the position to which it is set. (A compatible sensor is already under development for automatically recording the gate position on the same data-logger.) The means by which the Q/h relationships are used for calculating discharge rates are explained in standard texts such as Bos (Ref. 6), a description of the particular methods used in the Zimbabwe study is described by Lewis (Ref. 7).

3.2 INSTRUMENTATION

The capacitance gauge-board and the data-logger system are illustrated in the schematic diagram shown in Fig. 3. The system is based on a low-cost, solar-powered data-logger. This receives and stores the signal produced by the gauge-board indicating the water-level.

At regular intervals, eg weekly, a portable micro-computer is brought to collect the accumulated data from the loggers at all the field monitoring stations. The micro is then returned to the office where it provides either : (a) coarse processing of the data and print-out of results for local counterpart staff, or (b) dumping of the data onto micro-cassette so that it can be sent for further analysis on the mainframe computer at HR.

The logger has capacity for up to 4 channels and typically has space for up to 10 days of readings at hourly intervals.

The system comprises the following components although one should remember its design is versatile and that a range of compatible sensors can be used in place of the capacitance gauge-board.

3.2.1 Capacitance gauge-board

The capacitance gauge-board comprises a standard GRP gauge-board to which a tubular capacitance probe has been added along its whole length (see Plate 3). An integral control box at the top of the probe, which is powered up by the logger, sends back a signal indicating the level of water at the gauge-board.

The tubular probe is made from an aluminium cylinder coated by a layer of shrink-fit polyester which acts as the dielectric medium. The cylinder forms one pole of a capacitor, the water outside the polyester film acts as the other and forms an earth. The capacitance across the dielectric is directly proportional to the wetted area and therefore to the depth of inundation. Essentially the system operates by timing how long a constant current must flow in order to charge the variable capacitor. The capacitance of the probe is translated into a 15 Hz pulse train by a pulse generator in the control box. The length of the pulses is proportional to the capacitance and therefore the height of water. The pulses are themselves converted into an output signal by a frequency generator. A range of 0 - 10 Hz covers the full range of the capacitance probe. Thus the number of pulses produced is indicative of the water level. Since this is a digital signal, long cable runs are possible. Because the capacitance probe is connected to a gauge-board, visual checks of the output can be made any time. The output is displayed either via the micro-computer to be described in section 3.2.3, or via a specially-designed, portable gauge-board meter which gives a direct instantaneous reading.

3.2.2 Data-logger

The data-logger (see Plate 4) is designed on a low-cost concept with provision for recharge of batteries by a solar panel. Up to 4 channels are available, and with its 2K EPROM memory the data-logger can store approximately 250 2-byte records per gauge-board. CMOS integrated circuits are used to minimise current consumption.

To make and record a reading the data-logger counts the number of pulses received in a given time interval from the gauge-board. Typically the logger is set to switch on at a preset time and then to take a reading once an hour over a 5 minute measurement interval normally preceded by a 10 second 'warm-up'. The stored data has to be removed at regular intervals otherwise over-writing occurs when the memory allocated is full. The logger is reset automatically by the time of this data-collection

Two types of internal rechargeable battery are used, either Ni-Cad D-cells or lead-acid jelly sealed cells. Solar panels are normally used to provide continuous trickle recharge of the batteries during daylight hours. In the event of a solar panel or internal battery failure an external car battery can be used to operate the logger.

Although the data-logger is described here in use with the capacitance gauge-board, it can be connected to other sensors such as the float-level sensor and the ultrasonic current meter. In fact all the sensors developed for Overseas Development Unit use have been designed to be compatible with the logger and data retrieval system.

3.3.3 Data retrieval system

Data retrieval is based on a Husky portable microcomputer. The device was chosen for its physical robustness, water-proofing and durability in high temperatures - features which make it eminently suitable for being carried round hot and dusty irrigation schemes to carry out weekly interrogation of loggers at various measuring sites. The advantage of using a micro in the field is that as well as controlling the collection of data from the loggers, it also provides a number of options for handling and plotting the data.

Once data has been collected from a logger, it can be immediately reviewed on the micro's l.c.d. screen. This is done to check the correct functioning of the logger, gauge-board and/or any other sensor.

When all the data has been collected from the site and the micro is back in the local office, two operations are carried out. Firstly all the stage records are produced as individual figures using a graphical printer. In order to do this the micro has to perform some approximate data processing. The graphs produced (see Fig. 4) enable the counterpart staff quickly to review the functioning of all the logger/gauge-board systems so that any necessary maintenance can be planned. Also they can see very clearly how the water in the irrigation scheme has been distributed in the preceding week. This printed record is filed as a secure copy of the data. The second operation is subsequently to transfer the raw data files in the micro onto a micro-cassette. In doing this, the allocated memory area is reset. The micro-cassette is then sent back to the Overseas Development Unit at HR for further analysis.

The incoming micro-cassette is loaded onto a 256 K micro at HR where the data is quickly and conveniently listed, checked and, if necessary, corrected. It is then transferred to the mainframe computer where the sophisticated analysis required for research studies can be carried out.

3.3 SYSTEM PERFORMANCE

3.3.1 Field deployment overseas

The gauge-board/data-logger system has been successfully deployed at ODU research project sites in Zimbabwe, Sudan, Philippines and Thailand. Design modifications resulting from this field experience have already been incorporated (they are briefly reviewed in section 3.3.3).

The power and potential of the system is amply demonstrated by the stage record shown in Fig 4. Where previously observers were recording two instantaneous readings of stage every day at each structure (from which the flow regime was being deduced), a continuous and far more accurate picture is now being produced. This particular record is of a field-canal offtake and comes from the small-scale irrigation study in Zimbabwe. The aims of this study and how the gauge-board system is being utilised in it is described by Pearce (Ref. 8). The upstream record shows the daily pattern in the distributary canal, the downstream record demonstrates which days the field canal received water. This is the information that the local irrigation manager receives. For the research study, when the raw data is received it is processed on the mainframe to determine actual discharge rates on an hourly basis.

(The portable micro is not used to compute discharges because the workspace in its memory is already fully utilised.)

The main advantage of an automatic monitoring system over data-collection by observers is the large reduction in visits to each measurement location (typically once per week instead of twice per day) and the elimination of observer and transcription errors. However there are other advantages of the gauge-board/data-logger system to be considered over conventional water-level recorders. These are :- (i) researchers receive raw data quickly and can transfer it directly to a main-frame. This eliminates the need for digitising chart records; (ii) the system is sufficiently low-cost to enable large-scale monitoring exercises to take place that would otherwise be out of budget; (iii) there are no moving parts to cause maintenance problems, and (iv) local management receive a weekly summary of flows in the irrigation scheme which gives them a clear and virtually instantaneous overview.

3.3.2 Accuracy

The capacitance gauge-board is capable of an accuracy of +1 cm over a 1 m range. On longer (or shorter) gauge-boards the accuracy is still +1%. Response is linear over the whole range except for very low levels below 2 cm. Response to changing levels is adequate, the response time to a step increase in level is virtually instantaneous, but for a step decrease the gauge-board takes up to 10 minutes for a 95% response. However in its intended application where levels do not change rapidly this will not be a problem. Water surface disturbances can decrease achieveable accuracy since the gauge-board will record the maximum rather than the mean of short period fluctuations. In river and canal situations the small effect of capillary waves can be taken into account in the Q/h relationship, if necessary. A stilling well could be used in other circumstances but this would reduce the gauge-board's advantage of easy installation.

3.3.3 Difficulties encountered

The major difficulty encountered occurred with the earlier version of the gauge-board where the polyester dielectric was mounted flat on the board's back. Failure of the adhesive holding the polyester film on, led to the leading edge (against the current) starting to lift and allowing charge to leak to earth. The solution has been to use the tubular design which has no edges and is of similar surface area. In very small channels the 3 cm diameter does provide more of a constriction to flow than the earlier version but this has not been found to be significant.

Floating trash does occasionally get trapped around the sensor, but this is removed on a daily basis by the water-bailiffs. If caught above the water-level it causes the gauge-board to over-read. Trash-guards could be utilised if necessary.

Build-up of sedimentation around the base of the probe can occur and, if no regular cleaning were carried out, would cause misleading information to be produced when the channel was dry. However such an error would occur only at times of very low level and would not create a large systematic error in the data, in fact its occurrence should be caught by error-trapping procedures in the data-processing programs.

Penetration of moisture into both the gauge-board control box and the data-logger box has occurred during heavy rains. In equipment where precautionary potting of the electronic components was not carried out, some bridging across the p.c.b.s did occur. As a result cable-glands have been given extra protection, also higher specification boxes are now being used for the control boxes.

Occasional malfunction of the batteries has occurred. This has been attributed to the effect of high temperature in the logger box. High temperatures (40 - 50°C) cause the maximum acceptable recharge voltage of the batteries to be reduced below the protection threshold of the logger's voltage regulator. A suitable solution is being sought.

3.3.4 Further work

It is intended in the future to develop the data-handling capability of the data logger so that readings taken at a particular measurement location can be immediately processed. With an appropriate Q/h relationship programmed into it the logger could continuously output the instantaneous discharge rate to an LED. This would be of immediate application to irrigation schemes in general since no suitable low-cost means of measuring discharges in channels exists at present. In this context, such a device could be used to reassure small-holder farmers that they were receiving fair quotas of irrigation water.

Other development work already well advanced has been the design and building of alternative compatible sensors for measuring water level and current velocity; these have already been listed. The float-level transducer is of particular note since its use is parallel to that of the gauge-board. The device is intrinsically more accurate particularly in high velocity flow and is more suitable for wide ranges of water-level, but the installation of the necessary stilling well can be a major task.

A 'gate-opening sensor' has been designed and tested, and in the near future will be deployed at field sites where monitoring of the control-gate position is required. The sensor uses the same transducer as was developed for the float-level transducer and is therefore immediately compatible with the loggers already connected to the upstream and downstream gauge-boards.

4   FUTURE DEVELOPMENT

Both the Automatic Suspended Sediment Sampler and the Capacitance Gauge-board/Data-logger systems described in this paper are of immediate use in their own right to hydraulic measurements in overseas research studies such as those being carried out by the HR Overseas Development Unit. The fact that both systems are already deployed at several overseas sites is indicative of their suitability, and it is hoped that they will be of application to other practitioners involved in water resource management and research in both this country and the developing world.

The two systems are not so separate as they might seem, both being micro-processor controlled. It is well within the bounds of present hardware for one micro-processor to control both systems. The next logical step is to combine the two systems in this way. This would enable the sensor logging water-levels at a measurement location (be it a capacitance gauge-board or an alternative instrument) to also provide the triggering and event-marking required by the sediment sampler. Of course the laboratory analysis of bottle samples would still be required but sample collection could be conveniently combined with the regular visits of the field portable micro to abstract logger data. Then the type of output demonstrated in Fig. 2 could be straightforwardly and immediately produced by the simple expedient of inputing the results of the sediment analysis onto the same microcomputer processing the stage/discharge data.

5   ACKNOWLEDGEMENTS

Funding for the equipment development programme of the HR Overseas Development Unit is provided by the UK Government's Overseas Development Administration. Design and field work for the instruments described in this paper were carried out respectively in the Engineering Design Group headed by Mr I Shepherd, the Field Services Group headed by Mr C Waters, and the Overseas Development Unit headed by Dr K Sanmuganathan.

The authors are grateful for the many varied inputs from their colleagues at Hydraulics Research.

6   REFERENCES

1.   Amphlett, M B: "Measurements of soil loss from experimental basins in Malawi". Proc of Harare Symposium on Challenges in African Hydrology and Water Resources, IAHS Pub. 144, 1984.

2.   Sanmuganathan, K and Bolton, P: "Application of theoretical principles for improved sediment discharge measurement in rivers". Proc 4th Congress of Asian & Pacific Regional Division, IAHR, Thailand, 1984.

3.   Bolton, P: "Sediment discharge measurement and calculations - Techniques for use at river gauging stations". Hydraulics Research Report, OD/TN/2, Wallingford, 1983.

4. Crickmore, M J and Teal, C J: "Recent developments in pump samplers for the measurement of sand transport". Proc Symp on Erosion and Sediment Transport Measurements, IAHS Publ 133, Florence, 1981.

5. Kiff, P R: "Evaluation of a hydrocyclone for on-site sediment separation, J Sediment Petrol. 47 (3), 1977.

6. Bos M G (Ed): "Discharge Measurement Structures". ILRI Pub 20, Wageningen, Holland, 1976.

7. Lewis N S: "Flow measurement structures, Nyanyadzi, Zimbabwe". Hydraulics Research Report OD/TN 8, Wallingford, 1984.

8. Pearce G R: "Water flow analysis on a small holder irrigation scheme - initial results, Nyanyadzi, Zimbabwe". Proc African Regional Symposium on Smallholder Irrigation, University of Zimbabwe, 1984.

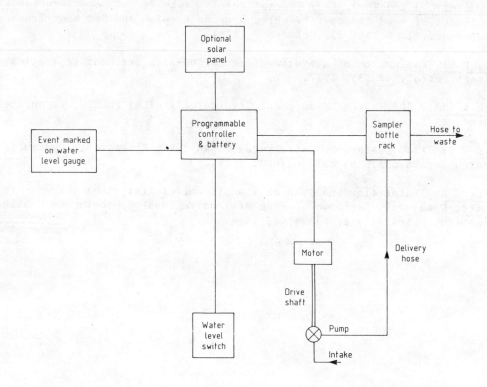

FIG 1   SCHEMATIC DIAGRAM OF THE AUTOMATIC SEDIMENT SAMPLER

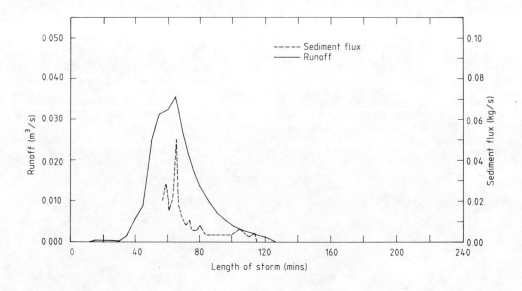

FIG 2   TYPICAL FIELD DATA FROM THE MALAWI SOIL EROSION STUDY

The runoff hydrograph from a study catchment during a storm is compared with the rate of soil loss calculated from the pumped samples.

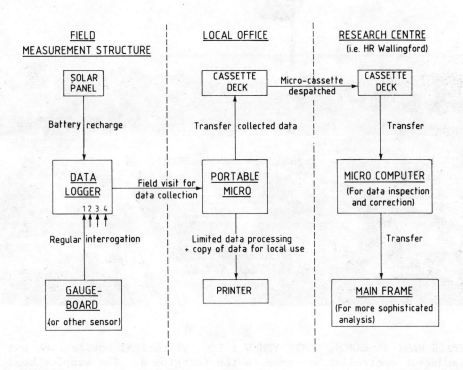

FIG 3   SCHEMATIC DIAGRAM OF THE CAPACITANCE GAUGE-BOARD/DATA-LOGGER SYSTEM

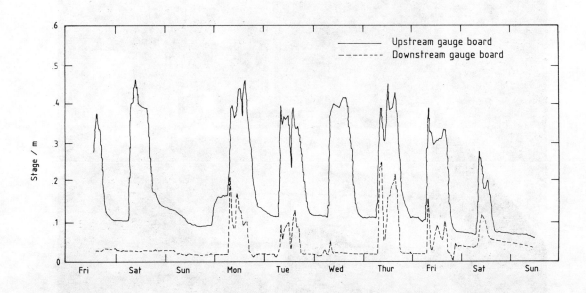

FIG 4   RECORD OF STAGE AT AN OFFTAKE CONTROL GATE,
ZIMBABWE SMALL-SCALE IRRIGATION STUDY

The upstream board record shows the daily cycling in the main canal. The
downstream board reveals which days the offtake was used. Actual discharges
are computed from the combined records.

PLATE 1 SPATE WADI RESEARCH, NORTH YEMEN : The cylindrical bottle rack and the
adjacent controller are seen in the foreground. The sampler head is located
in the offtake channel beside the weir. The solar panel is for battery
recharge.

PLATE 2 SOIL EROSION RESEARCH, MALAWI : A sediment sampler installation in the
mixing basin below an H-flume. The pump and sampler head are located at
the base of a rigid coupling from the 24V motor. A conventional water-
level chart recorder (and stilling well) is located to the right of the flume.

PLATE 3   FLOW MEASUREMENT SITE, ZIMBABWE : The Capacitance Gauge-
board is deployed to the right of the conventional
gauge-board and the data-logger protection box above it.
The discharge over the thin-plate weir in the foreground
is computed from the stage record.

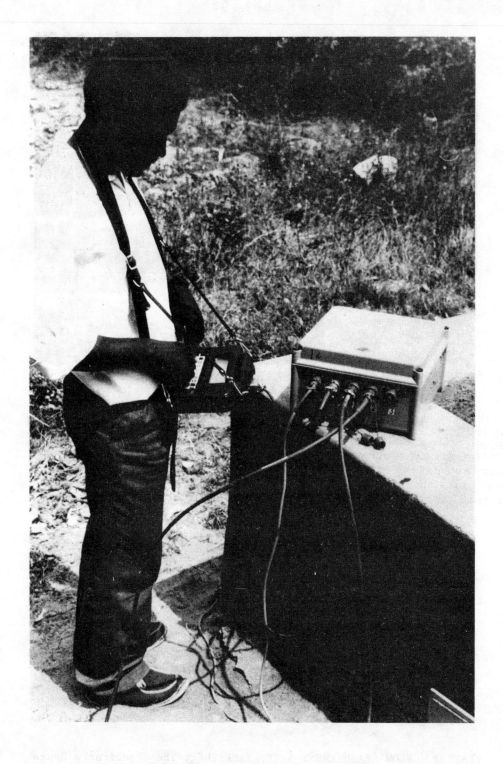

PLATE 4   FIELD VISIT FOR DATA-COLLECTION, ZIMBABWE : The week's
          data from this data-logger (removed from its box for the
          photograph) is being collected on the Husky portable
          microcomputer.

A NOVEL FIBRE-OPTICS FLOWMETER *

W.J. Easson and C.A. Greated

Physics Department
Edinburgh University
Edinburgh, U.K.

One of the most notable advances in fluid measurement has been the development of the Laser Doppler Anemometer. It combines accuracy, and high response and is non-intrusive to the measuring volume. Recently, the problem of positioning the laser and other bulky optical components in awkward flows (e.g. underwater) has been overcome by the use of fibre optics. However, the techniques are costly and lead to further complication rather than simplification of the optical system.

The authors have developed a new technique which provides accurate LDA signals using fibre optics. The simplicity of the new system means that two-dimensional flow measurements are as easily obtained as one-dimension and the low frequency signals obtained are easy to analyse. The optical heads have a diameter of less than 5 mm and are fixed to one probe therefore requiring no alignment. Tests have been performed against traditional LDA methods and the results are exceptionally good.

A further reduction in cost is envisaged by replacing the low power laser by a narrow spectrum L.E.D.

*Patent applied for.

Held at Imperial College of Science and Technology, London. Organised and sponsored by BHRA, The Fluid Engineering Centre. Co-sponsored by the American Society of Civil Engineers and the International Association for Hydraulic Research.
©BHRA, The Fluid Engineering Centre, Cranfield, Bedford MK43 0AJ, England 1986.

1.   INTRODUCTION

The measurement of fluid velocity by optical techniques has taken several forms, the most common being the two beam heterodyne arrangements (Ref.1). Less common are the time of flight method (Ref.2) and the grating method (Ref. 3).

Recently, attempts have been made to remove bulky and fragile parts of the L.D.A. system from the locality of the measuring fluid to enable Laser Anemometry to be applied to awkward or otherwise inaccessible situations. This varies from using a fibre to supply the laser light to the L.D.A. optics (Ref. 4) to a fully developed velocimeter probe using the heterodyne technique (Ref. 5).

It is very common for two perpendicular components of fluid velocity to be required simultaneously. Again many excellent methods have been devised of achieving this, commonly using two colour lasers and/or polarised pairs of laser beams. These are very expensive and often very difficult to align.

The proposed fibre-optic flowmeter will be capable of simultaneous measurement of two-dimensional fluid flow by means of an intrusive but very small probe head. The use of optical fibres to provide the light beam to the measuring volume and collect the scattered doppler signal removes the normal requirement of placing potentially dangerous electrical equipment and expensive optical equipment near the point of interest.

A one-dimensional form of the probe will be described first; with the results from an early bench model. Further developments will then be outlined, including 2-D measurements, back scattering probe heads and diode light sources.

2.   THE ONE-DIMENSIONAL PROBE

The system is based on the grating method (Fig. 1). Light from a low power laser is condensed into the transmitting fibre by a low power lens. The fibre is connected to the transmitting head of a rigid probe where the light may be either collimated or condensed to the measuring volume by the transmitting lens.

A receiving lens collects the scattered light from the measuring volume and focusses this on the end of the receiving fibre. A grid on this end of the receiving fibre modulates the intensity of the scattered light according to the rate of passage of the image of a scattering particle across the grid lines. This modulated signal is collected at the other end of the fibre by a diode in the usual manner and signal analysis can proceed by the best appropriate technique.

The fluid velocity is given by

$$u = \frac{\ell - F}{F} \, s \, f_D \qquad\qquad (1)$$

where u = velocity perpendicular to the grid
$\quad$ F = focal length of the receiving lens
$\quad$ ℓ = distance to the measuring volume (in air)
$\quad$ s = line spacing of the grid
$\quad$ f_D = the modulated frequency.

The size of the measuring volume is determined by the focal length of the lens and is nominally

$$\omega = \frac{(\ell - F)d}{F} \qquad\qquad (2)$$

wide, where d is the diameter of the fibre, and of length

$$r = \frac{2\ell(\ell - F)d}{aF} \qquad\qquad (3)$$

where a is the aperture of the receiving lens.

3.   BENCH TESTS

A bench model has been built and demonstrated for the purpose of testing the

system. The specifications of the components used were very low; a significant
improvement in performance is envisaged for better quality optics.

The fibres were of the multimode polymer type with an internal diameter of
1 mm. A 10 mw laser was used, the beam being condensed into the fibre by a single lens.
In the absence of suitable diameter transmitting and receiving lenses 10 mm microscope objectives were used, stopped to an aperture of 5 mm. Their focal length was
16 mm. The grid was produced by photography, using computer drawn parallel lines with
line width equal to half the separation of 1 mm. This was imaged on HP5 negative
film which was then used as the grid. The lines produced in this manner were very
'grainy' and an effective limit on signal to noise ratio with this system was found
at a spacing of s = 0.1 mm.

The head components were mounted in brass tubes with 0.5 cm outside diameter
and these were inserted in 2 cm diameter holders to allow manoeuverability on standard
optical benches. The grid/lens distance, V, was chosen as 3F such that the magnificat-
ion factor would be 2 giving a virtual grid spacing of half the spacing on the film.
Thus the measuring volume was 0.5 mm wide by 7 mm long and 2.4 cm from the lens.

The probe was calibrated on a spinning perspex disc, against a standard L.D.A.
The result are shown in Fig. 2. This gave an experimental virtual spacing of s'= 49μm
against the calculated value of 50μm.

Because of the small number of lines in the measuring volume (< 10) it was
necessary to use high pass filtering at $f_1 > 0.1\ f_D$. This is demonstrated in Fig. 3
where several filter settings are recorded for one speed of rotation. For $f_1 < 0.1\ f_D$
the correlogram is dominated by the light scattering over the beam width.

The probe was then tested on a steady fluid flow through a rectangular
section channel (Fig. 4). The standard L.D.A. and the probe were mounted 10 cm apart
with the measuring volumes at the centre of the tank. The flow was generated by a
recirculating motor and was found to be not quite steady. Fig. 5 shows the output from
a tracker analysing the signal from the standard L.D.A.. Upper bounds on the drift were
10% of the mean flow rate. The flow rate was varied by opening a valve on the recirculation pipe.

Correlograms of the probe signal were then collected for various rates of flow
measured on the standard L.D.A. A typical pair of correlograms is given in Fig. 6,
the frequencies are plotted against each other in Fig. 7 along with the conversion ratio
measured on the spinning disc. As can be seen the mean flow rates are in agreement to
well within the error bounds of the experiment,

Noise levels were high for the probe and fairly close filter settings were
required. This was mainly due to laser noise centred at ∿ 10ᴋHz.   However, a large
level of low frequency noise was also generated, the explanation for which is not
clear. The spectrum was gathered, without filtering and a schematic diagram
is given in Fig.8 . Several explanations are likely. Firstly, bad alignment of the
bench test reduced the number of lines in the measuring volume to an unacceptable level.
Secondly, the line spacing/seeding particle diameter ratio was too large giving a poor
signal/noise ratio. Thirdly, the measuring valve was too large and the seeding level
too high giving a poor signal/noise ratio (Ref. 7).

Each of the above explanations is easily tested and investigations are underway to refine the bench model to a useful prototype form .

4.   FURTHER DEVELOPMENTS

4.1   Two dimensional anemometer

The extension of the sytem to the measurement of two components of velocity is
simple and may be achieved by mounting a paired fibre in the receiving head of the
probe. Each fibre would have a separate grid and typically the grids would be mounted
perpendicular to each other (Fig.9).

4.2   Diode Laser

As the laser light is being fed directly to a fibre and the restrictions on
coherence are not so strict with the probe as with normal L.D.A. arrangements, the
possibility exists of reducing the cost of a L.D.A. system by replacing the laser with

a photo-emitting diode, possibly a laser diode. Apart from cost reduction this gives the possibility of straightforward modulation of the transmitting signal. High frequency modulation is one possible method of allowing velocity measurement around zero.

4.3   Backscattering optics

A backscattering probe would provide a further advantage for the fibre optic anemometer in necessitating only one head in the fluid at one side of the measuring volume. Naturally the power required in such instances would be higher than in the forward scattering mode but the probe would then be capable of measurement in more awkward flows.

5.   SUMMARY

The low cost fibre-optic anemometer has been primarily developed for new and routine research applications. In particular, the system is extremely practical in its application to: fundamental research in laboratories investigating fluid effects on shipping and offshore structures; routine measurement of fluid parameters in wave flumes and wind tunnels; and monitoring of offshore environments in the field. The main advantages in these situations are: the optics require no alignment; and electrical equipment may be removed from the locality of the water.

Further applications may include the monitoring of flow rates for semi-opaque and seeded flows, particularly where these involve flammable liquids.

6.   REFERENCES

1.   Durrani, T.S. and Greated, C.A.: "Laser Systems in Flow Measurement". New York, Plenum, 1977.
2.   Thompson, D.H.: "A Tracer-particle Fluid Velocity Meter Incorporating a Laser". J. Phys. E., 1, 1968, pp 929-932.
3.   Gaster, M.: "A New Technique for the Measurement of Low Fluid Velocities". J.F.M., 20, 1964, pp 183-192.
4.   Meakin, R.L., Koseff, J.R. and Street, R.L.: "A Fibre-Optical Link for Modular Two Component LDA Optics and Argon-Ion Lasers". In 2nd Int. Symp. on App. of Laser Anemometry to Fluid Mechanics, Lisbon, 1984, p 10.2.
5.   Durst, F. and Krebs, H.: "LDA Optics Development for Measurements in Internal Combustion Engines". In: 2nd Int. Symp. on App. of Laser Anemometry to Fluid Mechanics, 1984, p 6.5.
6.   Durst, F. and Zare, M.: "Optical Developments in Laser Doppler Anemometry". University of Karlsruhe, Report SF13, 80/E/65, 1975.
7.   Durst, F., Melling, A. and Whitelaw, S.H.: "Principles and Practice of Laser-Doppler Anemometry". London, Academic Press, 1976.

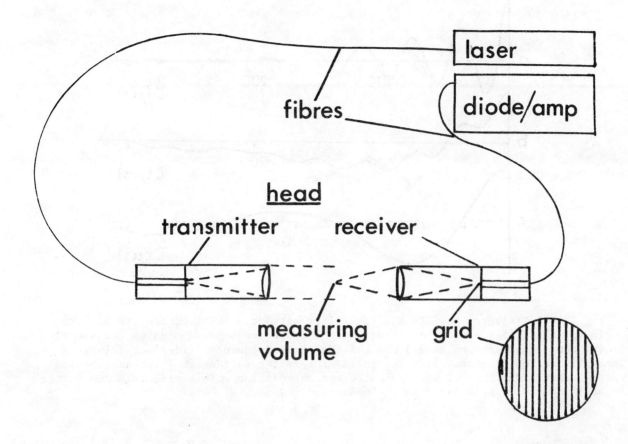

Fig. 1   Probe optics

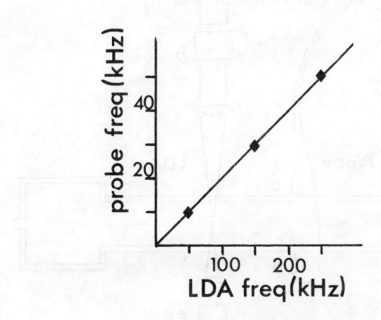

Fig. 2   Calibration of probe against two beam L.D.A. on a spinning disc.

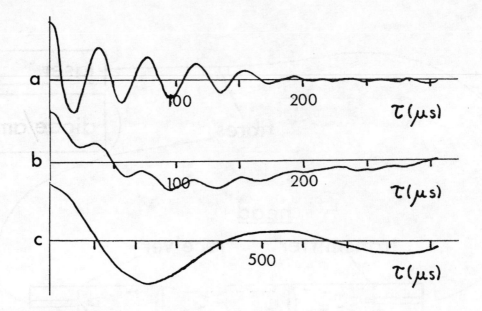

Fig. 3   Correlogram quality against filter setting. The frequency is 23.8 kHz .
In (a) the filter setting is 16kHz and the probe frequency dominates the signal.   In
(b) the filter setting is 3.3 kHz and the probe frequency 'rides' on a large
amplitude,low frequency signal.   In (c) the filter setting is 1 kHz and the low
frequency signal of $\sim$ 2 kHz dominates the correlogram.   This effect is due to the
small number of grid lines in the measuring volume.

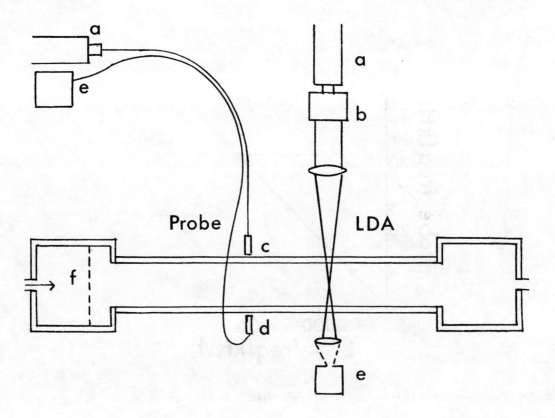

Fig. 4   Test set up for channel flow.   a.   lasers;   b.   beam splitter;   c,d.   probe
heads;   e.   diode/amplifiers;   f.   settling chamber.

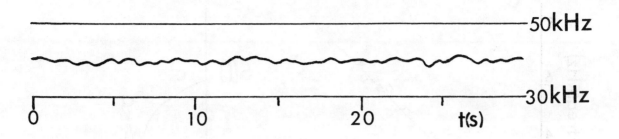

Fig. 5 Tracker output in channel test.
The flow rate was unsteady with an r.m.s. of about 5% of the mean as this record illustrates.

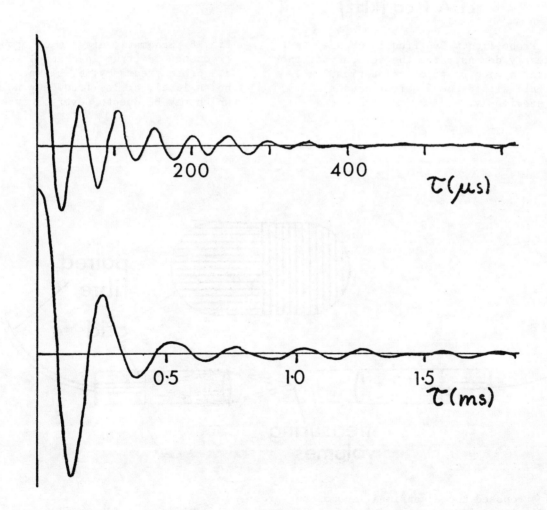

Fig. 6   LDA (upper) and probe correlograms for steady flow.

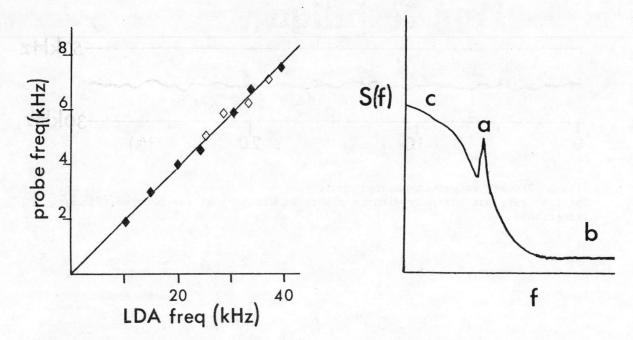

Fig. 7 Probe freq.v LDA freq.in channel test.
This clearly demonstrates the accuracy of the
probe over a wide range of velocities. All
points fall within the 5% error bounds discussed previously.

Fig. 8 Schematic diagram of probe
frequency spectrum in channel test.
a. probe frequency; b. background
noise level; c. low frequency noise
which may be due to several causes.

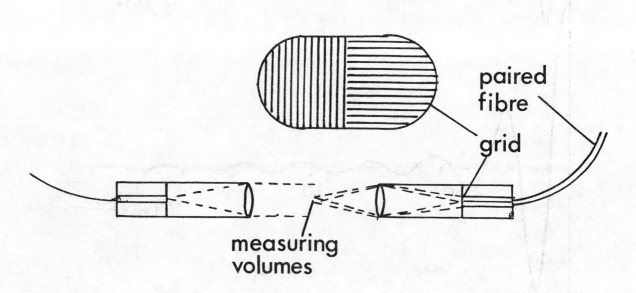

Fig. 9 Proposed two-dimensional anemometer probe.
Perpendicular grids are placed in front of a two-core fibre. Each fibre will then
carry information about one component of the velocity. The system then has two very
closely spaced measuring volumes which may be regarded as one measuring volume of
twice the diameter of the one component probe.

International Conference on

Measuring Techniques

of Hydraulics Phenomena in offshore, Coastal & Inland Waters

London, England: 9-11 April, 1986

MEASUREMENT OF DISCHARGE IN AN EXPONENTIAL
CHANNEL BY THE END DEPTH METHOD

K. Subramanya

Professor, Civil Engineering Department
Indian Institute of Technology
Kanpur-208016, India

SUMMARY

A general analytical method to predict the end depth in a free
overfall of an open channel is presented. This method is based on
the use of Boussinesq equation and the continuity of the water surface
at the brink. Application of this method to exponential channels is
presented. The role of the state of flow and the validity of the
method are also studied. Further, to use the end depth as a flow
meter, a procedure for estimating the discharge as a function of the
end depth in an exponential channel is indicated.

Held at Imperial College of Science and Technology, London. Organised and sponsored by BHRA, The
Fluid Engineering Centre. Co-sponsored by the American Society of Civil Engineers and the
International Association for Hydraulic Research.

<u>NOMENCLATURE</u>

A   =   area

C   =   a constant

E   =   specific energy

F_o =   Froude number at normal depth

K   =   a coefficient

S_c =   critical slope

$\bar{S}_f$ =   average energy slope

S_o =   bed slope

T   =   top width

V   =   mean velocity at a section

V_x =   horizontal component of velocity V

V_y =   vertical component of velocity V

Q   =   discharge

a   =   an exponent

c   =   suffix to indicate critical condition

e   =   suffix to indicate conditions at end depth

f   =   function

g   =   acceleration due to gravity

h_{ep} =   equivalent piezometric head

m   =   side slope of a trapezoidal channel

o   =   suffix to indicate normal flow

x   =   horizontal distance

y   =   depth of flow

y_c =   critical depth

y_e =   end depth

α =   kinetic energy correction factor

ϵ =   specific energy at end depth relative to critical depth

δ =   ratio of normal depth to critical depth

$\bar{\delta}$ =   ratio of end depth to normal depth

η =   ratio of end depth to critical depth

θ =   inclination angle of the channel

ρ =   radius of curvature

1. INTRODUCTION

1.1 General:

The depth at the end of a free overfall in an open channel is known as the end depth. The flow situation is similar to the flow over a sharp crested weir of zero height.

Fig. 1 shows a typical free overfall in a subcritical flow. Due to the continuity of the water surface profile the gravity effect of the free overfall extends a short distance upstream of the edge causing an acceleration of the flow. The minimum depth of flow y_e occuring at the brink of the overfall is termed the end depth. Due to the curvature of the water surface and boundary conditions the pressure at the end section departs considerably from the hydrostatic pressure distribution.

1.2 Experimental Studies:

Rouse (Ref. 1) is probably the first to recognise the interesting feature of the end depth at a free overfall. His measurements on the end depth y_e for subcritical flow in rectangular channels indicated $y_e = 0.715 \, y_c$ where y_c = critical depth. Since then a large number of experimental studies have been conducted on a variety of channel shapes and boundary conditions. Some important studies on subcritical flows are summarised in Table 1.

For supercritical flow cases, Delleur et al and others (Refs.3,4,5 and 8) showed by using their experimental data that the end depth ratio $\eta = y_e/y_c$ is expressible as

$$\eta = f \, (S_o/S_c, \text{ channel shape}) \tag{1}$$

where S_o = channel bed slope and S_c = critical slope. For a given channel the variation of S_o/S_c was shown to be a unique function of S_o/S_c. Considerable scatter of data of the order of $\pm$ 10%, especially at large values of S_o/S_c are observed in these studies. The factors affecting the end depth in a rectangular channel has been investigated in detail by Christodoulou et al (Ref. 9).

1.3 Analytical Studies:

The unique relationship for the end depth ratio for a given channel at a free overfall has given the end depth the status of a flow meter. The need for an accurate estimation of the end depth ratio under known flow conditions is evident. Most of the existing attempts to predict the end depth analytically are based on the momentum equation, with various assumptions regarding the velocity and pressure distribution at the brink. The solution invariably involve a pressure coefficient which has to be determined experimentally (Refs.3,4,7 and 9). An analytical one dimensional method of analysis based on the continuity of water surface curvature was first proposed by Anderson (Ref.10) for rectangular channels. Subramanya (Ref. 11) has given a description of Anderson's method and has indicated wider application possibilities.

This paper presents a general analytical solution, on the lines of Anderson's work, for solving the end depth ratio in channels of any shape. A particular application to exponential channels is presented in detail. The role of flow state is also studied. Further, a procedure for estimating the discharge as a function of the end depth is indicated.

2. THE GENERAL SOLUTION

2.1 Water Surface Curvature at the Brink:

In a free overfall, the water surface is a continuously falling curve. The water-surface profile starts in the canal somewhere upstream of the edge, passes through the brink and ends up as a parabolic trajectory of a gravity fall. In the present section, the expressions for the curvature of the water surface are separately derived for the canal flow and free overfall and matched at the brink.

2.2 Curvature of the Channel Flow:

The specific energy E at any section is given in one dimensional analysis as

$$E = h_{ep} + \alpha \frac{V^2}{2g} \qquad (2)$$

where h_{ep} = equivalent piezometric head, α = kinetic energy correction coefficient, V = mean velocity of flow at a section and g = acceleration due to gravity. For a free surface flow with a small curvature h_{ep} is given by the Boussinesq equation (Ref. 11) as

$$h_{ep} = y + \frac{1}{3} \frac{V^2}{g} \; y \; (\frac{d^2 y}{dx^2}) \qquad (3)$$

where y = depth of flow. This equation assumes that the radius of curvature $\rho = d^2 y/dx^2$ and that ρ varies linearly with depth. Thus, by assuming α = 1.0 for simplicity, Eq.(3) can be written by using the discharge Q and area A as

$$E = y + \frac{Q^2}{2g \, A^2} + \frac{1}{3} \frac{Q^2 y}{gA^2} (\frac{d^2 y}{dx^2}) \qquad (4)$$

Non-dimensionalising Eq. (4) with respect to the critical depth y_c and considering the brink section which is denoted by the suffix e,

$$\frac{E_e}{y_c} = \frac{y_e}{y_c} + \frac{Q^2}{2gA_e y_c} + \frac{1}{3} \frac{Q^2}{gA_e^2} \frac{y_e}{y_c} \frac{d^2 y/y_c}{d(x/y_c)^2} \Bigg|_{y=y_e} \qquad (5)$$

Denoting the critical flow conditions by the suffix c,

$$\frac{Q^2}{g} = \frac{A_c^3}{T_c} \qquad (6)$$

where T = top width.

Putting $y_e/y_c = \eta$, $\dfrac{A_c^3}{A_e^2 T_c y_c} = f(\eta)$

and $E_e/y_c = \epsilon$.

Equation (5) becomes

$$\epsilon = \eta + \frac{1}{2} f(\eta) + \frac{1}{3} \eta f(\eta) \frac{d^2 y/y_c}{d(x/y_c)^2} \Bigg|_{y=y_e}$$

The expression for the curvature of the channel flow at the brink is obtained as

$$\frac{d^2 (y/y_c)}{d(x/y_c)} \Bigg|_{y=y_e} = -\frac{3}{\eta f(\eta)} (\epsilon - \eta - \frac{1}{2} f(\eta)) \qquad (7)$$

2.2 Overflow Trajectory:

Referring to Fig.(1), V_x = x component of the velocity in the overflow trajectory is given by

$$V_x = V_e \cos \theta \qquad (8)$$

where V_e = mean velocity at the brink inclined at an angle θ to the horizontal. For a gravity fall

$$\frac{dV_x}{dt} = 0 \quad \text{and} \quad \frac{dV_y}{dy} = -g$$

where V_y = y component of velocity in the trajectory.

$$\frac{dy}{dx} = \frac{V_y}{V_x}$$

$$\frac{d^2 y}{dx^2} = -g/V_x^2 = -\frac{g\,A_e^2}{(V_e A_e)^2 \cos^2\theta}$$

$$= -\frac{g\,A_e^2}{Q^2 \cos^2\theta} \tag{9}$$

Using Eqs.(6) and (9) the curvature of the overflow, trajectory at the brink can now be written as

$$\left.\frac{d^2(y/y_c)}{d(x/y_c)^2}\right|_{y=y_e} = -\frac{1}{f(\eta)\cos^2\theta} \tag{10}$$

2.3 General Equation for End Depth Ratio:

The two expressions for the curvature of the water surface at the brink, viz. Eq. (7) and (10), must be identical and as such

$$-\frac{1}{f(\eta)\cos^2\theta} = \frac{3}{\eta\,f(\eta)}\left(\epsilon - \eta - \frac{1}{2}f(\eta)\right)$$

Simplifying

$$6\epsilon\cos^2\theta - 2\eta(3\cos^2\theta - 1) - 3f(\eta)\cos^2\theta = 0 \tag{11}$$

In the usual cases θ is small and $\cos\theta \approx 1.0$ and $\cos^2\theta \approx 1.0$. Eq.(11) simplifies as

$$6\epsilon - 4\eta - 3f(\eta) = 0 \tag{12}$$

This equation is a general equation relating the end depth ratio η with the non-dimensionalised specific energy at the brink. Depending upon the nature of the flow and the channel geometry and suitable assumption regarding the friction values of ϵ and $f(\eta)$ can be determined and η evaluated.

To illustrate the use of Eq.12 the end depth in exponential channels is studied in the following section.

3. END DEPTH IN EXPONENTIAL CHANNELS

3.1 Basic Relationship:

An exponential channel is one in which the area A is related to the depth y as

$$A = K y^a$$

where K and a are constants. It is easy to see that values of a = 1.0, 1.5 and 2.0 represent rectangular, parabolic and triangular channels respectively.

For an exponential channel

$$T = \frac{dA}{dy} = K \, a \, y^{a-1}$$

$$f(\eta) = \frac{A_c^3}{A_e^2} \cdot \frac{1}{T_c \, y_c} = \frac{1}{a} (y_c/y_e)^{2a} = \frac{1}{a \eta^{2a}}$$

Using these relations in Eq.(12)

$$6 \, \epsilon - 4 \eta - \frac{3}{a} \frac{1}{\eta^{2a}} = 0 \tag{13}$$

This is the basic relationship relating the end depth ratio, ϵ and a in an exponential channel.

Further, in an exponential channel the Froude number F_o at the normal depth y_o can be expressed, by using the suffix o to denote the normal depth conditions, as

$$F_o^2 = \frac{Q^2 \, T_o}{g \, A_o^3} = \frac{A_c^3 \, T_o}{A_o^3 \, T_c} = (y_c/y_o)^{2a+1} \tag{14}$$

or

$$\frac{y_o}{y_c} = \delta = (1/F_o^{2/(2a+1)}) \tag{15}$$

3.2   State of Flow:

3.2.1 Subcritical Upstream Flow:

If the flow upstream of the brink is subcritical, the critical depth must occur before the end depth. Hence, y_c exists at some location in the flow profile. Typically in a rectangular channel, y_c occurs at a distance of about $4y_c$ upstream of the brink. Now considering the specific energy E at the critical depth and neglecting the frictional losses in excess of that required to maintain uniform flow between the critical section and the end section (i.e.$(\bar{S}_f - S_o) = 0$)

$$\epsilon = \frac{E_e}{y_c} = \frac{E_c}{y_c} \tag{16}$$

Further it would be reasonable to assume the flow curvature to be negligible at the critical flow section and as such E_c can be written as

$$E_c = y_c + \frac{Q^2}{2g \, A_c^2}$$

i.e.

$$\epsilon = 1 + \frac{1}{2} (\frac{A_c}{T_c} \cdot \frac{1}{y_c}) \tag{17}$$

For an exponential channel Eq.(16) becomes

$$\epsilon = 1 + \frac{1}{2a} \tag{18}$$

It is seen that for a given channel shape (i.e. a = constant) ϵ is a constant for subcritical upstream flow.

3.2.2 Supercritical Upstream Flow:

If the flow upstream of the brink is supercritical the normal depth y_o is less than y_c and further the critical depth y_c does not exist in the profile between y_o and y_e. Considering an upstream section where the depth of flow is y_o and by assuming the frictional loss in excess of that required to maintain uniform flow between this

318

section and the end depth as negligible (i.e. $(\bar{S}_f - S_o) = 0$)

$$\in = \frac{E_e}{y_c} = \frac{E_o}{y_c} = \frac{y_o}{y_c} + \frac{Q^2}{2g \, A_o^2 \, y_c}$$

$$= \delta + \frac{F_o^2}{2} \, \frac{A_o}{T_o \, y_c}$$

$$= \delta + \frac{1}{2} \, F_o^2 \, f(\delta) \tag{19}$$

For an exponential channel

$$f(\delta) = \delta/a \tag{20}$$

From Eqs.(15) and (20)

$$\in = \frac{1}{F_o^{2/(2a+1)}} \, (1 + \frac{F_o^2}{2a}) \tag{21}$$

It is to be noted that in supercritical upstream flow $\in$ is a function of F_o for a given value of a.

3.3   End Depth Ratio:

Using Eqs.(18) and (21) the values of $\in$ for three channel shapes viz. a = 1.0, 1.5 and 2.0, were calculated for both subcritical and supercritical flow conditions i.e. as $f(F_o)$. Substituting the appropriate value of $\in$ corresponding to a given shape in the end depth equation (Eq. 13) the value of η for a given flow situation was obtained. It is easy to see that in a given exponential channel η is a constant in subcritical flow and is a function of F_o in supercritical flows. The variation of the end depth ratio η with F_o for various values of a obtained as detailed above is shown in Fig.2.

3.4   Results:

Table 2 shows a comparison of the available experimental data on subcritical flows with those obtained analytically by using Eq.(13). It can be noted that the results obtained in the present study are for a frictionless channel. These values are seen to be generally less by about 5% of the mean experimental values. The difference is due to the neglect of friction and other assumptions. Considering the extent of variation of the experimental values (Table 1) the prediction of η by Eq.(13) is considered to be satisfactory.

Detailed experimental data are not available for the verification of the variation of η with F_o completely. Computations were done by using the reported values of η for a triangular channel (Ref. 3) to present the data in the form of η vs F_o. These data had considerable scatter often as much as 12%. A comparison of the mean curve of experimental data with those obtained analytically by the use of Eq.(13) is shown in Table 3. It can be seen that the results are once again satisfactory. The trend of variation of η with F_o is predicted very well and the present analytical method generally overestimates the value of η . Similar results were obtained when certain experimental data on rectangular channel (Ref. 6) were compared.

3.6   Estimation of Discharge:

One of the main uses of the end depth is in flow measuring. Relationships for the estimation of discharge for known end depth in an exponential channel under subcritical as well as supercritical flow conditions is presented below.

For subcritical flows in exponential channels, it was shown earlier that $y_e/y_c = $ a constant. Denoting the constant of proportionality = C,

$$y_e = C \, y_c \tag{22}$$

Expressing y_c in terms of K and a, Eq.(22) leads to

$$Q = K\sqrt{g/a} \; C^{a+0.5} \; y_e^{a+0.5} \tag{23}$$

Values of C for different a values are as in Table 2.

In supercritical flows, however, the end depth ratio has been shown earlier to be a function of Froude number, i.e.

$$y_e/y_c = f(F_o) \tag{24}$$

Since by Eq.(15) for an exponential channel

$$y_o/y_c = 1/F_o^{2/(2a+1)}$$

Values of $\delta = y_e/y_o$ for various values of F_o in different channels are calculated by using Fig.(2) and Eq.(15) and is shown in Fig.(3).

Noting that $F_o = \dfrac{Q}{\sqrt{g \, A_o^3/T_o}} = \dfrac{Q}{\sqrt{g(K^2/a) \, y_o^{(2a+1)}}}$

Fig.3 can be used for quick estimation of the discharge for known y_e and y_o.

It is thus seen that while only one depth measurement viz., the end depth measurement is enough to estimate the discharge in subcritical flows in exponential channels, two depths viz., the normal depth y_o and end depth y_e are needed for the supercritical flow discharge estimation. The normal depth y_o can either be directly measured or estimated from other measurements and uniform flow equations.

4. CONCLUSIONS

Using the continuity of the water surface profile at the brink section and the Boussinesq equation for the specific energy, a general equation has been developed for estimating the end depth ratio. This equation simplifies considerably for an exponential channel. The role of the states of flow viz. subcritical and supercritical flows are identified and the end depth ratios for three channel shapes are obtained analytically. It is shown that for a given exponential channel the end depth ratio is a constant in subcritical flow and is a function of the normal flow Froude number in the supercritical flow. Comparison of the results of analytical prediction with the available experimental data indicate satisfactory agreement. Thus the present method which does not involve any experimental coefficient affords a convenient analytical method for predicting the end depth in an exponential channel for both subcritical and supercritical flow states. Further, a procedure for the estimation of the discharge by knowing the end depth in exponential channels is indicated.

5. REFERENCES:

1. Rouse, H:"Discharge Characteristics of the Free Overfall", Civil Engineering, ASCE, April 1936, pp. 257-260.

2. Boss, M.G., (Ed): "Discharge Measurement Structures". Int. Inst. for Land Reclamation and Improvement, Wageningen, The Netherlands, Pub No. 20, 1976.

3. Rajaratnam, N. and Muralidhar, D.: "End Depth for Exponential Channels", J. of Irr. and Dri. Div., Proc. ASCE, March 1964, pp.17-36.

4. Rajaratnam, N. and Muralidhar, D.: "End Depth for Circular Channels", J. of Hyd. Div., Proc. ASCE, March 1964, pp.99-119.

5. Rajaratnam, N. and Muralidhar, D.: "The Trapezoidal Free Overfall", J. of Hyd. Resch, IAHR, Vol.8, No.4, 1970, pp.419-447.

6. Rajaratnam, N. and Muralidhar, D.: "Characteristics of a Rectangular Free Overfall", J. of Hyd. Resch., IAHR, Vol.6, No.3, 1968, pp.233-258.

7. Diskin, M.H.: "The End Depth at a Drop in Trapezoidal Channels", J. of Hyd. Div., Proc. ASCE, July 1961, pp.11-32.

8. Delleur, J.W. et al: "Influence of Slope and Roughness on the Free Overfall", J. of Hyd. Div., Proc. ASCE, Aug. 1956, pp.30-35.

9. Christodoulou, G.C. et al: "Factors affecting Brink Depth in Rectangular Overfall", Proc. 1st Int. Conf. on 'Channels and Channel Control Structures, Southampton, England, April 1984, pp.1-3 - 1-17.

10. Anderson, M.V.: "Non-uniform flow in front of a Free Overfall", Acta Polytechnica Scandinavia, Ci 42, Copenhagen, 1967, pp.1-24.

11. Subramanya, K.: "Flow in Open-channels", Vol. II, Tata McGraw-Hill, New Delhi, 1982.

Table 1 :   Experimental Results on End Depth in Subcritical Flow in
            Horizontal Channels

No.	Shape	Ref.No.	y_e/y_c	Variation (Approx. percent)
1	Rectangular	(2)	0.715	$\pm$ 2.0
		(3,6)	0.715	$\pm$ 3.5
2.	Parabolic	(3)	0.772	$\pm$ 5.0
3.	Triangular	(3)	0.795	$\pm$ 2.5
4.	Circular	(4)	0.725	$\pm$ 3.5
5.	Trapezoidal	(5,7)	$f(my_c/B)$	

Table 2:   Comparison of the End Depth Ratios in Subcritical Flows
           Obtained by Eq.(12)

a	Channel shape	Mean experimental value of η (Table 1)	η by Eq.(12)	Percent under esti-mation
1.0	Rectangular	0.715	0.694	2.9
1.5	Parabola	0.772	0.735	4.8
2.0	Triangle	0.795	0.762	4.2

Table 3 :   Comparison of values by Eq. (13) for Supercritical Flow
            in Triangular Channels

S_o/S_c	F_o	η by eq.(13)	η by experiment (Ref. 3)	Percent overesti-mation above mean Expt. Value
2	1.384	0.755	0.73 $\pm$ 3%	3.4
3	1.673	0.726	0.70 $\pm$ 7%	3.7
4	1.915	0.700	0.67 $\pm$ 3%	4.5
5	2.126	0.684	0.636$\pm$ 5%	7.5
6	2.316	0.675	0.62 $\pm$ 6%	8.9
8	2.650	0.655	0.60 $\pm$ 6%	9.2
10	2.942	0.625	0.58 $\pm$ 12%	7.8

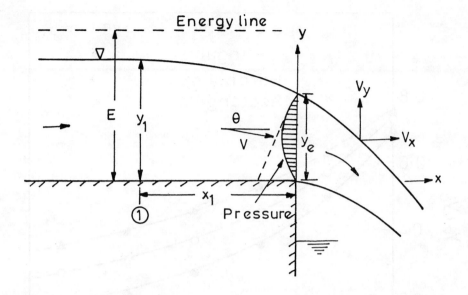

Fig.1   Definition sketch of flow at a free overfall.

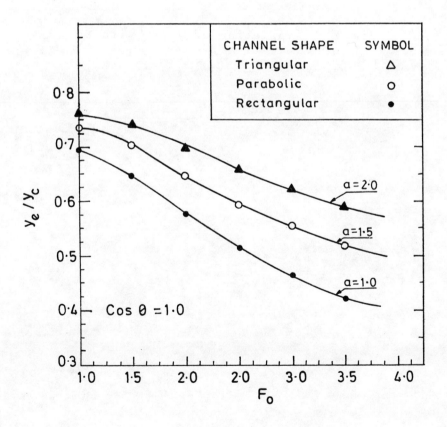

Fig.2   Variation of end depth ratio in supercritical flows.

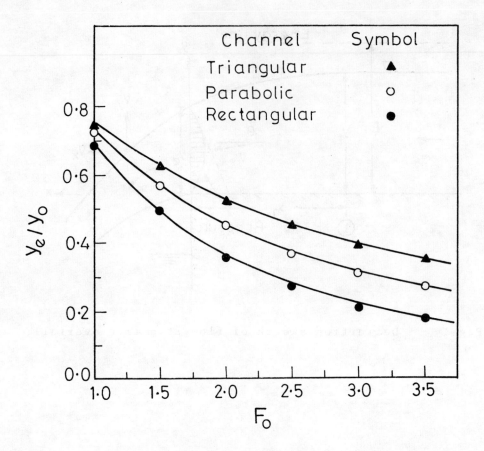

Fig.3   Variation of y_e/y_o with F_o.

International Conference on

Measuring Techniques
of Hydraulics Phenomena in offshore, Coastal & Inland Waters
London, England: 9-11 April, 1986

Monitoring cohesive sediment transport in estuaries

T N Burt, B.Sc, J R Stevenson, B.Sc

Hydraulics Research Limited
Wallingford
Oxfordshire
OX10 8BA
England

SUMMARY

A good knowledge of the cohesive sediment transport regime of an estuary is
fundamental to making sensible predictions of siltation rates in harbours, at jetty
berths and in dredged channels. It also has considerable bearing on determining the
fate of heavy metal pollutants which attach themselves to the sediment. Yet in many
cases available data on this aspect is very sparse, often being the random sampling by
an unspecified method at an unrecorded depth and without regard to tidal cycle and
seasonal variations.

The paper uses case studies to demonstrate the need for high frequency and carefully
located monitoring. A method of achieving this using optical monitors and data
loggers is presented. The problems of calibration are acknowledged and discussed.
Some principles of interpretation of the results are outlined again using case studies
as examples.

The paper concludes that there is no substitute for long term monitoring if long term
siltation predictions are required but that much can be learned from detailed
observations during a single tide.

Held at Imperial College of Science and Technology, London. Organised and sponsored by BHRA, The
Fluid Engineering Centre. Co-sponsored by the American Society of Civil Engineers and the
International Association for Hydraulic Research.

1.   INTRODUCTION
A good knowledge of the cohesive sediment transport regime of an estuary is
fundamental to making sensible predictions of siltation rates in harbours, at jetty
berths and in dredged channels. The accuracy of such predictions becomes more and
more important with increasing dredging costs; significant sums of money can be saved
if such information is readily available, particularly when large estuarine projects
are being planned. Sediment transport rate also has considerable bearing on
determining the fate of heavy metal pollutants attached to the suspended material.
This aspect too, is becoming increasingly important as we begin to take a greater
interest in preserving our environment.

Despite the evident importance of good measurements, in many applications the
attention given to the collection of meaningful data is far from sufficient. Civil
engineers, water authorities and consulting engineers sometimes rely on random
sampling by unspecified methods at an unrecorded depth with little regard to tidal
cycle and seasonal variations.

A comprehensive survey of methods of measurement is given by McCave in Ref 1. These
are:-
    i)   water sampling and filtration. The problem with this method is the large
         quantity of samples to be analysed and the consequent high cost.

    ii) optical methods which offer convenient and relatively cheap means of
         continuous measurement.

    iii) gamma ray measurements which work on the same principle as optical systems
         but are appropriate only to very high concentrations ($>$150 000 g/m^3).

    iv) remote sensing. The problems of interpretation for these techniques have
         not yet been solved.

The authors have found that an optical system usually provides the information
required most efficiently and economically.

McCave points out that there have been few studies which have made detailed analyses
of the temporal variations in suspended solids concentration and it is hoped that the
authors' experience in many UK estuaries will make a useful contribution in focusing
attention on the need for a rational approach to silt monitoring in estuaries.

2.   MEASUREMENT CRITERIA
There are several criteria to consider when a knowledge of suspended sediment
concentrations in an estuary is required, whether it be for a fundamental research
project or to answer specific questions related to a siltation problem.

2.1 Time variations
One of the most important decisions to make is the frequency at which measurements
should be made. The movement of sediment in a tidal situation is very complex. At
any one time in a tidal cycle the concentration at a point may depend on the addition
of (i) material eroded locally and diffused upwards by turbulence and (ii) material
already in suspension being transported laterally by the current.

To investigate first of all whether short term fluctuations in suspended sediment
concentrations were significant a field exercise was carried out by HR in the Thames
in 1974 (Ref 2). Every half hour, for 10 hours, 20 instantaneous optical turbidity
measurements were taken at 3 second intervals and later converted to suspended solids
concentrations. The band of results obtained is shown in Fig 1. It is evident from
this that the greater fluctuations occur at the higher concentrations. The average
variation expressed as a percentage of the mean value is $\pm$ 22%. Therefore an
instantaneous reading could be about 20% different from a value averaged over 1
minute. These very short term variations are evidence of clouds of sediment passing
the observation position.

We have also seen from measurements of suspended solids concentrations through the
tidal cycle in the River Thames (Ref 2) that concentration peaks and troughs can come

and go in periods of well under 1 hour. These events can be quite independent of velocity variations; the trough, 40% into the ebb in Fig 2, for example, occurs whilst the current velocity is still rising. Nevertheless, in this case there is a general trend which does relate to water velocities. Low values are recorded around the times of slack water as sediment settles towards the estuary bed and high values as the current velocities accelerate.

Water velocity which varies throughout the tidal cycle also varies with tidal range, the highest velocities coming during the fortnightly spring tides. Generally, tidally averaged suspended solids concentrations will be greatest during the highest range tides. This is illustrated in Fig 3 which compares mean suspended solids concentration with tidal range. The data is from Avonmouth in the Severn (Ref 3) and is a summary of 1 year's measurements. The concentration axis in the figure represents the mean of the tidal mean concentrations for tides of similar range. Moving to an even larger time scale we would expect to see some seasonal variations in suspended solids concentrations and Fig 4 demonstrates such a variation with fresh water flow in data from Gravesend Reach in the River Thames (Ref 4).

Thus it is apparent that the concentration of sediment in an estuary can vary with time ranging from seconds to the long term seasonal fluctuations.

2.2 Site variations

In most estuaries the suspended solids concentration varies through the depth with more sediment moving close to the bed than at the surface. Fig 5 shows a comparison of measurements at 0.6 and 5.5m above the bed at the southern end of Long Reach in the Thames. The very high early ebb peak close to the bed is not seen at all at the higher level (Ref 2).

Large variations can also occur between different positions across a river. A comparison between 3 positions across the River Thames at Halfway Reach (Ref 2) showed that although the general shape of the suspended solids time characteristic throughout a tidal cycle was similar at each of the sites, the magnitude did vary significantly (Fig 6).

A good example of the misleading effects of variations due to height above bed and location across the channel occurred in the River Parrett, Somerset. Here, values of suspended solids concentration 0.6 metres above the bed at the river's edge showed high concentration peaks occurring during the flood and ebb tides and, in between, a period of almost zero concentrations at slack high water. Mid channel measurements at the same height above the bed showed a large peak at slack water, Fig 7. Higher up in the water column (2.5m above the bed) concentrations did decrease. So it was reasoned that the decrease at slack water at the bank was due to suspended material falling below that level and concentrating near the bed in the deeper water in the channel.

The above examples illustrate that, when a knowledge of suspended solids moving in a river estuary is required, careful consideration should be given to the frequency and location of measurements.

3. MEASURING SYSTEM

A monitoring programme should be designed to meet the particular needs of the job in hand. A small amount of data collected over one or more tidal cycles can be useful and may indeed be entirely satisfactory for some projects. For others, however, data may be required to cover the full range of tidal and seasonal variations. For example, for accurate siltation predictions or with regard to siting a cooling water intake. In this case practical and financial considerations almost invariably mean that the data collection system must be automatic and require as little day-to-day maintenance as possible. This effectively rules out sampling the water and requires some sort of direct reading transducer.

In most cases, HR has used Partech optical suspended solids monitors, consisting of a simple white light source and photocell (Ref 5). HR have also developed an infra-red system which is used for rapid profiling of the water column but not as yet for unattended operations (Ref 6).

The optical suspended solids transducer operates on the photo-extinction principle. The sensing head has a light source, a gap through which water passes and a photo cell which measures the amount of light passing through the gap. In some applications the sensor may have two gaps of unequal length and two photo cells to compensate for uniform clouding of lenses and ambient diffused light.

It should be stressed that the optical instrument does not directly measure suspended solids concentration but rather turbidity which is a measure of the amount of light which passes through the suspension in the gap. Two suspensions having the same turbidity do not necessarily have the same concentration of suspended solids, due to differing light diffusing properties. The principal factor is particle size.

In laboratory tests at HR (Ref 7) it was discovered that in a silt and sand mixture the Partech system almost totally ignored the sand content (Fig 8). However, in a silt environment the use of optical suspended solids monitors can yield very good results provided care is taken to use proper calibration procedures.

If measurements of sand concentrations are required an optical system is not satisfactory and resort must be made to regular sampling of the water and subsequent laboratory dry weight analysis. A possible alternative has been developed by the Institute of Oceanographic Services, Taunton, in the form of an instrument which measures the concentration of suspended sand movement as a function of the impact of the moving grains on a transducer. This has been used in the field and shown to give reliable results (Ref 8).

Calibration of optical instruments can be effected using either suspensions of the indigenous material of predetermined concentration or by using standard turbidity suspensions such as Formazin. The latter was originally developed by the brewing industry for quality control in their vats. If Formazin is used then it is still necessary to obtain samples of the water being monitored for laboratory dry weight analysis in order to get a factor relating the actual concentration to the equivalent formazin concentration. This factor may or may not be linear with concentration.

4. MONITORING

The information measured at the transducer then has to be relayed onto some form of data logger. If high frequency measurements are required over a short time period a chart recorder is useful as it records continuously. However, this form of output is difficult to analyse in detail and HR have normally opted for a magnetic tape storage system. It has been illustrated that the frequency at which measurements are taken is very important, the shear bulk of data produced is also an important consideration from a financial point of view. After carefully weighing these two factors, HR decided that, for the original monitoring project in the Thames, a 15 minute time interval would be suitable. It was accepted that this would lead to some loss of definition because of shorter term fluctuations but in the context of indentification of tidal cycle and seasonal trends average values over a tide are relevant. These average values in practice were largely unaffected by fluctuations of less than 15 minute period.

Perhaps a better approach now, with the advent of more sophisticated logging techniques would be to record a time averaged value every 15 minutes.

5. SITE SELECTION

Apart from careful choice of recording interval and meticulous calibration procedures serious consideration also has to be given to the location of the instruments. The obvious choice for much short term survey work is an anchored vessel in the main flow of an estuary, but in situations where variations in suspended solids concentrations caused by tidal range or seasonal fluctautions are required a jetty mounted monitor is more practical. The examples above showed that measurements taken from arbitrarily chosen sites could yield completely uncharacteristic, and in some cases, meaningless results. Therefore, great care must be taken in the choice of such sites and any limitations in their usefulness understood and allowed for before conclusions are drawn.

Within these limitations a well chose jetty site can yield good representative data. Characteristics of a good site are:-

i)   river flow at the site is outside the influence of adjacent structures and is typical of the river channel

ii)   a convenient site for the logger is available

iii)   there is sufficient water at the site at low water to avoid drying out

iv)   access is available throughout the tide

v)   reasonable site security is provided against theft, damage or vandalism

The first of these conditions is most critical unless knowledge about a very small site close to a jetty is specifically required. There must be a convenient site for the logger (condition ii) but 'convenience' can vary from suspended underneath an unsafe road bridge in Portland (Ref 9) to comfortably situated in a centrally heated office on Tate and Lyles' jetty on the Thames (Ref 10). Sites which meet all of the requirements are rare and in many instances second best has to be accepted.

If more than one site can be used in an estuary then careful selection of the relative positions can enhance the value of the data considerably. Suspended solids concentrations recorded at two sites within a tidal excursion can provide useful supplementary data. This advantage is lost if the sites are too close together.

6.   INTERPRETATION
The measurements in the Thames Estuary in the early 1970's soon produced large amounts of data, and serious thought had to be given to a sensible and economic method of analysis (Ref 2).

Visual inspection of the data showed that the form of the graph of suspended solids concentrations against time for each tidal cycle was fairly constant and characteristic for a given site. Even profiles recorded at sites known to have an in-erodible bed displayed these repeating peaks of suspended concentration and the peaks were not necessarily in phase with velocity peaks. This lead to the hypothesis that at least some of the sediment observed at a particular location represented material from a remote source being transported by the tide past the recording position.

A good example of this is from the River Crouch in Essex, where HR monitored suspended solids concentrations for 4 months in 1974. The concentration rarely exceeded 200 g/m^3 (Ref 5) and the tidal cycle profile showed peaks of concentration occurring only around HW (Fig 9). This shows material arriving from, and beginning its journey back to, a remote source at the other end of the tidal excursion. The existance of the source was confirmed by bed samples.

Although the patterns were seen to be tidally repetitive the magnitude of the concentration peaks did vary with tidal range. These two facts lead to the following method of analysis. For each period of data collection, which was normally one or two spring-neap-spring tidal cycles, the values of suspended solids concentration for each cycle were normalised by dividing each by the mean concentration for that tide. The time basis was also standardised so that each tide was separated into the same number of sub-divisions. A single graph of normalised concentration against time for each period of record was then produced by combining the normalised records and averaging. A further graph was produced of mean concentration for a tidal cycle against tidal range. Computer programs were written to facilitate this method of analysis and are documented in Ref 5.

Both of these graphs, summarising a single recording period, were observed to vary between different periods. To simplify the comparison between mean concentrations from one period to the next a single value representing the whole of the record was calculated. Plots of this value against time then gave a measure of seasonal variation. Using this in conjunction with the normalised concentration time profiles the relative height of peaks in the latter indicate how much material is coming from a particular source.

Thus, careful interpretation of long term silt monitor data, particularly from 2 or more sites, in conjunction with relevant velocity data, can be used to build up an overall picture of sediment sources and 'clean' areas in an estuarine environment.

7. CONCLUSIONS

This summary of silt monitoring techniques used and developed by HR over a period of about 15 years shows that there is no substitute for long term monitoring if long term siltation predictions are required or if the results of long term engineering projects are to be detected. At the other extreme there is often little or no value in taking arbitrary samples in an undocumented way, and resources of time and money may be wasted in this way. However, detailed observations during a single tide at a well chosen location can be of great value and in some cases sufficient to satisfy the requirements of a particular project.

8. ACKNOWLEDGEMENTS

This paper draws on experience gained from a large number of field studies carried out over a 15 year period. The data was made available from Hydraulics Research records and reviewed under Contract Number PECD 7/6/59, funded by the Department of the Environment. The DoE nominated officer was Dr R P Thorogood. The work was carried out by Mr T N Burt and Mr J R Stevenson on behalf of Mr M F C Thorn, Head of the Tidal Engineering Department at Hydraulics Research Limited.

9. REFERENCES

1. DYER K R (Editor) Estuarine hydrography and sedimentation Cambridge University Press, 1979.

2. THORN M F C and BURT T N. The silt regime of the Thames Estuary. HR Report IT 175, January 1978.

3. HYDRAULICS RESEARCH. The Severn Estuary. Silt monitoring, April 1980 – March 1981. HR Report EX 995, August 1981.

4. HYDRAULICS RESEARCH. Thames flood prevention investigation. Analysis of suspended solids regime prior to barrier construction. HR Report EX 934, June 1980.

5. THORN M F C and BURT T N. Transport of suspended sediment in the tidal River Crouch. HR Report IT 148, November 1975.

6. SHEPHERD I E. A silt, salinity, depth profiling instrument. Journal of the Institute of Measurement and Control. Vol 12, No 5, May 1975.

7. HYDRAULICS RESEARCH. Partech turbidity monitors. OD/TN 1, September 1985.

8. SALKIELD A P, LE GOOD G P and SOULSBY R L. An impact sensor for measuring suspended sand concentration. Conference on Electronics for Ocean Technology, Birmingham, September 1981. The Institution of Electronic and Radio Engineers.

9. HYDRAULICS RESEARCH. Ferry Bridge, Weymouth. Portland road A 354. Effects on hydraulic regime of resiting the bridge. HR Report EX 1039, January 1982.

10. HYDRAULICS RESEARCH. Thames flood prevention investigations. Continuous silt monitoring 1970 – 1973. HR Report EX 671, July 1984.

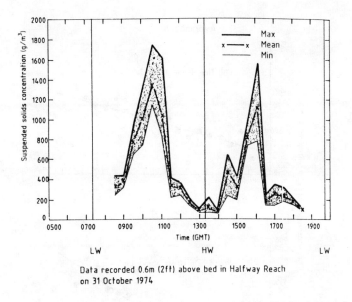

Data recorded 0.6m (2ft) above bed in Halfway Reach
on 31 October 1974

Fig.1   Short term variations in suspended solids concentration

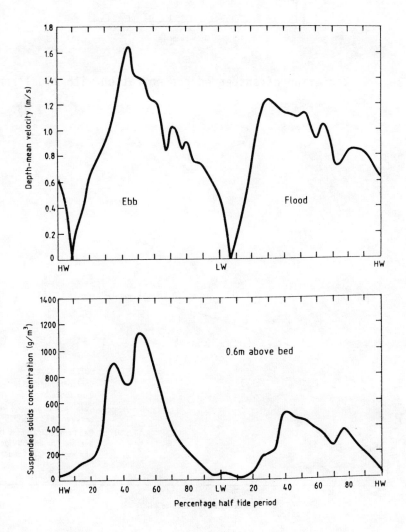

Fig.2   Variation of suspended solids concentration with velocity

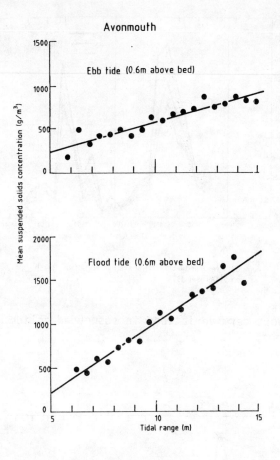

Fig.3   Variation of suspended concentration with tidal range

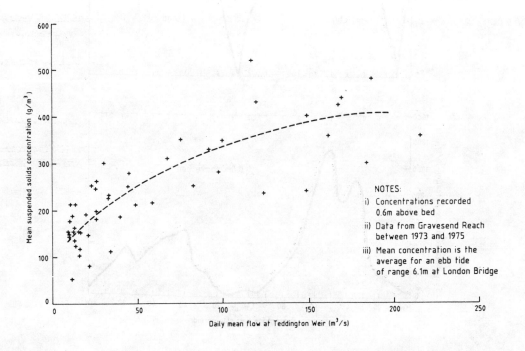

Fig.4   Variation of suspended solids concentration with freshwater flow

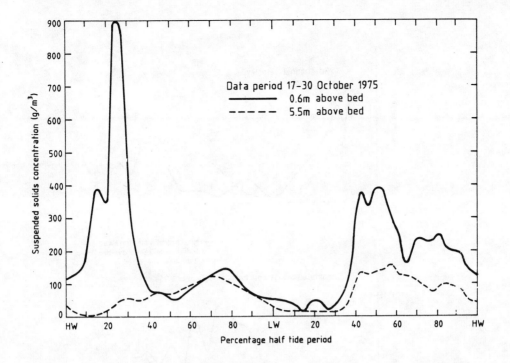

Fig.5   Variation of suspended solids concentration with height above bed

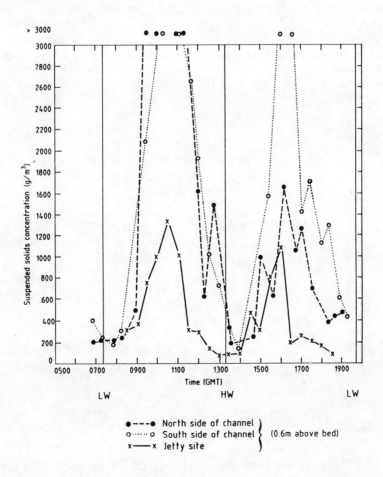

Fig.6   Cross-river variation of suspended solids concentration

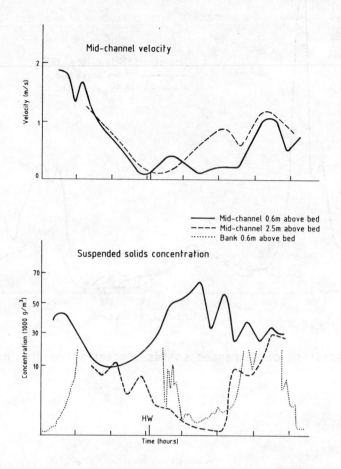

Fig.7   Variation of suspended solids concentration with height above bed and position across river

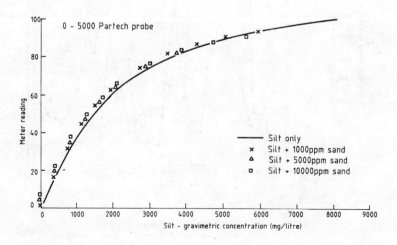

Fig.8   Effect of sand on the calibration of an optical silt monitor

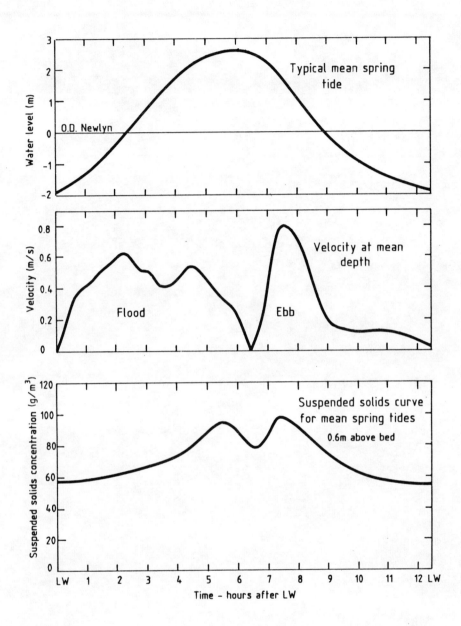

Fig.9   Suspended sediment from a remote source

International Conference on

Measuring Techniques

of Hydraulics Phenomena in offshore, Coastal & Inland Waters

London, England: 9-11 April, 1986

PAPER H2

EVALUATION OF ESTUARINE MIXING MECHANISMS
USING FLUORIMETRIC TECHNIQUES.

I. Guymer,BSc.,DipIS.,PhD.*
J.R. West,BSc.,PhD.,MICE,MIWES.**

* Dept. of Civil Engineering, Heriot-Watt University,
Edinburgh, EH14 4AS.

** Dept. of Civil Engineering, University of Birmingham,
Birmingham, B15 2TT.

Summary

The further development of mathematical models describing estuarine mixing mechanisms is at present limited by an incomplete understanding of the physical processes involved. Careful application of fluorimetric techniques using synchronous measurements at three levels in the flow to monitor a continuous surface point source injection, combining with detailed synoptic velocity and density measurements at up to 24 points throughout a cross-section and an E.D.M. position fixing technique, have helped to elucidate the flow structure and mixing processes.

Field measurements have been performed in a narrow reach of the Conwy estuary, North Wales, in the presence of a longitudinal salinity gradient, where the techniques have been successfully employed to detect transverse shear induced secondary flow effects, vertical density effects and both the short and longer term stochastic and temporal nature of the flow. Estimates of the diffusion coefficients have been made and are shown to be dependent on flow direction.

Held at Imperial College of Science and Technology, London. Organised and sponsored by BHRA, The Fluid Engineering Centre. Co-sponsored by the American Society of Civil Engineers and the International Association for Hydraulic Research.
©BHRA, The Fluid Engineering Centre, Cranfield, Bedford MK43 0AJ, England 1986.

<u>NOMENCLATURE</u>

$$
\begin{array}{ll}
c & = \text{solute concentration} \\
s & = \text{salinity} \\
t & = \text{time} \\
u, v, w & = \text{turbulent mean velocity components relating to } x, y, z \\
& \quad \text{directions in a Cartesian coordinate system} \\
x, y, z & = \text{longitudinal, transverse \& vertical coordinate directions} \\
& \quad \text{in a Cartesian coordinate system} \\
x_c & = \text{injection rate} \\
\varepsilon_{sx}, \varepsilon_{sy}, \varepsilon_{sz} & = \text{turbulent transport coefficients relating to } x, y, z \\
& \quad \text{directions in a Cartesian coordinate system} \\
\sigma_m^2 & = \text{mean transverse variance of twenty five traverses} \\
\sigma_n^2 & = \text{normalised transverse variance} \\
\sigma_y^2 & = \text{individual value of transverse variance} \\
f_A & = \text{area mean value of } f \\
f_d & = \text{depth mean value of } f
\end{array}
$$

1. INTRODUCTION

A knowledge of the factors which contribute to the overall transport of momentum, solutes and suspended solids within an estuary is important in several aspects of engineering. In particular the transport of solutes in the tidal reaches of estuaries is of interest in determining factors such as the limit of saline intrusion,needed to site water abstraction schemes and the mechanisms affecting pollution transport,needed for the sucessful management of water quality.The transport of these solutes is dependent on the complex interaction of parameters such as : tidal range, river flow, channel geometry, bed topography, position in the estuary, suspended solids concentration, density variations and bed material composition.

Due to scaling problems it is rarely possible to accurately physically model the transport processes of an estuarine environment. Consequently mathematical models have gained a wide acceptance as a tool for studying the flow field and solute mixing in estuaries. These mathematical models require a knowledge of the relative importance of various mixing mechanisms before the governing equations can be formulated in a computationally efficient form for a given study. Predictive formulae are then required for the appropriate forms of the diffusion and dispersion coefficients in the solute mass balance equation. For estuaries both the transport mechanisms and the mixing coefficients are poorly understood at present, but recent advances in field observation techniques should lead to significant advances in the understanding of these phenomena.

This paper describes the application of a dye tracing technique to the Tal-y-Cafn reach of the Conwy estuary in North Wales and gives examples of the results obtained. The technique has proved successful for studying the effects of the interaction of fluid shear and longitudinal solute induced density gradients on the turbulent mean flow field and for investigating the spreading effects of bed generated turbulence in the vertical and transverse directions.

2. SOLUTE TRANSPORT

In turbulent open channel flow, the transport of a solute may be given by

$$\frac{\partial c}{\partial t} + u\frac{\partial c}{\partial x} + v\frac{\partial c}{\partial y} + w\frac{\partial c}{\partial z} = \frac{\partial}{\partial x}(\varepsilon_{sx}\frac{\partial c}{\partial x}) + \frac{\partial}{\partial y}(\varepsilon_{sy}\frac{\partial c}{\partial y}) + \frac{\partial}{\partial z}(\varepsilon_{sz}\frac{\partial c}{\partial z}) \qquad (1)$$

where c = solute concentration,
         t = time,
    x,y,z = longitudinal, transverse & vertical coordinate
            direction in a Cartesian coordinate system,
with u,v,w = turbulent mean velocity components and
   $\varepsilon_{sx}, \varepsilon_{sy}, \varepsilon_{sz}$ = turbulent diffusion coefficients respectively
corresponding to the three coordinate directions. Mathematical models used to investigate solute transport require the solution to this equation or a simple derivative in conjunction with the momentum transport equation (Ref.1,p.35).

The solution of equation (1) or a one or two dimensional spatially averaged form of it, is complicated by density and bend induced secondary flows which can lead to the terms $v\frac{\partial c}{\partial y}$ and $w\frac{\partial c}{\partial z}$ being significant. Also the diffusion coefficients, ε_{sy} & ε_{sz} are affected by the influence of acceleration and vertical density gradients on the structure of the turbulent eddies and their capacity to transport momentum, solute and suspended solids.

The field investigation of the turbulent mean and the turbulence fields may be undertaken by Eulerian or Lagrangian measurements. The former at a fixed point, offers the advantage of investigating the temporal variation of the vertical and transverse structure of the flow with fairly good spatial resolution. A limitation is that the flow field can be influenced by velocity and density fields generated up to several kilometres upstream of the measurement point.

Thus a Lagrangian approach can complement fixed station data. Drogue studies are useful, but very time consuming if a range of depths are to be investigated or ensemble means are to be obtained. Aerial photographs and satellite images can provide very useful information on the evaluation of plumes but are limited to the surface plane. The plumes generated by tracers injected continuously from a fixed point source are able to add two dimensional (f(y,z)) information on the integrated effects of secondary flow and turbulence between the injection and measuring points which cannot realistically be obtained by other methods. The method is also useful for investigating some aspects of turbulence in highly stratified flows where fluctuations are too small for most transducers to measure or where the flow is too deep or fast flowing for a bed frame to be safely deployed.

3. FIELD TECHNIQUE

Rhodamine dye diluted to approximately 5 g/l was injected at a constant rate of about 1 x 10-3 l/s onto the surface of the flow from an unmanned 3.5 m inflatable boat using a 240v A.C. Watson Marlow peristaltic pump powered from two 12v accumulators through an inverter. The pulsating effect of the pump was removed by running it at full speed and tapering the 3m length of 3 mm I.D. tube to a 1 mm outlet using a brass nozzle. The rapid mixing and spreading of the dye plume over the first few metres of travel indicated the absence of any adverse effects caused by the density difference between the injected and ambient fluids.

The concentration of the dye plume at three different levels in the flow as it mixed throughout the test reach was logged on chart recorders connected to three Series 10 Turner Designs Fluorimeters. This model fluorimeter may operate in two modes, either discrete or continuous sampling and has a total range selection in excess of 1:3000. It has four ranges which may be selected either manually or automatically, with an additional range multiplier of x100, which may only be selected manually.For all the tests described in this paper the fluorimeters were operated in the continuous mode to allow the varying dye concentrations to be recorded.

High density 35 mm I.D. P.V.C. tubing was used for all the pipes throughout these experiments to minimise dye absorbtion to the inner pipe walls. This tubing was connected to both the inlet and outlet fittings of the instruments with the first metre of tubing blacked out to prevent external light reaching the sample . The inlet pipes were attached to perspex inlet nozzles suspended at different levels from the bows of the tracking boat. While the outlet pipes from the fluorimeters were connected to 12v Jabsco Water Puppy pumps so that a stream of fluid containing dye could be passed through each instrument. After the concentration of dye had been measured the samples were pumped through orifice meters , to measure the flow rate and then discharged over the stern of the tracking boat. All equpiment was powered from 12v accumulators.

During the initial laboratory tests of the fluorimeters it was found that the response time of the instruments was quite slow. It took approximately 4 seconds to register around 90% of the final reading, which previous studies (Ref.2) had shown could cause serious attenuation of the peak concentration value unless the dye plumes were traversed at inconveniently slow speeds. After consultation with the manufacturers a minor modification was performed which produced a significant improvement in the response time. A better than 90% reading was achieved in less than 0.5 second at the expense of only a slight increase in background noise.

The auto-ranging facility available on the Series 10 Turner Designs Fluorimeter could take over 15 seconds to select the correct range. Even after adjustment to the rapid response time more than 2 seconds were required. If used in the field this could cause the loss of significant data ,thus on each traverse of the plume one scale was manually selected and retained for the full traverse to ensure that a complete record would be obtained on the chart rolls.

The position of the tracking boat as it traversed the dye plume was recorded from a shore station using a Zeiss Th2 theodolite and Geodimeter 112 with a horizontal array of six corner cube prisms. These prisms were mounted on a mast close to the bows of the tracking boat and were constantly directed towards the shore station. This allowed the polar coordinates, azimuth and distance, of the tracking boat to be recorded by the shore station approximately once every twenty seconds. Radio contact was maintained between the shore station and the tracking boat and marks were simultaneously made on each of the three chart rolls as a fix of the position was obtained.

Knowing the lengths of the inlet pipes to each fluorimeter, the flow rates through each instrument and the chart speed it was possible to apply a correction to the fixes on the chart rolls and hence locate the actual position of the dye plume relative to the shore station. This assumes that the tracking boat maintains a constant velocity between consecutive position fixes. An accuracy estimated at better than 1m was obtained.

To assist with the interpretation of the results from the tracer studies, during each of the experiments up to five stations were positioned transversely at the landward end of the test reach. These fixed stations recorded the depth variation of primary velocity and salinity at up to twelve positions over the total flow depth at intervals of twenty minutes.

At the three inner stations arrays of four Braystoke velocity meters were used to record the mean primary velocity during 100 seconds over a 1m depth, while at the two outer stations single meters were used. Each station performed three sets of velocity readings, at the surface, bed and mid-depth of flow. The vertical salinity variations were obtained by simultaneously pumping one litre samples over approximately one minute from various depths spaced throughout the flow. The inner stations sampled at six positions and the outer at three. These samples were stored and later the salinity determined using an MC5 salinometer. In addition a constant record of the water level in the test reach was maintained during each of the experiments.

4. RESULTS

In order to determine the relative magnitude of the stochastic variation in dye concentration and tidal variations and to determine a suitable number of traverses to define an ensemble mean concentration distribution, twenty five surface traverse profiles were measured on an ebb tide between 120 and 150 minutes after high water.

The actual distributions recorded during the experiment are presented in Fig. 1 where the individual values of the transverse variance, $\sigma_{\bar{y}}^2$ have been normalised ($(\sigma_{\bar{y}}^2 - \sigma_m^2)/\sigma_m^2$) with respect to the mean variance of the twenty five traverses, σ_m^2. These results show considerable variations from the mean with two distributions, traverse numbers 4 and 10, exhibiting a value of the normalised variance, σ_n^2 more than twice the mean value.

In an attempt to determine the optimum number of traverses required, calculations were performed to determine the variance and associated standard errors of an average distribution produced by considering running ensemble means of various sample sizes between three and nine. The individual variances together with those obtained for the extreme cases considered are reproduced in Fig. 2. Comparing the standard errors (Ref.3,p.704) produced from averaging successive variances shows similar magnitudes when either three or nine distributions are considered, Fig. 3.

The nature of these results shows that both a tidal variation, from 0.3 to -0.2 over approximately a 15 minute period during which the flow depth reduced by 0.25m and the velocity may be assumed constant and larger stochastic variations, probably mainly due to turbulence were detected. It appears that an ensemble mean of three or four traverses may be considered to be a reasonable compromise between minimising stochastic errors and defining the longitudinal variation of plume concentration over several cross sections in a period short enough for the tidal variations to be relatively small.

Fig. 4 presents velocity and salinity distributions recorded at the landward limit of the Tal-y-Cafn test reach 89 minutes before high water, at 14-46 hrs on 15 th July 1983. The overall distribution of salinity, increasing with depth below the surface and distance from the banks, together with vertical profiles of velocity exhibiting peak values below the surface, are consistent with the presence of a density induced twin vortex type secondary flow structure as postulated by Smith (Ref.4). This is further supported by surface velocity measurements performed on the same test reach by Nunes (Ref.5) and similar observations by West et al (Ref.6) on the Great Ouse estuary.

Results from fluorimetric dye studies performed between 14-19 and 14-48 hrs on the same tide are presented in Fig. 5. At section 1, nearest to the injection boat, the results from the upper sampling level, I, show the dye transported towards the west bank. This relative transverse movement of the dye plumes between different sampling levels is increased at the second section and suggests either bank effects or a secondary current as the cause. At the third section, nearest to the section recording the flow structure, more dye has been recorded at the lowest sampling level, suggesting that the vertical velocity, w is significant.

The indication that the dye distributions result from a twin vortex type secondary circulation are further supported by visual observations of the axial convergence line (Nunes,(Ref.5)) during the tests and the flow structure records presented in Fig. 4, which although not recorded at the exact time of the dye studies, do show the presence of the necessary structure.

The results from the fluorimetric tracer studies have been used to estimate values of the mixing coefficients (Guymer,(Ref.7)) and mean values for the transverse mixing coefficients of 270 cm^2/s and 840 cm^2/s for ebb and flood tides respectively were obtained.

5. CONCLUSIONS

The fluorimetric dye tracing technique has proved a useful method for elucidating the complex flow structure and dominant physical mechanisms found in estuarine flows. Improvements could be made to the method of recording the dye distributions on chart rolls, which gives a visual representation of the results and is advisable during the development stages however the use of data loggers and automatic position fixing equipment is suggested.

6. REFERENCES

1   McDowell,D.M. and O'Connor,B.A.: "Hydraulic Behaviour of Estuaries", London, MacMillan Press Ltd., 1977, pp.292.

2   Cotton,A.P.: "On Mixing Coefficients in an Urban Stream and a Tidal River", Ph.D. Thesis, University of Birmingham, England, 1978, pp.240.

3   Bajpai,A.C., Mustoe,L.R. and Walker,D.: "Engineering Mathematics", London, John Wiley & Sons, 1974, pp.793.

4   Smith,R.: "Longitudinal Dispersion of a Buoyant Contaminant in a Shallow Channel" J. Fluid Mech., 78, pt. 4, 1976, pp.677-688.

5   Nunes,R.A.: "The Dynamics of Small Scale Fronts in Estuaries", Ph.D. Thesis, University of Wales, 1982, pp.175.

6   West,J.R., Knight,D.W. & Shiono,K.: "A Note on Flow Structure in the Great Ouse Estuary", Est.,Coastal & Shelf Science, 19, 1984, pp.271-290.

7   Guymer,I.: "Some Aspects of Solute Transport Processes in the Conwy Estuary" Ph.D. Thesis, University of Birmingham, England, 1985, pp.167.

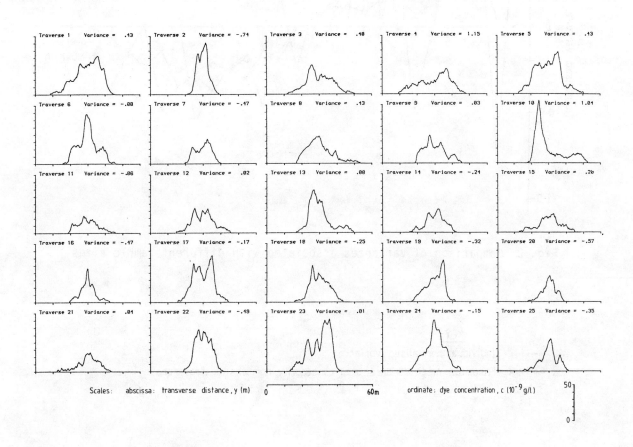

Fig. 1 Individual dye distributions for multiple traverse experiment

343

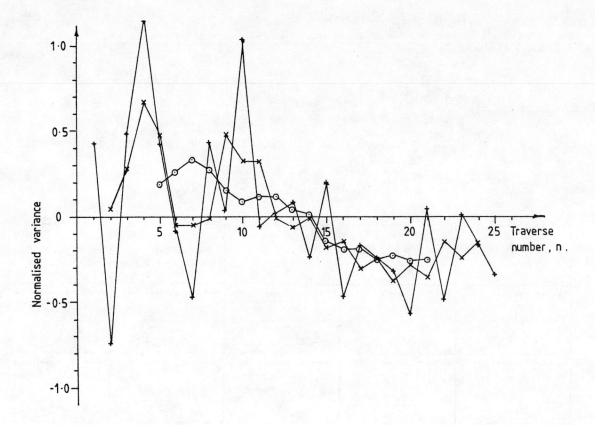

Fig. 2 Comparison of variances associated with different sample sizes

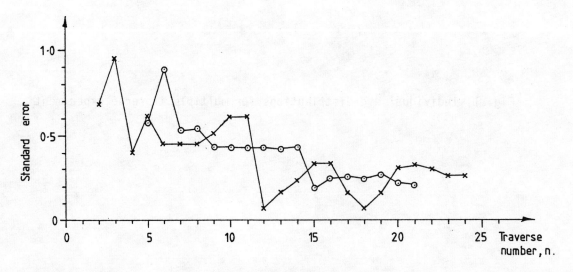

Fig. 3 Comparison of standard errors associated with different sample
        sizes

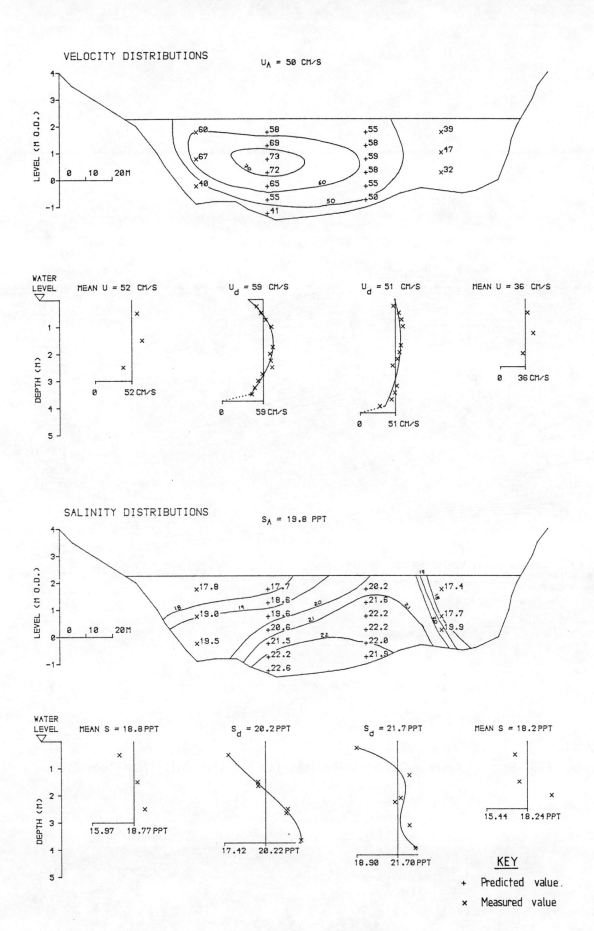

Fig. 4 Velocity and salinity distributions measured at 14-16 hrs on
15th July 1983 flood tide

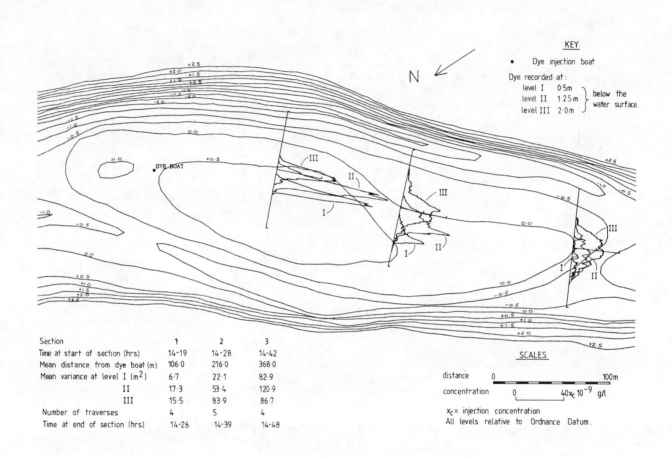

Section	1	2	3
Time at start of section (hrs)	14·19	14·28	14·42
Mean distance from dye boat (m)	106·0	216·0	368·0
Mean variance at level I (m²)	6·7	22·1	82·9
II	17·3	53·4	120·9
III	15·5	83·9	86·7
Number of traverses	4	5	4
Time at end of section (hrs)	14·26	14·39	14·48

SCALES

distance   0 — 100m

concentration   0 — $40 x_c \cdot 10^{-9}$ g/l

x_c = injection concentration
All levels relative to Ordnance Datum.

FIG. 5   Average Dye Distributions for the 15th July 1983 Flood Tide.

International Conference on

Measuring Techniques
of Hydraulics Phenomena in offshore, Coastal & Inland Waters
London, England: 9-11 April, 1986

PAPER H3

THE DEVELOPMENT OF A SILT MONITORING RIG

Dr. P. Tosswell (Research Fellow)

and

Mr. N.B.Webber (Senior Lecturer)

DEPARTMENT OF CIVIL ENGINEERING

UNIVERSITY OF SOUTHAMPTON

SUMMARY:

The experience of four years of field investigation of the
siltation charactertistics of Port Hamble Marina, in the
estuary of the River Hamble, indicated the need for a more
fundamental study of the relationship between the behaviour
of the water column and silt collecting at the bed in a
situation of quiescent settling. The development of a
fully instrumented monitoring rig was essential to such an
evaluation.

This rig has recently been installed at one of the marina berths
at Port Hamble where siltation is known to occur. It comprises
a base tripod supporting a vertical stainless steel spar on which
are mounted two electro-magnetic current meters and four suspended
solids sensors. The sensors operating in the infra-red, have
been specially designed to meet the requirements of low concentration measurement in conditions where ambient light is of
varying intensity. Tidal levels and salinity/temperature are
measured conventionally. Provision has also been made for the
collection of silt at various levels.

Information is recorded by a 20-channel Orion Delta data logger,
enabling the data to be processed during recording and ensuring
the identification of the bursts of suspended sediment activity
which were found to be a feature of the earlier research.

Held at Imperial College of Science and Technology, London. Organised and sponsored by BHRA, The
Fluid Engineering Centre. Co-sponsored by the American Society of Civil Engineers and the
International Association for Hydraulic Research.
©BHRA, The Fluid Engineering Centre, Cranfield, Bedford MK43 0AJ, England 1986.

1. INTRODUCTION

This paper describes the background to, and development of a silt monitoring device for deployment in a research project directed to investigating the settling process of cohesive sediments in an estuarine environment.

This current project stems from an earlier field investigation (Ref.1), commenced in 1979, which was aimed at establishing the siltation characteristics at Port Hamble Marina (location map:Fig.1). The earlier research entailed, inter alia, the establishment of a recording station on the outer periphery of the Marina, equipped with an electro-magnetic current meter and a suspended sediment monitor, the information being registered on chart paper in a hut ashore; water level variation was monitored by a pneumatic tide recorder installed nearby.

Continuous recording of these parameters was undertaken over a period just exceeding one year. The data was digitised and processed computationally with a view to establishing the factors conducive to siltation, such as tidal range, wave action and rainfall.

The principal findings (Ref.2) were as follows:-

(a) Suspended sediment concentrations were generally very low ($\leq$ 100 mg/l), but there were bursts of much higher activity, generally occurring during the flood tide, and thus implying an external source. Continuous monitoring was therefore essential if these peaks were to be identified. (see also Ref.3)

(b) In spite of exhaustive examination of all relevant data it was concluded that the siltation process was extremely complex, and that a number of sources, including those well outside the estuary were contributory.

It was recognised that this was a problem that was widely encountered in ports and harbours and the overriding conclusion, therefore, was that further research of a basic nature was justified. This should be aimed at establishing the correlation between the silt accumulating at the bed in a fairly quiescent area, such as marina or harbour berth, with the characteristics of the fluid flow above it, including the degree of turbulence. It would be necessary to concentrate scientific effort at one site of relatively small extent so that comprehensive knowledge of ambient conditions could be readily acquired.

The research project as envisaged was commenced in August 1984. Since it was the deposition behaviour that was of particular interest a site (Figs. 2 and 3) was selected where conditions were known to be conducive to siltation and where re-suspension would be minimal. An additional attraction was the recent dredging of the area so that more active deposition might be expected. Furthermore, there was an overhead beam which could be adapted as a lifting device to facilitate the frequent haul-outs so vital in an operation of this kind.

One of the first tasks was to develop the monitoring rig, the information from which would be registered by a data logger located in a hut nearby. It is this instrumentation aspect which is the topic of the present paper.

2. FRAME

The philosophy behind the design of the rig was that it was sufficiently versatile to be installed in any typical estuarine situation.

The main member consists of a vertical tubular spar (Fig.4) that functions as the main attachment point. This member is supported by a tripod constructed of scaffold tubes with base pods so as to prevent excessive settlement.

A 316 grade stainless steel has been used for all members except the base tripod. From previous experience, the authors have found this grade to be extremely resistant to corrosion by sea water. However, it is worth noting that most marine instruments are fabricated from stainless steel of the 321 grade, which is less expensive but of inferior quality.

As many of the existing oceanographic instruments still contain brass in their construction the authors also consider it necessary to provide sacrificial anodes, Zinc anodes are normally employed, the main proviso being that an electrical connection must be made between the anode and the metal it is protecting. Mechanical metal to metal contact has been shown to be satisfactory but care must be taken with plated brass units for which a direct connection to the brass is recommended.

The frame has been treated with an aluminium/bitumen primer and coated with a hard wearing marine anti-fouling paint.

To facilitate periodic inspections and calibration checks of the instrumentation a block and tackle gear is retained between the frame and the overhead beam.

3. GENERAL INSTRUMENTATION

It was required to record the physical parameters of tidal level, tidal current (magnitude, direction, and turbulence intensity), salinity, temperature, and most importantly suspended and deposited sediment.

Whilst, for the most part, instruments were found to be commercially available for these measurements, adaptations have in several cases been required. The more conventional instrumentation is outlined in this Section.

An exception has been in regard to suspended solids monitoring, where, because of the rather special circumstances of a low silt concentration and a need for high sensitivity and good resolution it was decided to develop a purpose-made instrument. There is a detailed account in the subsequent Section.

A most time-consuming aspect of the previous research was the laborious digitising of the chart records. The avoid this in the present investigation an Orion Delta data logger was acquired, an additional advantage of which is its ability to perform cross-channel computation and decisions. By processing the data during recording, with raw and processed data being logged, considerable time-saving in analysis may be achieved. Like many of the more sophisticated data loggers it has the potential for replacing part of the electronic circuitry in an instrument.

3.1 Tidal Level

In the previous study a Neyrpic pneumatic tide recorder was employed and found to be extremely reliable. This unit was installed in the conventional manner with the buffer chamber attached to the quay wall above the estuary bed. However, the chart recorder unit was replaced by a pressure transducer connected to the manometer in the unit, the pressure changes (indicating tidal state) being recorded as a voltage in the data logger. If desired, the chart recorder can be operated in tandem.

3.2 Tidal Current

For velocity measurements, two electro-magnetic current meters have been installed. Experience with this instrument in the previous investigation was very favourable, only occasional cleaning of the sensor head being required. The Orion data logger is capable of sampling at a rate of up to 140 channels/sec which is ample for the derivation of turbulence intensity from the velocity recordings. This parameter is of considerable significance in the settling process.

3.3 Salinity and Temperature

A digital salinometer and temperature recorder is employed. The salinity, which is an important parameter in the flocculation of cohesive sediments, is in the range 28 to 32 parts per thousand.

3.4 Silt Deposition:

Silt mats constructed of 'Astro Turf' appear to be the most promising means of recording deposition. The advantage of this surface over other similar plastic mats is the greater rigidity of its fibres. Once the sediment has settled on to the mat the fibres will resist the potential erosion currents.

Some interference with the natural regime is inevitable but these mats offer minimal resistance and are readily raised for inspection.

Each sensor has a purpose-made mounting bracket and a universal fitting, enabling it to be attached to the rig at any position in the water column

4. PROBLEMS ASSOCIATED WITH CONVENTIONAL SUSPENDED SOLIDS SENSORS

To obtain as near an undisturbed measurement of suspended sediment as possible it is normal practice to measure the concentration in-situ. This is usually achieved by optical methods, either by beam attenuation (transmittance) or scatterance (nephelometry).

In the case of the transmissometer, light is shone directly at the sensor and the amount reaching it is related to the suspended solids concentration.

With the nephelometer, the sensor measures the light scattered (reflected and refracted) by suspended particles.

The main limitation of the transmissometer as compared to the nephelometer is that some dissolved substances absorb light, especially the dissolved organic material · so common to river water, coastal areas, sewage etc. Therefore increased amounts of these organic dyes could be mistaken for suspended material. This is especially relevant when dealing with very low ($\leqslant 100$ mg/l) concentrations.

The problem with an sensing system that uses visible light as a measuring device is its response to ambient light. Various commercial optical suspended sediment sensors are available. However, the authors are unaware of any commercial nephelometer that is unaffected by ambient light. Most commercial transmissometers use a twin path system (Fig.5) to compensate for lens clouding and ambient light; in shallow waters the latter is seldom completely successful and light shields are necessary. The light shield introduces a further impediment to the natural flow, and in particular the vertical motion, since it is normally mounted in a horizontal plane, and in an extreme case a layer of silt has been found to be deposited on the shield covering the sensor.

Previous experience showed that the performance of optical sensors was prone to calibration drifting and poor reliability. Some improvements in these respects have been effected in the present study by dispensing with the commercial electronics and linking the sensor directly to the data logger. However, this still left much to be desired; for example, preliminary analysis of early data has shown a tendency for negative values to occur (Fig.6), Calibrations (in a darkened container) either side of the negative readings have confirmed that the sensor is functioning correctly, and the problem evidently arises from ambient light reaching the photocells in intensities which are not in the same ratio as the path lengths (which the design concept for compensation assumes).

It is some confirmation that sensors at greater depth have been found to be less affected. The site is well lit and there is no noticeable difference between night and day. Although hydrodynamically undesirable, larger light shields have been tried but these have had little effect due it is though to bottom reflections in the relatively clean and shallow water of the marina berths.

The effect of ambient light is at its most adverse in conditions low concentration in shallow water. At higher concentrations - the type of field conditions in which these sensors are more commonly used - it is quite possible that this feature may be occurring unnoticed, even if none of the readings are negative there is no guarantee that a negative component is not included.

5. AN ALTERNATIVE INFRA-RED SENSOR

Since suspended sediment measurements are generally required in relatively shallow water and because it is in this situation that ambient light has its greatest intensity, there is a particular need for a sensor that is insensitive to visible light. The need becomes even greater if it is low concentrations that are to be measured, as in the present case.

A pumped sampler offers one possibility but apart from the logistical difficulties of siting the unit for continuous recording, there are also the almost insuperable problems of inlet orientation, sampling rate, and the disruption of any flocs that may be present.

The use of non-visible light, in the form of infra-red enables the desired specification to be very largely fulfilled. A prototype suspended solids monitor, incorporating this principle, has recently been developed. It comprises a Ga As infra-red emitting diode spectrally matched to a high photo sensitivity silicon photodiode (Fig.7).

The photodiode is encapsulated in black plastic which reduces the effect of ambient light. Ambient light is readily compensated for by a second photodiode, fitted back to back with the infra-red emitting diode, recording ambient light. Thus there is no requirement for the twin gap compensating arrangement that is a rather cumbersome feature of the conventional systems. The author are of the opinion that the use of twin path lengths to compensate for lens clouding is not effective, and is unnecessary since the instrument is inspected and calibrated (before and after cleaning) regularly - at present at not more than fortnightly intervals.

The prototype has a transmission gap of 50 mm and has been specifically designed for measuring low concentrations ($\leqslant 150$mg/l). But there is the facility for reducing the gap so that the same sensor may be used to measure any higher concentration.

At present, the data logger provides the measurement electronics for the infra-red sensor, but the design of a purpose-made control unit is in hand. Also, a subsequent prototype will be constructed as a nephelometer because as indicated earlier this gives a truer indication of the inorganic suspended sediment proportion.

Careful attention has been directed to ensuring watertightness because this is a defect that has been a source of frustration with the more complicated housing of the previously used optical units. Notwithstanding these precautions, there is ready access to the electronics.

The instrument has been calibrated in the conventional manner by immersion in know proportions of Formazin (Fig.8). It is seen that it responds well over the desired range ($\leqslant 150$ mg/l), but, in common with all types of sensor, has diminishing sensitivity with increasing concentration. But as indicated earlier this can be adjusted for by narrowing the transmission gap. Two sets of plotted points are show in Fig.8, representing the contrasting cases of calibrating in a darkened container and then in direct sunlight. The stability exhibited is a remarkable testimony to the insensitivity to ambient light of the infra-red sensor.

The unit has been operating in the field, attached to the rig, for about one month and is performing satisfactorily. Fig. 9 is typical of the data obtained, with sampling at 15 min intervals; for comparision with the tidal behaviour, the curves of tidal level and tidal current are included.

6. CONCLUSIONS

The siltation of harbour berths in estuaries is a common problem and is generally a costly one to deal with. Greater knowledge of the settling process of cohesive sediments is needed and requires continuous monitoring of suspended sediment, with the influencing parameters, in a field situation.

A monitoring rig has been developed and is currently deployed at Port Hamble Marina, where there are siltation problems in spite of the fact that the suspended sediment concentration is generally very low.

Accurate and reliable measurements of concentration are essential to this research and the conventional type of suspended solids optical sensor has been found to have serious deficiencies. However, an infra-red sensor that has been developed has performed encouragingly well in early field trials; moreover it is of simpler structural form and has an adjustable range of operation.

The use of a sophisticated data logger, in this case an Orion Delta, has reduced the effort of data collection and processing, and greatly assisted in the field trials of the infra-red sensor without the more usual delay for control electronics development.

7. ACKNOWLEDGEMENTS

The authors are indebted to the Science and Engineering Research Council for funding this research. They would like to thank Capt. H. Hamilton of Rank Marine International Limited for making available the test rig site, and for practical assistance and encouragement.

8. REFERENCES

1. Tosswell,P.: "A Study of the Siltation of Marinas with Particular Reference to Port Hamble". Ph.D. Thesis, Dept of Civil Eng., University of Southampton 1984

2. Tosswell, P. and Webber, N.B.: "Siltation in Marinas", Dock and Harbour Authority, 64, 760, April 1984, pp.263-264

3. Reinemann, L., Schemmer, H and Tippner, M.: "Turbidity Measurements for Determining Suspended Sediment Concentration". (Trübingsmessunen zur Bestimmung des Schwebstoffgehalts). Deutsche Gewasserkundliche Mitteilungen, 26, 6, pp 164-174 (In German)

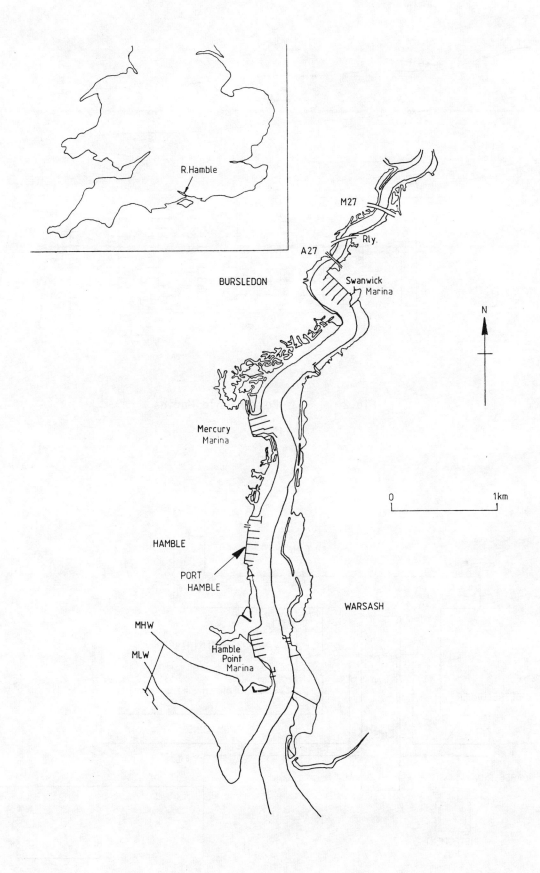

Fig.1   The Lower Hamble

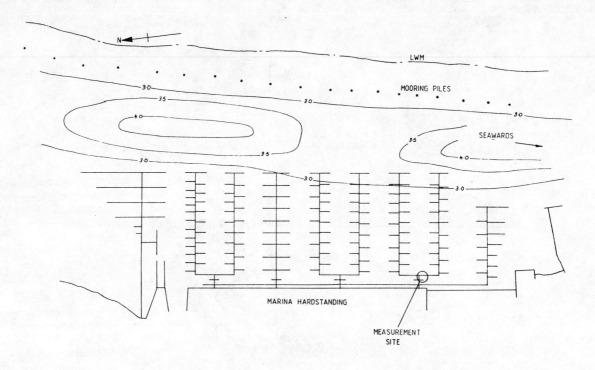

Fig.2   Port Hamble Marina

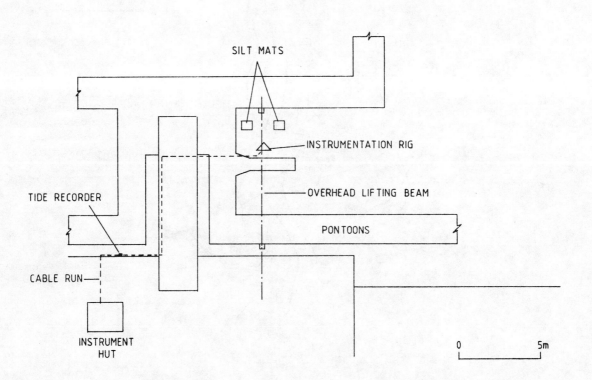

Fig.3   Measurement Site

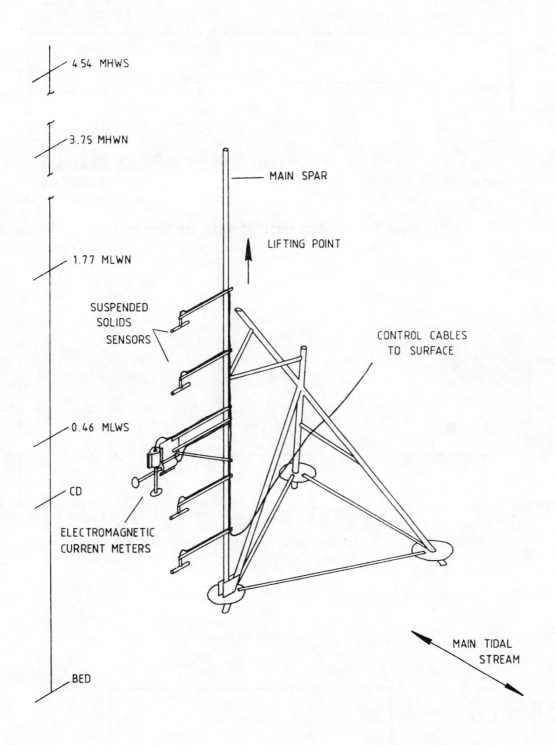

4.54 MHWS

3.75 MHWN

1.77 MLWN

0.46 MLWS

CD

BED

MAIN SPAR

LIFTING POINT

SUSPENDED
SOLIDS
SENSORS

ELECTROMAGNETIC
CURRENT METERS

CONTROL CABLES
TO SURFACE

MAIN TIDAL
STREAM

Fig.4   Instrumentation Rig

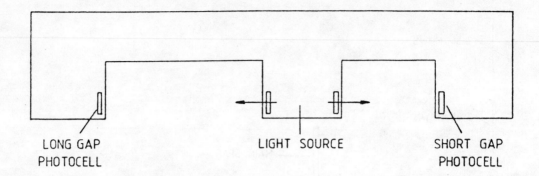

Fig.5   Schematic of Twin Gap Sensor

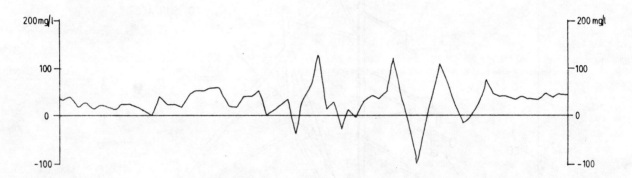

Fig.6   Typical Data from Standard Photocell Unit

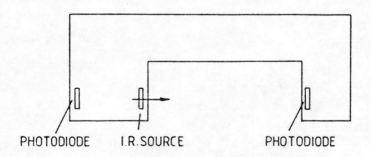

Fig.7   Schematic of Infra-Red Sensor

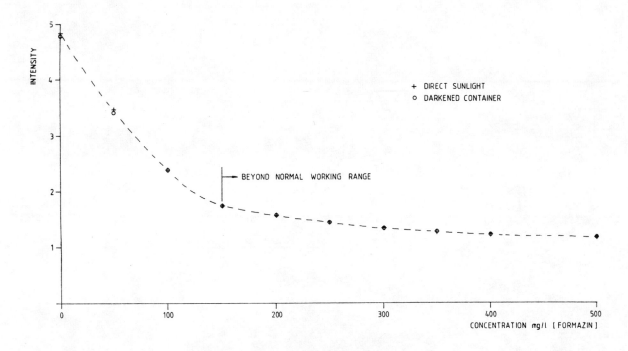

Fig.8   Graph showing Calibrations on Infra-Red Sensor (Low Range Unit)

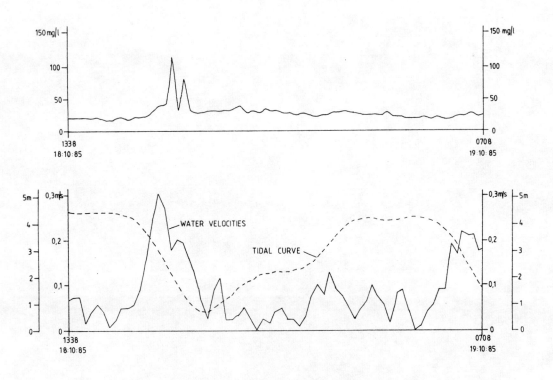

Fig.9   Typical Data from Infra Red Unit

International Conference on
Measuring Techniques
of Hydraulics Phenomena in offshore, Coastal & Inland Waters
London, England: 9-11 April, 1986

PAPER J1

AN INTEGRATED SYSTEM FOR MEASURING ESTUARINE TURBULENCE

by

R.W. Brierley, MBE
K. Shiono, BSc, MSc, PhD
J.R. West, BSc, PhD, MICE

Dept. of Civil Engineering,
University of Birmingham,
Birmingham.

Summary

A system has been developed for measuring up to 12 channels of data from transducers capable of detecting most of the energy spectrum of the turbulent perturbations of fluid velocity, salinity and suspended solids within 5 m of the bed in velocities of up to 1 m/s.

The system for deploying the instrumentation is self contained, portable and can be handled by two men. It has been designed to minimise the interference to the fluid flow. The data are transmitted by UHF telemetry to a shore based tape recorder and microcomputer which can provide an immediate analysis of several turbulence parameters. This permits close control of the field investigations and efficient use of field survey resources.

Held at Imperial College of Science and Technology, London. Organised and sponsored by BHRA, The Fluid Engineering Centre. Co-sponsored by the American Society of Civil Engineers and the International Association for Hydraulic Research.

1. INTRODUCTION

Numerical models of the movement of water, solutes and suspended solids are widely used for investigating the effects of Civil Engineering works and for the management of water quality in estuaries. In many cases the transport of momentum, solutes and suspended solids is greatly influenced by turbulence generated either at the bed or by the breaking of surface or internal waves. The complexity of the temporal and spatial variations of estuarine density and velocity fields has led to slow progress in the development of an understanding of estuarine turbulence phenomena. Due to the conflicting requirements of the scaling laws governing physical hydraulic models, the field investigation of estuarine turbulence is desirable.

Sporadic attempts to measure turbulence in estuarine channels over the last two decades have met with limited success due to only short lengths of record being obtained, lack of spatial resolution and the cost and physical difficulties of obtaining measurements in the field. This paper firstly describes a purpose built system for the investigation of estuarine turbulent transport phenomena and then gives some examples of data from observations of turbulence, fronts and internal waves.

2. TRANSDUCER DEPLOYMENT

The estuarine flow field is influenced by tidal range, fluvial discharge, channel width, depth, cross-sectional shape and sinuosity, and bed sediment size, composition and bed form configuration. Thus a wide range of the flow fields can be found. Previous experiments (Shiono (Ref. 1), West et al (Ref. 2)) have shown that a 5 m mast could be deployed in velocities of upto 2 m/s on a bed frame weighing ~ 250 kg over the side of a 12 m long vessel. Major disadvantages of this system were the cost of the vessel if a range of flow conditions were to be sought and studied sytemmatically and its interference to the flow in narrow channels, particularly if the flow was stratified.

The concept of a 5 m long vertical support mast for the transducers was maintained. This covers most of the flow depth in the upper limits of saline intrusion in many British estuaries. In the deeper, lower reaches the stratification could lead to turbulent eddies having length scales above 5 m from bed which cannot easily be resolved by existing transducers. The mast and frame are deployed from a 7 m purpose-built, self-propelled catamaran which can be transported on a trailer and deployed by a 2-man team. A 4 m rubber dinghy is left attached to the bed frame. This dinghy contains navigational warnings, instrument control boxes, transmitters and power sources. Plate 1 shows a part of a bed frame and mast fitted with inclinometers and a compass to determine alignment. Examples of the surface electronics units are also shown.

3. TRANSDUCERS

Laboratory studies of wall turbulence in well mixed flows show that the macro-scale of the turbulence is of the same order as the distance from the wall. Marine data have been used to show that high frequency losses from measured energy spectra due to transducer size may be estimated by fitting empirical formulae to the observed data. Thus commercially available transducers for measuring velocity, salinity and suspended solids concentration which have a measurement path of approximately 50 mm are suitable for investigating most of the 5 m depth permitted by the support mast. The transducers may be located within about 50 mm of each other.

Two components of velocity are measured using a 50 mm disc shaped electro-magnetic current meter (ECM) (Colnbrook Instrument Development Ltd.) based on an Institute of Oceanographic Sciences design. The head is energised by a 40 Hz squareware and its two velocity signals (of the order of microvolts) successively amplified, filtered and sampled to produce final outputs of approximately 300 mV/m/s. The commercially available units have been slightly modified to synchronise four transducers to avoid mutual interference by using a common clock generator.

For salinity measurement, four Conductivity Cells Type 2105 were obtained (from Aanderaa Instruments, Norway) and suitable electronics developed to enable them to be used as a multi-depth array. Physically, the cells are toroidal in form, with an O.D. of 37 mm, a bore of 12 mm and 28 mm thick. Electronically, they comprise two toroidal cores, one having a primary winding, the other both a secondary and a tertiary winding.

Primary-induced current in the encircling water couples with the secondary and produces a voltage proportional to conductivity. For the present application, primary energisation is a 2.56 kHz pulsed wave form and, repetitively, current in the tertiary winding is adjusted by a 10-step successive approximation to counteract the current in the water-loop and so reduce the secondary voltage to zero or near zero. In this way, conductivity is obtained as a bridge measurement with a 10-bit digital output. The resulting theoretical resolution is ~ 0.03 p.p.t. System timing is such that all four cells are read sequentially with each cell producing 40 readings per second.

Because these particular cells require short, fast rise-time pulses for their energisation, connecting cables need to be short. Thus, most of the associated electronics are housed in a submersible container. Cables to the cells (4.5 m long) are run within nylon tubes and both they and the container are pressurised to greater than the working hydrostatic pressure. The great advantage of pressurisation lies in its proof of watertightness; seals which maintain air pressure (verifiable from an integral pressure gauge prior to submersion) can be relied up on to exclude water.

Suspended solids concentration is measured using a Partech Electronic Ltd., siltmeter (Type 7000 3RP) with either, a type S-100 or type SDM 10 transducer. Both transducers employ a photoelectric principle with the former having a notional 0-250 mg/l range and the latter a 0-7000 mg/l range. The low concentration range instrument has a twin gap and one gap is filled with a perspex insert so as to maintain the reference light path and to restrict turbulence fluctuations to a single gap.

4. DATA TRANSMISSION

A range of strategies are available for recording and analysing the large volumes of data involved with turbulence studies. Although both magnetic tape and hard disc have sufficient capacity, tape recorders are at present generally more rugged and reliable for field conditions. A Racal Store 4FM recorder with 4 tape channels has been found to be ideal for this task. Real time analysis of the data is important for monitoring equipment performance and for the efficient management of expensive field survey work. A microcomputer is satisfactory, as standard production equipment can operate under field conditions and has the capacity and speed to undertake preliminary turbulence data analysis. The longer recursive calculations needed for spectral analysis studies require that the data is stored and fed into a main frame computer.

The requirement of real-time computer analysis called for the data to be got ashore from the small boat moored at the bed-frame. This has been accomplished by radio telemetry (see Fig. 1), using three UHF frequencies with four channels of data multiplexed on each.

All transducers (salinity, ECM and suspended solids) have been arranged to produce 10 - bit digital data words. The system multiplexes four such words sequentially at 50 ms intervals (20 Hz) and a complete reading of each four-transducer array is thus a 40 - bit block of data. In order to make use of standard integrated circuits (I.C.'s), each data block is first split into five 8 - bit words. Standard UART'S (Universal Asynchronous Receiver/Transmitter) can then accept each 8 - bits in turn, add START, STOP and PARITY bits and produce serial outputs suitable for asynchronous transmission. The next stage is conversion of these outputs to audio tones suitable for frequency - modulating a radio transmitter. Configured to the CCITT V.23, Mode 2 signalling standard, modem I.C.'s (AM 7910) readily perform the conversion, binary '0' becoming 2,100 Hz binary '1', 1,300 Hz. Commercial printed-circuit modules (Micro Modules Ltd) transmit and receive the data at 1200 baud on three frequencies in the band 458.5 - 458.8 MHz, which is that allocated for low - power telemetry in the U.K.

On shore, the received tones are first recorded (for subsequent full analysis) and, simultaneously, replayed into three electronic units, one for each of the telemetry channels. Each unit performs demodulation demultiplexing and digital-to-analogue conversation and provides four 0 - 1 volt analogue outputs, representing the readings of four transducers in an array.

5.   IN-SITU DATA ANALYSIS

In order to monitor the progress of the turbulence observations and to analyse the data, it is desirable to be able to calculate mean values, turbulent intensities after linear trend removal, Reynolds fluxes, correlation coefficients, spectra and cospectra. A versatile system has been developed which can achieve these results.

The total of twelve analogue channels transmitted via telemety are fed into a Zenith Z160 microcomputer (640 K memory 2 $5\frac{1}{4}$" disc drives) fitted with an additional board, the Metrabyte Dash 16 sixteen channel high speed A/D interface with direct memory access. This versatile expansion board is easily controlled by BASIC software which writes the data into blocks, into the upper part of the memory registers. Each channel is sampled at 100 ms intervals which leads to one block of data covering 273 seconds of data. This operation can be handled as a background job by the microprocessor on the expansion board but has yet to be implemented.

The calculation of the turbulence parameters is undertaken by a FORTRAN program stored on the same disc as the BASIC data collection programme. At present the two programs are run sequentially with 4.55 minutes of data collection and 5 minutes of data analysis. The results of the analysis are displayed on a VDU, output on a line printer and stored on disc for subsequent plotting using a BASIC program. The utilisation of a micro-computer for the analysis of 32,000 readings in five minutes has required some careful software development. The computer may be run off a 240 volt AC supply from mains or from two 12V 60 amp-hr accumulators, using a 500 W transverter (Valradio Ltd, Type D24/500s). Plate 2 shows a telemetry receiver, the Zenith computer and the Racal tape recorder.

6.   EXAMPLES OF RESULTS

Two examples of temporally and spatially varying results obtain during the development of the salinometer system are shown in Fig. 2 and 3. Fig. 2 shows transverse profiles of a salinity measured at Tal-y-Cafn in the Conwy estuary with three transducers suspended (at 0.2, 0.7 and 1.2 m below the surface) from the bows of a small boat. The 13 traverses from the east bank to the west bank show that the vertical stratification betweem 1631 and 1706 hrs BST decreases as the flood tide reaches its maximum velocity at about 1700 hours. A distinct front is shown near to the east bank (near the main channel in this reach) between 1631 - 1640 hrs. This front then weakens towards 1700 hrs. Between 1654 and 1706 hrs a predominantly surface feature of low salinity occurs near to the west bank.

Fig. 3 shows for ~ 6 minutes long salinity records obtained at distances of 0.05, 1.25, 2.0 and 2.7 m above the bed during an ebb tide at Cotehele in the Tamar estuary. The data show the temporal and spatial variability of salinity field perturbations. The magnitude of the perturbations is generally of the order of $\pm$ 0.5 kg/m^3 and the time scales vary from turbulent perturbations (~ 1 second) through wavelike motions (~ 10 seconds) to longer period effects of the order of 300 secs.

7.   CONCLUSIONS

A system has been developed for the field investigation of the turbulent perturbations of fluid velocity, salinity and suspended solids in shallow estuaries. Novel features of the system are the ease of deployment, relatively little inter-ference to the flow and real time analysis of the data. Preliminary results obtained for salinity during the development of the system show the potential of the system for investigating the spatial and temporal characteristics of the unsteady stratified turbulent flows found in estuarine channels.

ACKNOWLEDGEMENTS

The authors gratefully acknowledge the support of the Natural Environment Research Council for this work.

REFERENCES

1. Shiono, K. Vertical turbulent exchange in stratified flow. Ph.D. thesis. 1981. University of Birmingham.

2. West, J.R., Knight, D.W. and Shiono, K. Turbulence measurements in the Great Ouse estuary. (submitted for publication).

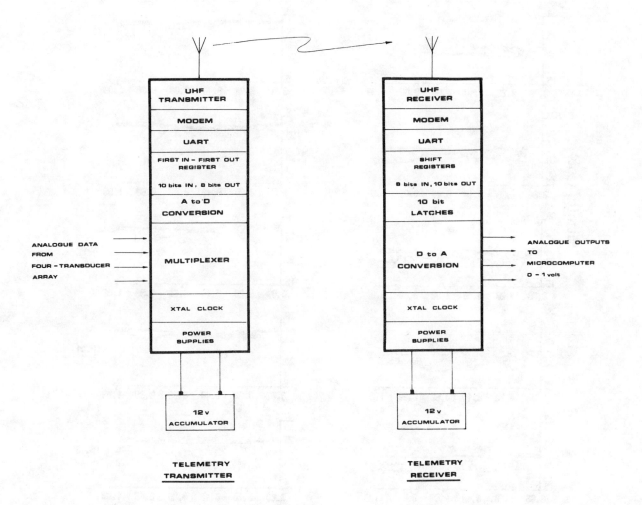

Figure 1   Typical Telemetry Channel   (for ECM and Suspended Solids)

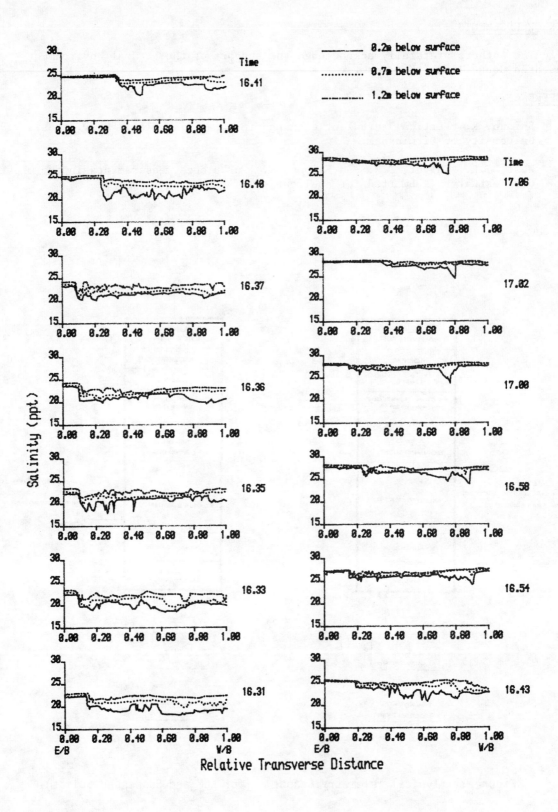

Figure 2   Transverse Salinity Distributions at Three Depths
R. Conwy, Tal-y-Cafn, 6.7.84

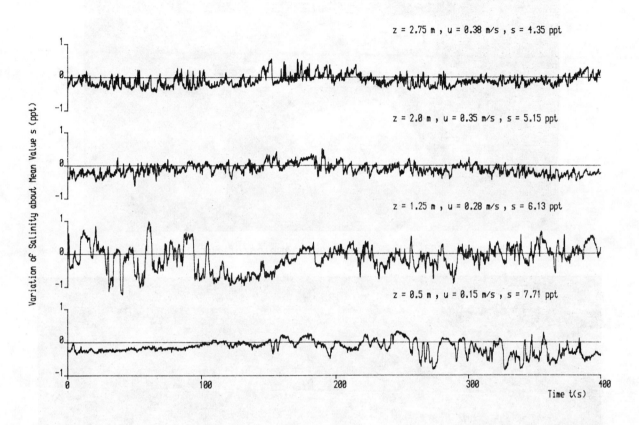

Figure 3   Example of Salinity Fluctuations, R. Tamar, Cotehele, 13:47 hrs, 8.7.85

Plate 1. Bed frame, Mast and Examples of Surface Units

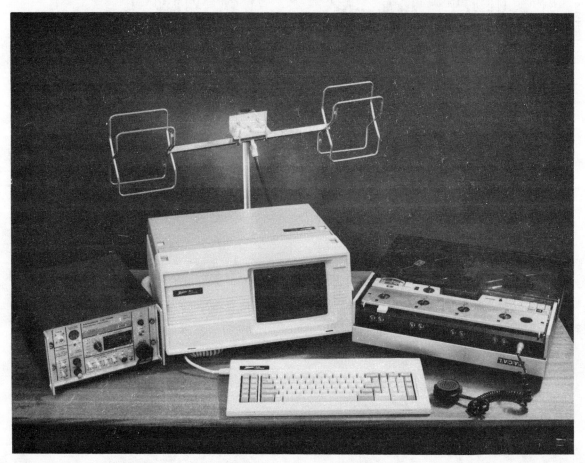

Plate 2. Telemetry Receiver, Computer and Tape Recorder

A FLUORESCENT TRACER FOR COHESIVE SEDIMENT

C.J. Louisse [1]

R.J. Akkerman [2]

J.M. Suylen [1]

Rijkswaterstaat

Tidal Waters Division

[1] The Hague, Hooftskade 1

[2] Delfzijl, Duurswoldlaan 2

The Netherlands

SUMMARY

In studies concerning the transportation of cohesive sediment in
estuarine or coastal waters, tracer methods are rarely used. One reason
may be the difficulty of finding a suitable tracer, i.e.: representative for the behaviour of the cohesive material (silt), well detectable, harmless for the environment.
At the Netherlands organisation for Applied Scientific Research (TNO) a
tracer material has been found, satisfying at least a part of the
demands. It is a fluorescent powder with very small particle sizes and
a rather high fluorescing capacity, which were detected by a fluorescence microscope.

The physical behaviour of labelled cohesive sediment was investigated
in laboratory conditions. Also environmental aspects of the use of the
tracer in natural conditions were studied (toxicity, mutagenicity).
Operational techniques were developed for labelling, injecting and
sampling, for laboratory analysis of samples and detection of the

Held at Imperial College of Science and Technology, London. Organised and sponsored by BHRA, The
Fluid Engineering Centre. Co-sponsored by the American Society of Civil Engineers and the
International Association for Hydraulic Research.

tracer. A field experiment in the Ems-Dollard estuary in the Netherlands was set up with two purposes: testing the tracer method in a natural system and answering a question on the spreading of dredge material dumped at a subaqueous dump site in the estuary. The applied techniques were proven to be suitable for tracing the labelled silt on relatively large time scales (up to several weeks). Limitations encountered were: by not being able to measure on short time scales initial conditions for the spreading on large time scales were not known; for the same reason the development of the patchy character of the tracer distribution in the estuary, observed at large time scales, could not be analysed. Because of this patchy distribution mass balance closure was not practicable, despite the conservative character of the tracer.

In general it will not be easy for a tracer method for cohesive sediment to obtain quantitative results in terms of total amount of transport in a certain direction. However, it will allow for observations of some basic features of sediment transport like residual transport direction, time scales of relevant transportation and tranportation characteristics (suspension or near-bottom transport). To better fulfil requirements, a method for real-time detection of the tracer is being developed.

1.   INTRODUCTION

Transport and accumulation of silt is one of the major problems in
estuaries and coastal waters. The problems encountered traditionally
concern the morphologic aspects of silt, but more recently the environmental impact of silt behaviour also requires the attention of public
authorities.

One way of studying silt behaviour is using a tracer. The most
characteristic feature of tracing methods is the possibility to track
the (labelled) silt through a water system from the place of release to
(more or less) final accumulation sites. Application of a tracer-method
is therefore especially useful for studies related to management
problems, such as control of dredge material disposal. However,
application is not restricted to this field as has been shown by
several authors (Ref. 5, 6, 14), who indicate that tracer methods are
used for all kinds of studies of sediment, such as sand-, silt- and
gravel transport problems.

Most sediment tracer investigations in the past were carried out by
means of radioactive tracers. This holds especially for silt transport
studies. Although usually no special reasons are given for chosing
between different types of tracers, it is rather clear why radio-
activity was often preferred as the detection principle: it is easy
detectable (in situ) and apart from silt tracers based on neutron
activation (e.g. Ref. 8) no practicable alternatives were available.
For sand tracer studies also fluorescent tracers were applied in the
past. Especially when direct impact or human activities is likely
(e.g. beaches) fluorescent tracers are preferable (Ref. 3). The main
advantage of these fluorescent tracers is the absence of safety problems, both during injection and detection, in contrast to radioactive
tracers. The use of fluorescent tracers also enabled investigating
vertical mixing of labelled sand in the sea bed (Ref. 9).

In the Netherlands sedimentological studies with radioactive tracers
were carried out until 1971. They mainly concerned sand transport
problems (Ref. 11), but also silt transport studies took place. The
latter were to study the spreading of dredge material from a dump site
Ref. 2). This silt tracer study turned out to be not very successful,
mainly because of the (unexpected) strong dispersion. After 1971 it has
become more and more difficult, if not impossible to use this kind of
tracer for sedimentological studies, mainly for reasons of public
hazard. Another type of artificial silt tracer was not immediately
available and it lasted until 1981 before new perspectives came in

sight. A new development started when the Netherlands Organisation for Applied Scientific Research (TNO) got on the track of a fluorescent powder with potentially interesting properties for silt tracer applications and developed a prototype detector, the first steps for new developments in this field were made (Ref. 7).

Even before possibilities for large scale applications were fully developed it was decided to carry out an experiment in a natural system: the Ems-Dollard estuary. A tracer experiment could be a suitable means to investigate the spreading of material dredged from the harbour of Delfzijl and dumped in the Dollard entrance (Fig. 1). It was hoped that the tracer experiment would give insight in the distribution over accumulation areas, like possibly the Emder Vaarwater being the entrance channel to the harbour of Emden.
The principal aims of this experiment were:
. to investigate possibilities of the fluorescent powder as a silt tracer in prototype conditions
. to gain insight in the spreading and accumulation of dredged silty material dumped in a selected spot in the estuary

2. THE TRACER METHOD

In general two practicable methods of sediment tracing can be distinguished. With the spatial integration method (Ref. 6) the spreading behaviour of a momentary discharge of a certain amount of labelled material is investigated as a function of time (discrete time steps) by observing the distribution of the tracer concentration in space. With the steady dilution method (Ref. 6) a (quasi)continuous release of labelled material is maintained and the tracer concentration is (quasi) continuously observed at a limited number of representative places. The last method can only be applied when the ways of travelling of the labelled material are rather well known. This is usually not the case in rather complex estuarine areas as the Ems-Dollard. So the method to be used for the relatively short time scales of the experiment (several months) had to be the spatial integration method.

To obtain conditions representative for the usual dredge disposal program, it was decided to carry out six discharges of 800 m^3 (being the volume of the barge) labelled dredge material, spread over the entire tidal period. Because of the sequence of labelled discharges over such a period, detection of the spreading on comparable short time scales (in the order of 1 day) was not useful. Besides sample analysis and consecutive proper tracer detection required at least 20 minutes for each sample. On time scales in which important transport of silt

370

takes place in this stage of the experiment ($\leqslant$ 6 hours) this frequency
of sampling was too low to allow for real-time tracer detection.
This led to the choice of large scale field sampling with consecutive
analysis on shore. A major disadvantage of this method was the
inability to adapt the observation program as time proceeded. This was
mainly caused by the laborious handling procedure of the samples.
To minimize the sampling effort, a bottom sampling method was chosen.
The sampling was carried out around high water slack. Measurements on
various locations in the estuary showed the suspended silt concentrations to be minimal during this tidal phase over the entire column.
This was supposed to be caused by settling of the greater part of the
suspended silt which probably formed a thin layer near the bottom.
Since this silt is very easily eroded it was necessary to use a
sampling device which would be able to properly sample the transition
area from water column to bottom. Two devices were developed and
simultaneously applied: a corer and a pistonsampler.

Various sampling programs were executed, the most important being the
synoptic observation program (T-observations), and the monitoring
program (M-observations). Synoptic observations at one, two and four
weeks after the release of the labelled dredge material, were executed
by 12 boats, sampling at the same time at locations selected in advance
(Fig. 1). In the monitoring program a more frequent survey of several
locations in the inner and outer estuary was realised. Furthermore
several supplementary observations were made.
The collected samples were analysed in a mobile laboratory, especially
equipped for this purpose. The distribution of the tracer material over
the estuary was expressed in load per unit area ($[mg/m^2]$).

3. TRACER QUALITIES

For an investigation of this kind the tracer material must satisfy
several requirements: it must be representative for silt behaviour, it
should meet environmental impact regulations, it has to be detectable,
and so on. These requirements among other things will be further
discussed in the following paragraphs.

3.1 The fluorescent powder

The tracer material is a powder consisting of synthetic particles which
contain a fluorescent pigment. The pigment accounts for only 3% of the
weight of the powder. As to particle sizes electron microscopic photographs show a very fine material with nearly all particles smaller than
10 μm (Fig. 2). This is expected to give the powder favourable properties with respect to coagulation with silt particles.

The powder can be obtained in ten different colours, but up till now only two colours are seperately detectable by the detector. These two dyes are fluorescent in green and orange when excited with blue light (Fig. 3). Both colours were simultaneously used in the Ems-Dollard experiment.

The pigments of the selected powders were found not to be mutagenic. The orange pigment appeared to be toxic in concentrations at least three orders of magnitude higher than the typical tracer concentrations. For the green pigment no toxic level could be determined.

A somewhat unfavourable property of the powder is its low degradability, due to which the natural system may be loaded up, eventually causing high background concentrations for future investigations. A favourable aspect of the tracer is its low price.

3.2 Detection

Detection of the tracer is obtained by counting the number of fluorescence pulses, caused by fluorescent particles passing the detection cell. The detection cell is built in a fluorescence microscope. The principle of the detection is schematically drawn in Fig. 4. Light from the microscope lamp is filtered to obtain a proper excitation wavelength for the tracer particles and is led through the suspension of silt and tracer in the detection cell. The light emitted by the suspension (scattered and fluorescent light from silt and tracer particles) is again filtered, transmitting the fluorescent light only. Counting takes place after amplification and modification of the pulse signal. A discrimination level is set to minimize background counts and optimize counting results. Particle counts are related with tracer concentrations by means of a calibration curve (Fig. 5).

The lower detection limit is determined by the duration of the counting period, the natural background of the suspension (number of pulses counted in a sample without tracer material) and the requirements with regard to accuracy. This is further elaborated in the appendix. For the Ems-Dollard experiment the detection limit is about 30 ng/l in a counting interval of 5 minutes and a minimum accuracy of 50%. There are no limitations on maximum detectable tracer concentrations, since for higher concentrations corrections for masking effects are introduced (see appendix) and for very high concentrations dilution of the suspension offers a practicable solution.

The masking effects of silt particles in the suspension on the tracer particles becomes inconvenient for silt concentrations above 400 mg/l (Fig. 6). Therefore siltconcentrations of the suspension samples are measured before the detection and the samples are diluted when the concentration exceeds 400 mg/l.

3.3. Labelling procedure

To obtain a homogeneous distribution of the tracer material over the dredge hopper content (800 m^3) a method was developed to bring the powder into an emulsion of high stability making it suitable for injection. The most important demand made upon the emulsion was that the particle-character of the tracer had to be maintained. Furthermore injection needed to be performed without interruptions, so the emulsion had to be ready for use on deck and therefore needed to be prepared several days in advance. The tracer emulsion obtained in this way contained 167 gram dyed powder per liter and was injected by means of a centrifugal pump into the supply line during the entire period of filling the dredger.

After the filling procedure 6 small samples (30 ml) were taken to check the homogeneity of mixing in the barge. On this small volume scale, mixing was not entirely homogeneous, but the average concentration was within 15% of the expected value. Since handling of the tracer material is absolutely safe, mixing of the tracer in a large volume is simple compared to using radioactive tracers (Ref. 5 and 13).

3.4 Bottom samplers

Two devices for bottom sampling have been developed, a corer and a pistonsampler. A schematic picture of the corer is shown in Fig. 7. The plastic dismountable sampling tube is driven into the bottom by a weight consisting of lead discs. Maximum length of a core is about 25 cm. The sample is kept in the tube against gravity by skin friction of the wall and by reducing the pressure difference between bottom and top of the tube with a sealing ball of stainless steel on a rubber ring.

On board the cores are stored in vertical position to prevent the sediment from mixing. The cores are qualitatively described and photo-graphed on shore before freezing them with liquid nitrogen. Freezing was necessary to be able to investigate distinct slices of the cores (by sawing the cores) and it was established that quick freezing was needed to prevent severe texture distortions, which were observed after slow freezing.

The pistonsampler consists of two main parts: the sample-box, built in a frame, and the air-circulation unit. This unit regulates the air flux through the sample-box when lowered through the water column (Fig. 8a). This air flux is kept constant independant of waterdepth and keeps the sample-box free from water. Reaching the bottom, part of the bottom surface is locked in (Fig. 8b). By removing the overpressure from the sample-box with respect to the surrounding water the hydrostatic pressure pushes the sample obtained into the sample-box (Fig. 8c). A vertical tube prevents the sample from being lost again when pressure differences are removed; the bend in the pipe prevents the sample material from penetrating into the air tube. Back on deck the sample-box can be disconnected from the frame and shut off by plastic caps and thus stored to wait for analysis.

3.5 Sample-analysis

Sample-analysis is divided into two procedures: preparation and detection. The preparation procedure will not be described in detail here. The main considerations are the following: at first the necessity of dilution of high concentrated samples was faced, while also low concentrated (piston) samples had to be handled. Besides silt suspension concentrations had to be measured over a wide range (200-4000 mg/l). If necessary, further dilution to a maximum concentration of about 400 mg/l was carried out to enable undisturbed counting of the tracer particles. Administration of the sample properties and regulation of the analysis procedure was executed by means of a HP-9825 Desk computer.

4. REPRESENTATIVENESS OF THE TRACER FOR SILT BEHAVIOUR

Before applying of the tracer in the Ems-Dollard field experiment, laboratory experiments were carried out to verify whether the behaviour of the labelled material would be representative for silt behaviour. Since the most suitable experimental facility, i.e. a column with oscillating grids to create a (homogeneous) turbulence field, was not yet available at the time (Ref. 11), experiments were carried out in a vertical tube without turbulence grids. The main process represented in this way is flocculation of suspended material due to differences in settling velocities and the vertical transportation of silt. Floc break-up due to high turbulent velocities and erosion/resuspension processes could not be studied in this way. The experiments were carried out by the Delft Hydraulics Laboratory (Ref. 10).

The experimental set-up is drawn in Fig. 9 showing the two sample intakes. After homogeneous mixing by pumping around and subsequently terminating the circulation, a series of samples was taken in the course of time to determine silt and tracer concentrations. Two types of silt material were tested, viz. suspensions of kaolinite powder clay and suspensions of Ems-Dollard silt, each in two concentration levels, (50 and 1000 mg/l) and two tracer/silt ratios (1:1.000 and 1:10.000).

The results are presented as the ratio between the measured concentrations of silt and tracer at time t and the initial concentrations at t=0. For the Ems-Dollard system Fig. 10 represents the most representative situation: rather high concentrations (1000 mg/l) and silt material derived from the region itself. Silt and tracer material are found to behave largely identically when brought together in the settling tube. The tracer material in itself shows much lower fall velocities in concentrations comparable to the combined experiment of silt and tracer. This indicates that the greater part of the tracer material is embedded in the silt flocs which are formed during the settling process and thus behave in the same way as the flocculating silt. No significant influence of the tracer material on the fall velocities of the silt was observed. For radioactive labelled silts (Gold 198 and Chromium 51) similar conclusions were drawn from laboratory experiments (Ref. 4), although results from settling velocity tests were less convincing. Representativeness of the radioactive tracers for silt tracer behaviour however has not been tested by these experiments.

Although the test results can not be regarded as a complete proof for the suitability of the tracer, it was shown to possess one of the main properties for representing the transport of the mud flocs and was therefore suitable for testing in a field experiment.

5.   THE EMS-DOLLARD ESTUARY

The Ems-Dollard estuary is a tidal system located on the border of The Netherlands and Germany (Fig. 1). The principal fresh water input is supplied by the river Ems and its tributaries (mainly the river Leda) which amounts to 110 m^3/s (average annual discharge). Other fresh water discharges are relatively small. The outer part of the estuary (seaward from Delfzijl) is generally well-mixed; the same applies to the Dollard area. The Emder Vaarwater and the more landward part of the Ems are partially mixed in the region where the salt can intrude the system. In this area a turbidity maximum occurs where silt concentrations increase to values beyond 2000 mg/l at 1 m above the bottom. The seaward part of the estuary as well as the Dollard consist of sandy channels, separated by large tidal flats. They are also mainly composed of sand and bordered by marshes.
Both Emden and Delfzijl have made a great effort to maintain the accessibility of the harbours for ships. Dredged material from Emder Vaarwater and Ems is partly released into the estuary and partly stored on land sites. The greater part of the dredged material from the harbour of Delfzijl is dumped into the estuary on various dump sites,

from where it is being transported by the tidal currents. One of these dump sites is by some authorities believed to be located unfavourably with respect to the probability of saddling the Emder Vaarwater with this silt (Fig. 1). This was the main reason for aiming the tracer experiment on this problem.

6.   THE FIELD EXPERIMENT

The release of six labelled hopper loads took place on 5 April 1984. Three orange labelled loads were released during the flood and three green ones during the ebb, with the purpose to distinguish between differences between these tidal phases. The result of the experiments will not be dealt with in detail. Only the main experiences with respect to techniques, methods and physical processes will be presented.

The experimental set-up of the field experiment may be called successfull in some respects and susceptible for improvements in others. The technical means and methods - e.g. the bottom samplers, the injection technique, most of the sample analysis procedures - were on the whole quite satisfactory.

The methodological concepts did not always satisfy. As had been expected the sampling method did not offer a sufficiently high resolution for the spatial distribution of the labelled silt in spite of using a large number of survey vessels. The distribution over the selected area generally appeared to be very patchy (Fig. 11, 12, 13), and on much smaller scales (5-50 m) compared to the network (0.5-2 km). Obtaining reliable samples from the estuary bottom was not always easy either and sometimes even impossible. In general it can be established that, for very soft bottoms with a considerable thickness ($\geqslant 0.3$ m), as they seem to occur in the Emder Vaarwater and in the harbours, both samplers were not adequate, because they proved not to be able to collect correct samples. On the other hand the corer (Fig. 7) appeared to sample more consolidated silty soils quite well and even worked smoothly on the sandy channel bottoms. Nevertheless a problem arose in the latter case: when the silt material is encountered in the main (sandy) channels of the estuary during the first days after release it is confined to a rather thin layer of soft material close to the bottom. This layer will be easily blown up by the pressure wave of the frame of the corer and will therefore not (entirely) be collected. The pistonsampler (Fig. 8) proved to reach better results in this case. When after some more days the tracer material had for the greater part vanished from the sandy channels and had been transported to accumulation areas, the core samples were more reliable, both on sandy and on

376

silty bottoms. As to sampling, it can be stated that both samplers were useful for certain applications, but neither of them worked satisfactorily in all conditions.

The rather patchy character of the measured tracer concentrations makes it impossible to interpret the results in a quantitative way as for instance suggested by Courtois and Tola (residual transport velocities; Ref. 5 and 12). There is hardly any sense in estimating the quantity of tracer recovered. If tried the estimates range from 6 to 20% with a tendency to increase as time proceeds. So the mass balance does not close.

Some qualitative results are worth mentioning. From the observations (e.g. Fig. 11-14) a rather good impression of transport directions and time scales was derived. To indicate the order of magnitude of the tracer concentrations shown in these figures, it is pointed out that a uniform distribution of the labelled silt over the selected area (Fig. 1) would have resulted in tracer concentrations of about 1 $mg/m2$ in each sample point. After releasing the labelled silt rather strong tidal velocities cause considerable displacement and dispersion of the labelled silt. Due to the succession of the labelled disposals during the whole tidal period, both Dollard and Emder Vaarwater are loaded with labelled silt within a few tidal periods. A part of the labelled material is kept in the inner estuary and is moved back and forth, and is eventually trapped in Emder Vaarwater, the harbour of Delfzijl, the Dollard or other accumulation areas. Under normal conditions this seems to happen within a week after the release of the labelled silt: observations on the 4th day revealed this behaviour, while on the 8th day (T1, Fig. 11) mobility of the silt in the inner estuary seemed to have decreased. Accumulation can take place on shallow tidal shoals or in the marshes. Labelled silt can be brought into circulation again from the shoals due to wave forces which erode the top layer(s). This probably happened at location 63 (Fig. 1) before T2. Fig. 14 indicates that such an event has occured one or two days before T2; wave- and wind-observations indeed show a remarkable change around that period.

Although some of the labelled material has been recovered in Emder Vaarwater and Ems there are indications that an important part of the labelled silt in this area has not been tracked. From other measurements (continuous silt registration at Pogum, Fig. 1) it was concluded that the main silt circulation area, the turbidity maximum, was located landward from Pogum at high water slack. Since tracer measurements were rather scarce there, a main sink of labelled material may have been

missed. Also the Dollard appeared to belong to the area where accumulation of the labelled silt occured. Inward silt transportation could
also be established in this area from silt concentration and current
velocity measurements carried out in the same period (17/18 april
1984). This was caused by tidal asymmetry.
Tracer material occasionally found in the outer region of the estuary
was believed to be transported that way by dumping dredged material
from the Emder Vaarwater in the middle of the estuary and consecutive
tidal displacement. Systematic large scale natural transportation of
labelled silt in seaward direction was not supposed to take place. Only
a small amount of tracer material was sometimes recovered in the see-
ward part of the estuary. Also transport of labelled silt landward from
the turbidity zone was not observed.

7.   DISCUSSION

The tracer experiment in the Ems-Dollard estuary has shown that fluorescent (powder) material provides a means for silt transport studies.
The advantages of this tracer material above radioactive tracers are
connected with the harmlessness of the fluorescent powder, which enables comparatively simple mixing procedures, sample-handling, and so
on. Furthermore radioactive tracers and tracers based on neutron activation often require rather severe measures to connect the tracer with
the silt, like heating the sediment and tracer in an oven (Ref. 4).
This labelling procedure might also affect the properties of the labelled sediment. For Gold and Chromium, Caillot (Ref. 4) shows this to be
true for the settling behaviour of silt suspensions; the influence on
the representativeness of the tracer for silt behaviour was not reported in that study. The fluorescent tracer was shown not to affect the
settling properties of the silt and in fact appeared to represent the
falling process of the silt fairly well (Ref. 10).

An obvious disadvantage of the fluorescent tracer with respect to
radioactive tracers is its "primitive" detection method. Radioactive
tracers can be detected real-time and continuous by a towing system
from a vessel (Ref. 14). This enables detection on short time scales,
immediately after the release of the labelled silt, and provides a much
more detailed description of the distribution of the labelled silt.
Besides, sample analysis was very laborious and it was not possible to
evaluate results in such a short time as to adapt the next sampling
program. However improvement in this situation for the fluorescent
tracer is expected, since a new detection apparatus is being developed,
which will enable real-time detection.

The Ems-Dollard experiment did not succeed in answering the question concerning the quantitative distribution of dumped silt over the estuary. Insufficient methods, e.g. the poor detection method, were partly responsible for this. However, even in case of continuous detection of tracer concentrations, mass balance closure can be difficult. This can be concluded from field experiments by Smith and Parsons in the Firth of Forth (Ref. 13), where only 30-40% of the released material was recovered using a continuously measuring system. Especially on larger time scales proper closure of the mass balance becomes more difficult since the initial cloud is torn apart in a number of smaller clouds, which are difficult to track. In the Ems-Dollard estuary this problem seemed to be dominant as well. General occurence of this feature in natural systems may be the reason why comperatively little information is usually presented in quantitative description of tracer experiments. Exceptions concern dispersion of mud material on short time scales (several hours) (Ref. 3).

An interesting, although expected feature is the property of the system to trap the silt in inner basins. Transport of silt by natural causes either landward from the turbidity zone or seaward does not seem to occur unless it is brought there by dredge hoppers. This was also indicated by the tracer experiments in the Gironde (Ref. 1).

It may be stated that quantitative description of the distribution of fairly mobile fine sediments on large time scales (months) by means of tracer experiments with a singular (set of) injection(s) will often encounter problems in the field of interpretation, since mass balance closure is achieved with great difficulty. Other tracer methods ((quasi) continuous injection) may provide better results for this kind of problems.

The Ems-Dollard experiment has shown the fluorescent tracer to be a good means for (at least) qualitative studies of the spreading behaviour of silt in natural systems. Developments in detection techniques, which are foreseen, will contribute to improve field detection.

8.   ACKNOWLEDGEMENTS

The research reported here was realised by the efforts of people of several institutions. The authors acknowledge the members of the research group electro-optical instrumentation from the department of Analytical Chemistry in the Netherlands Organisation for Applied Scientific research (TNO) P.M. Houpt, R. Tadema Wielandt, A. Draaijer, J.W. König who initiated research into silt tracers and are now developing a real-time detection method. The technical equipment for the Ems-Dollard experiment was developed in a very short time by the members of the Physics Division: L.B.A. van Riet, P.J. Lagerwey and L.C.M. Jonathans (who also did a large part of the sample analyses) and the group of J. Vegter (Rijkswaterstaat Delfzijl). We are grateful to many people from Rijkswaterstaat Delfzijl who contributed to the various activities of the experiment, e.g. injection of the tracer, field sampling, sample analysis, etc.). Special thanks to R. Jungcurt who made the drawings. G.C. van Dam and P. Schreurs made many suggestions for the improvement of the manuscript.

9.   APPENDIX

For a unit volume (1 liter) of suspension of labelled silt the number of fluorescent particles, present in a unit time interval (1 minute), is determined by:

$$N = (1+P)N_r - N_a$$

$$P = \frac{0.01 * N_r}{60}$$

in which:

N_r = total number of pulses counted

N_a = number of pulses counted in a "particle-free" suspension

P = correction factor to account for the particles that may have been "masked" by others

Passage of a fluorescent particle in the area of focus in the cuvet causes a light pulse which is counted to measure tracer concentration. Since the pulse has a duration of about 10^{-2} s it is to be taken into account that more particles can pass the focus area during the time one particle is counted. With a circulation velocity of 0.7 m/s the minimum distance between two particles must be more than 7 mm to prevent them from being counted as one. The chance that a particle is not counted in a detection period of one minute, is expressed by P.

If the number of pulses counted in a suspension is too small the inaccuracy of the observations becomes too large. The relative error in one observation of N is:

$$\frac{\Delta N}{N} = \frac{N_r . \Delta P + (P+1) . \Delta N_r + \Delta N_a}{(P+1)N_r - N_a}$$

$\Delta N_r = \sqrt{N_r}$

ΔN_a = standard deviation of all measurements of N_a

$\Delta P = \Delta N_r$

A treshold level for this error must be chosen to be able to discriminate between relevant and non-relevant data.

10. REFERENCES

1. Allen, G.P., Sauzay, G., Castaing, P. and Jouanneau, J.M.:
    "Transport and deposition of suspended sediment in the Gironde
    estuary, France". Estuarine Processes; Proceedings: 3rd Int. Est.
    Res. Conf.(Galveston USA Oct.1975), Volume 2, 1977, pp. 63-81.

2. (Anonymus): "Report on the measurements with radioactive labelled
    silt on "Loswal Noord" in the North Sea" (Verslag van de metingen
    met radioactief gemerkt slib op de "stortplaats Noord" in de
    Noordzee), Rapport W.69.009 (in Dutch).

3. Bastin, A.L., Caillot, A. and Malherbe, B.: Zeebrugge port extens-
    ion, Sediment transport measurements on and off the Belgian Coast
    by means of tracers. In: Proceedings 8th International Harbour
    Congress, Antwerp, 13-17.06.1983.

4. Caillot, A: "Les méthodes de marquage des sédiments par des
    indicateurs radioactifs." La Houille Blanche 25, 1970, pp.
    661-674.

5. Courtois, G.: "La dynamique sédimentaire et les traceurs
    radioactifs, Point de la situation en France." La Houille Blanche
    25, 1970, pp. 617-628.

6. Crickmore, M.J.: "Tracer techniques for sediment studies; their
    use, interpretation and limitations". In: Proceedings Diamond
    Jubilee Symp. on Modelling techniques in Hydraulic Engineering,
    vol 1, Central Power and Water Research Station, 1976, Paper A13,
    19 pp.

7. Draaijer, A., Tadema Wielandt, R. and Houpt, P.M.: "Investigations
    on the applicability of fluorescent synthetic particles for the
    tracing of silt" (Onderzoek naar de toepasbaarheid van fluoresce-
    rende kunstofdeeltjes bij het traceren van slibbewegingen),
    Netherlands Organisation for Applied Scientific research, Report
    No: R 84/152, Nov. 1984, (in Dutch).

8. Ecker, R.M., Sustar, J.F. and Harvey, W.T.: "Tracing estuarine
    sediments by neutron activation". In: Proceedings 15th Coastal
    Eng. Conf., 1976, Honolulu, Vol II, part V, Ch 117, pp. 2009-2026.

9. Lees, B.J.: "Sediment transport measurements in the Sizewell-
    Dunwick Banks area, East Anglia, U.K.". Spec. Publs. Int. Ass.
    Sediment (1981) 5, pp. 269-281.

10. Van Leussen, W.: "Investigations with the fluorescent silt tracer Day glo". (Onderzoek met de fluorescerende tracer Day glo), Delft Hydraulics Laboratory, report WL-WGSL-84-17, (in Dutch).

11. Van Leussen, W.: "Laboratory experiments on the settling velocity of mud flocs". Proceedings 3rd International Symposium on River Sedimentation, Jackson, Mississippi, USA, 1986. (submitted).

12. Pilon, J.J.: "Mésures par traceurs radioactifs du movement des sable aux Pays-Bas 1957-1962". C.R. Réunion, Bruxelles 3-5 oct. 1963, Eurisotop, Cahier d'information n° 8, communication n° 15, (1965), pp. 201-314.

13. Smith, D.B. and Parsons T.V.: "The investigation of spoil movement in the Firth of Forth using radioactive tracers". In: Proceedings Symp. on dredging, London, 1967, pp. 47-54.

14. Tola, F.: "The use of radioactive tracers in dynamic sedimentology". Note CEA-N-2261, Centre d'Études Nucleaire de Saclay.

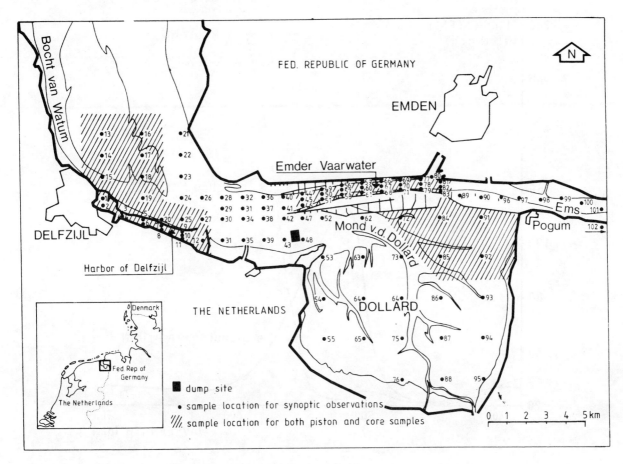

Figure 1: The Ems-Dollard: the part of the estuary in which the tracer material was expected to be mainly accumulated.

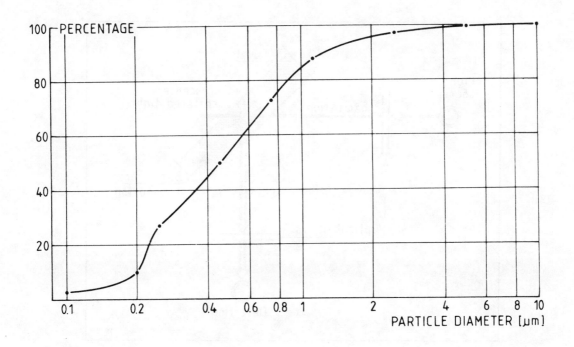

Figure 2: Particle size distribution of the fluorescent powder.

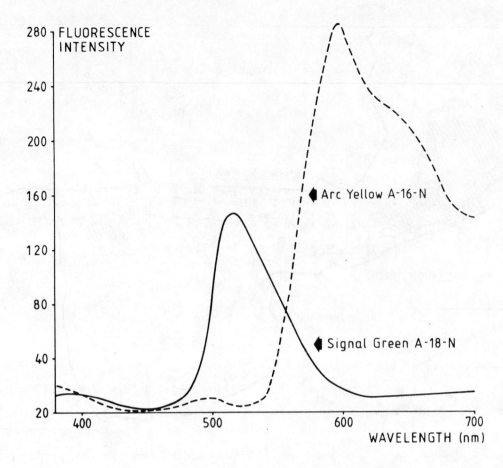

Figure 3: Fluorescence spectra of the two colours applied in the experiment: green and orange.

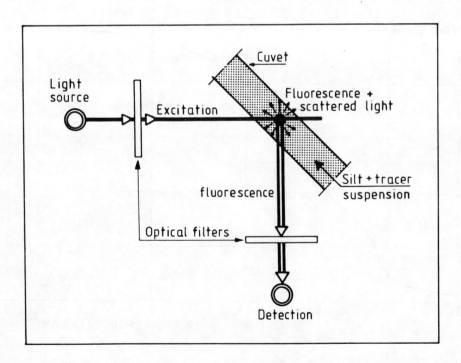

Figure 4: Schematic presentation of the detection principle.

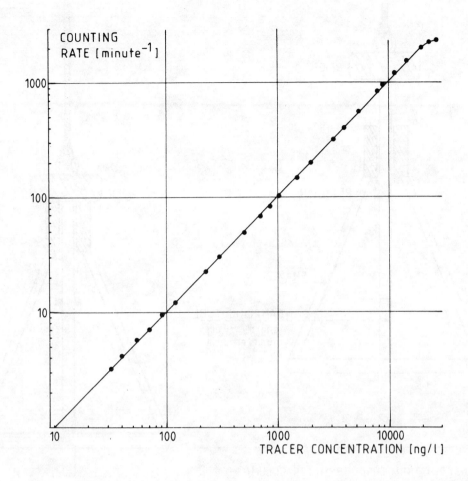

Figure 5: Calibration curve for the orange tracer.

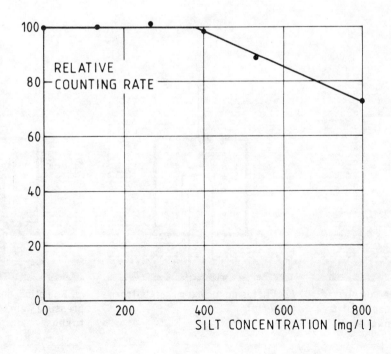

Figure 6: Influence of siltconcentration on fluorescent particle counting-rate.

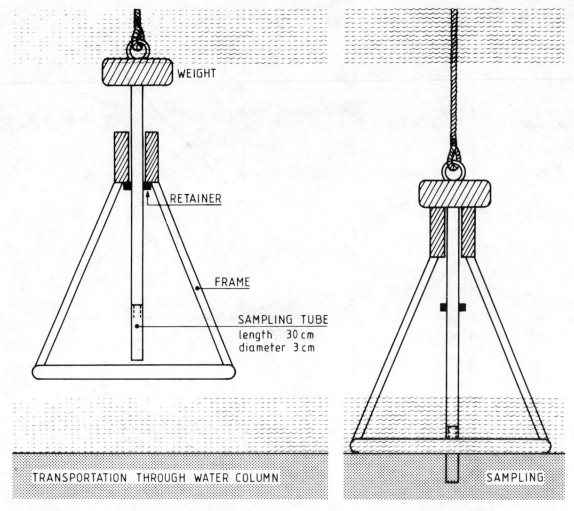

WEIGHT

RETAINER

FRAME

SAMPLING TUBE
length 30 cm
diameter 3 cm

TRANSPORTATION THROUGH WATER COLUMN

SAMPLING

Figure 7: Schematic impression of the corer, both during
transportation to the bottom and sampling.

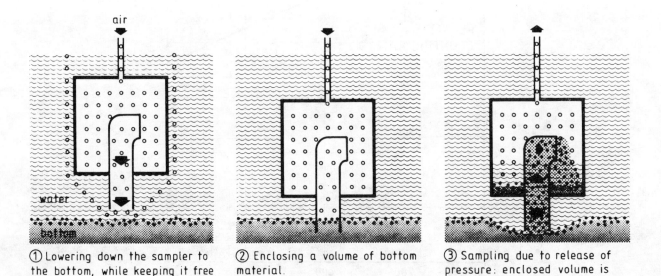

air

water

bottom

① Lowering down the sampler to
the bottom, while keeping it free
from water by downward air
circulation.

② Enclosing a volume of bottom
material.

③ Sampling due to release of
pressure: enclosed volume is
taken.

Figure 8: Principal operation procedure of the pistonsampler in three stages.

386

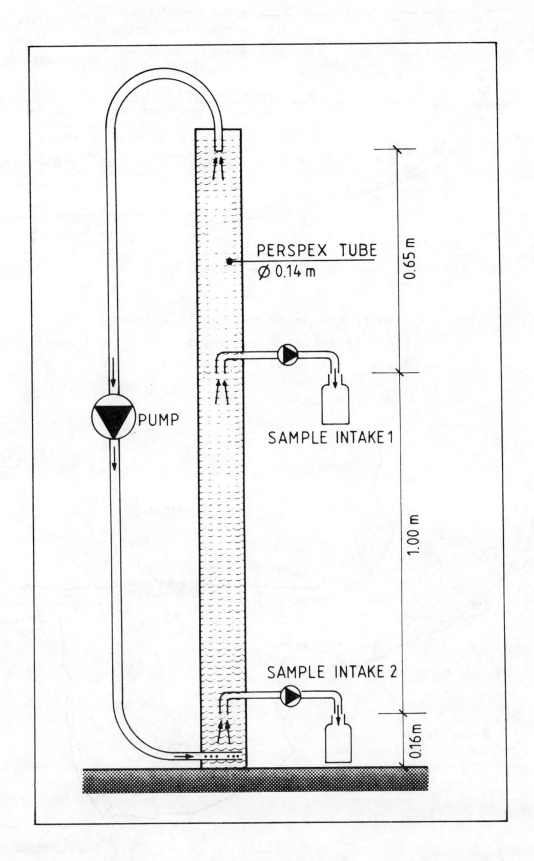

Figure 9: Schematic picture of the laboratory facility used for physical testing of the tracer.

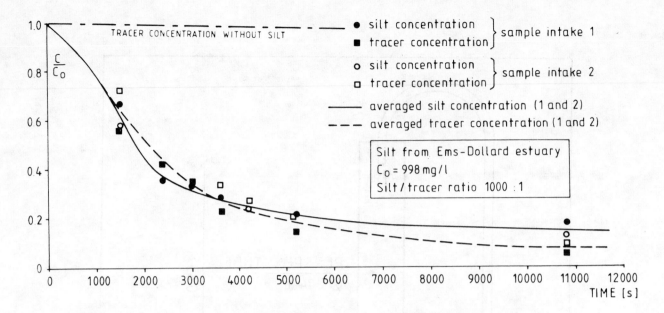

Figure 10: Results of one of the settling column tests:
Ratio of measured and original silt and tracer
concentrations as a function of time.

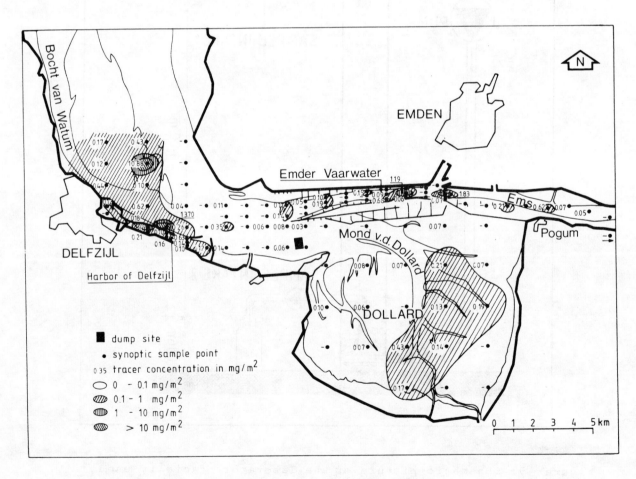

Figure 11: Quantitative tracerconcentrations (core samples)
for the synoptic observation program (T1) a week
after the release of the tracer.

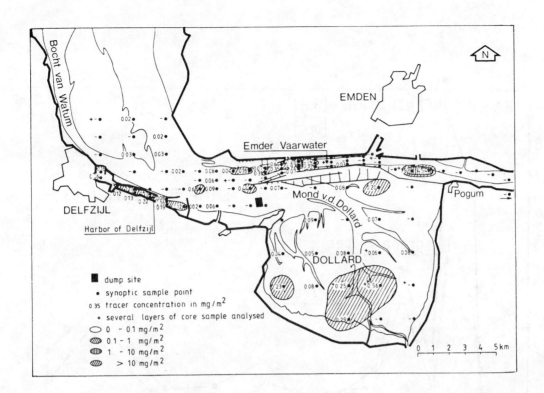

Figure 12: Quantitative tracerconcentrations (core samples) for the synoptic observation program (T2) two weeks after the release of the tracer.

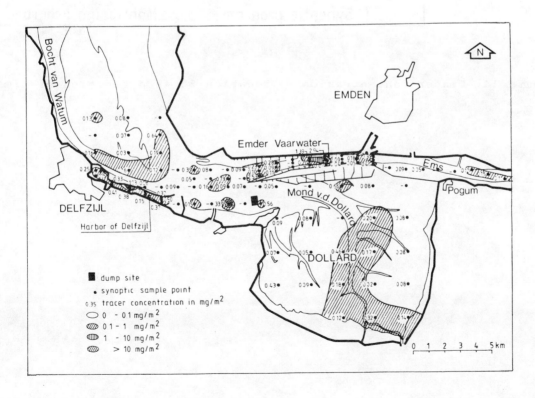

Figure 13: Quantitative tracerconcentrations (core samples) for the synoptic observation program (T3) two weeks after the release of the tracer.

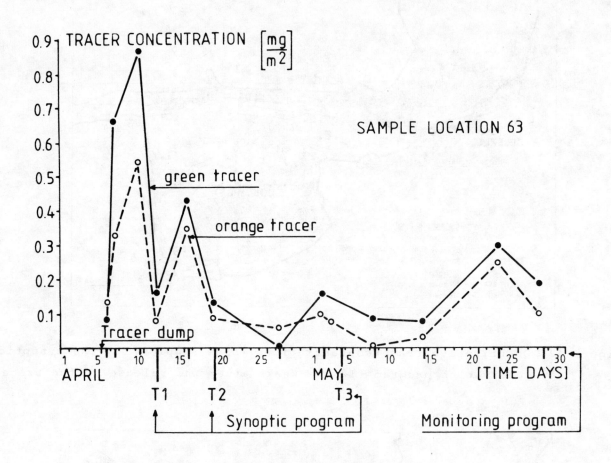

Figure 14: Tracer concentration at location 63 as a function of time.

TV TECHNIQUE USED FOR BOTTOM PROFILE MEASUREMENTS

P.Georgiev, N.Khadzhiisky, M.Stephanov, M.Mikhailov

Measuring Equipment Laboratory

Bulgarian Ship Hydrodynamics Centre

Varna 9003

Bulgaria

Summary

Successful erosion protection requires, as a rule, detailed investigation of the lithodynamic processes in the coastal zone. Field and model measurements are necessary to monitor the bottom profile and its variations due to sediment transport.

The paper presents a novel technique for contactless measurement of the bottom profile near the coast line. The theoretical background of the measurement and the influence of different error sources are discussed. For experimental verification this technique is used to measure the bottom profile dynamics during hydraulic model tests in a wave flume.

The experimental set-up works as follows: A laser beam is steered in such a way that a vertical plane, parallel to the longitudinal (X) axis of the flume, is formed. A colour TV camera is used to observe the intersection line of this plane with the bottom. The measurements are performed at daylight, so special processing of the video signal is necessary to eliminate the background light. As a result a video image is obtained, containing only the investigated bottom profile. This image is digitized and fed into a microcomputer where the X and Z coordinates of the points of one longitudinal bottom profile are calculated. By moving the laser plane across the flume, a set of curves Z = f(X), corresponding to different Y values, is obtained. The zoom lens of the TV camera allows optimum resolution of the system to be obtained.

Discussions of the experimental results and of the field measurement application are also presented.

Held at Imperial College of Science and Technology, London. Organised and sponsored by BHRA, The Fluid Engineering Centre. Co-sponsored by the American Society of Civil Engineers and the International Association for Hydraulic Research.

NOMENCLATURE

x,y,z   = coordinate system of the TV camera's image

X,Y,Z   = coordinate system of the object

$\left.\begin{array}{l} a \\ b \end{array}\right\}$ = constants

f   = focal distance of the TV camera's lens

$\left.\begin{array}{l} N_x \\ N_z \end{array}\right\}$ = numbers corresponding to the (x,z) coordinates of an image point

$\left.\begin{array}{l} K_x \\ K_z \end{array}\right\}$ = coefficients

$\left.\begin{array}{l} X_o \\ Z_o \end{array}\right\}$ = constants

ABBREVIATIONS

BSHC   = Bulgarian Ship Hydrodynamics Centre

FIFO   = First in - first out

He Ne   = Helium-Neon

TV   = television

WL   = water level

1. PROBLEM DESCRIPTION

The activities on coastal protection from erosion are preceded, as a rule, by detailed studying of lithodynamic processes in the coastal zone. The bottom profile changes induced by sediment transport are investigated, both in full scale and by model experiments.

In the modelling of a movable bed usually statistic data on full scale relief, accumulated for a long period of time, are used.

The correct modelling of historical hydraulic conditions requires the precise measurement of model topography in the different stages of its construction and investigation (Ref. 2 and Ref. 3).

At the Bulgarian Ship Hydrodynamics Centre (BSHC), Varna, model investigations of coastal protection and coast fortification hydrotechnical facilities are carried out since 1981. For bottom profile measurement in the investigation of movable bed models, the classical method is used - by stepwise drainage of the model and photographing of the suitably marked waterlines.

The main disadvantage of this method is the low bottom stability at stepwise draining of the model, when artificial sediment materials are used (Ref. 1).

That is why a new, contactless method for bottom profile measurement was developed and experimented on at the BSHC Measuring Equipment Laboratory. It is the subject of the present paper.

2. ESSENCE OF THE METHOD AND EQUIPMENT USED

The method is based on the possibility for computerized processing of TV images.

With the aid of a fast rotating optical head, a focused beam is rotated in a plane perpendicular or parallel to the hydrotechnical model's bottom. The intersection line of this plane and the bottom relief represents the sought profile - in the first case a vertical section, and in the second - a horizontal one (isobate).

The thus visualized section is recorded by the TV system and processed by a specialized microcomputer.

In Fig. 1 is presented schematically the experimental set-up for measurement of a vertical profile and the coordinate system used. The location of the profile being measured along the Y axis is determined by the optical head's location on this axis, and it is specified by the experimental conditions. The TV system, consisting of a TV camera and a videoprocessor, enters into the microcomputer data for the X and Z coordinates of the different points of the profile being measured.

The TV camera's orientation relative to the beam's plane is selected so as to observe in the best way the intersection line with the bottom relief. It is necessary only to know the relative orientation of the coordinate system (x,y,z) connected with the camera and the object coordinate system (X,Y,Z). The coordinates' transformation is simplified if the directions of both coordinate systems' axes coincide and if the beam plane is perpendicular to the Y axis.

In this case we have:

$$(1) \qquad \begin{pmatrix} X \\ Z \end{pmatrix} = - \frac{a+b}{f} \begin{pmatrix} x \\ z \end{pmatrix} + \begin{pmatrix} X_o \\ Z_o \end{pmatrix}$$

where (X,Z) are the coordinates of a point of the profile being investigated, (x,z) are the coordinates of this point's image on the TV tube's target. $(X_o, -b-f, Z_o)$ are the coordinates of the origin of (x,y,z) relative to (X,Y,Z), f is the focal length of the TV camera's lens. The beam plane is described in the coordinate system (X,Y,Z) with the equation $Y = a$. The values of a, b, f, X_o, Z_o are all known.

Instead of the coordinates (x,z), into (1) can be substituted their corresponding values, obtained by digitizing the TV image of the profile being investigated. Then we obtain:

$$(2) \quad \begin{pmatrix} X \\ Z \end{pmatrix} = - \frac{a+b}{f} \begin{pmatrix} K_x & N_x \\ K_z & N_z \end{pmatrix} + \begin{pmatrix} X_o \\ Z_o \end{pmatrix}$$

where N_x and N_z are the values computed by the microcomputer, as coordinates of the respective point, and K_x and K_z - coefficients. For each profile point, the microcomputer computes N_x and N_z. The coefficients $-\frac{a+b}{f} K_x$ and $-\frac{a+b}{f} K_z$ can be determined experimentally by physically locating in the beam plane of reference points with known coordinates.

If a window is present between the TV camera and the profile measured, the coordinates (X,Z) calculated after equation (2) are incorrect, because the window shifts the images of the object points:

$$(3) \quad \begin{pmatrix} X' \\ Z' \end{pmatrix} = \begin{pmatrix} X \\ Z \end{pmatrix} - \begin{pmatrix} \Delta X \\ \Delta Z \end{pmatrix}$$

where X', Z' are the real object coordinates, and X, Z are the object coordinates calculated according to (2).

If we assume that the window is flat, with thickness t and refraction index n, and is parallel to the (XZ) plane, then:

$$(4) \quad \begin{pmatrix} \Delta X \\ \Delta Z \end{pmatrix} = \frac{n-1}{n} \frac{t}{Y} \begin{pmatrix} X-X_o \\ Z-Z_o \end{pmatrix} \quad .$$

This equation gives acceptable results if $X/Y \leqslant 0.1$, $Z/Y \leqslant 0.1$.

Equations (2) and (4) can be combined so that the above correction is included in the K_x and K_z coefficients.

In this case, if a window is present between the camera and the profile measured, the object coordinates are calculated according to equation (2), but the K_x, K_z coefficients must be replaced with:

$$(5) \quad \begin{aligned} K'_x &= K_x (1 - \frac{n-1}{n} \frac{t}{Y}) \\ K'_z &= K_z (1 - \frac{n-1}{n} \frac{t}{Y}) \end{aligned}$$

As a source of light a He Ne laser with 10 mW output power is used; it is mounted in a common case with the optical system for creation of a light plane.

The TV system for bottom relief registration consists of a TV camera and a videoprocessor (see Fig. 2), which performs the following operations:
- elimination of the light background from the image and obtaining of the bottom relief as a light line on a dark background;
- digitizing of the TV image into 512 x 512 elements;
- generation of signals for connection with the microcomputer.

A personal microcomputer IBM PC has been used for information processing.

Since the rate with which data are transmitted by the videoprocessor exceeds the microcomputer's data acquisition capabilities, a quick FIFO type buffer memory with a size of 128 20-bit words has been used.

The buffer memory is connected to the microcomputer bus via a specialized interface.

The digitized image (512 x 512 elements) is processed by the microcomputer according to the following algorithm:
(a) The addresses of the beginning and the end of the light sectors in each TV line are determined.
(b) The middle of these sectors is computed. In this way the N_x and N_z values for the profile points are obtained. The thus reduced information is stored for further processing.
(c) The data stored are processed by an application program package which gives several outputs - a smoothed profile, calculation of the real object's scale, etc. - depending on the experimenter's requirements.

(d) The end result from the processing is output on a plotter, hardcopy   or printer. Output to a higher level computer is also envisaged.

3.   DESCRIPTION OF THE EXPERIMENTAL SET-UP. EXPERIMENTAL RESULTS

The new method for bottom relief measurement was experimented in the   investigation of models of coastal protection and coast fortification facilities with novel construction in the BSHC wave flume.

With this method, the influence of each structure type on underwater coastal slope lithodynamics, and the stability of ground foundation around the facility   under storm waves, were investigated.

The measuring device was calibrated only once in operation conditions, before the experiment start. The procedure included suitable location in the camera's field of view of a reference grid with step interval 0.010 m along the X and Z axes. The grid was drawn on a photographic plate with the aid of a precise photoplotter. The plotting accuracy was 1.10^{-5} m. After processing the grid image, the calibration coefficients along the X and Z axes were obtained, taking into account the nonlinear distortions in the TV scan, and the influence of the windows.

The wave flume dimensions (see Fig. 1) are: length 27 m; breadth 0.8 m   and depth 0.8 m. One of the working section walls has nine windows with dimensions 1.000 x x 0.700 x 0.014 m.

The laser with the optical head and the TV camera were mounted on a mobile carriage, moving along levelled rails above the flume. The fixing of the camera's coordinate system relative to the object's coordinate system was realized by means of a rigid frame connecting the TV camera and the optical head. The light beam's plane was perpendicular to the channel bottom and parallel to its longitudinal axis. In this way, as a result of the measurement, a longitudinal cross-section of the bottom profile in the region in front of the hydrotechnical structure's model was obtained. The measurement was performed with dry bottom. The TV camera observed the object through one of the windows and was located at a distance 2 m from the beam's plane. A longitudinal section sector with about 0.7 m length was seen in the frame. With this set-up the measuring device resolution was about 0.00135 m along the X and Z axes.

The results from TV image processing were output in graphical form on a plotter, and a sample can be seen in Fig.3a,b,c(curves 1).Three profiles are shown measured at three longitudinal sections of the bottom in the region being investigated.

In the same figures (curves 2) are shown profiles calculated for Y = 0.00 m, Y = -0.2 m and Y = 0.2 m from isobates obtained by the stepwise draining method.   The good coincidence is seen of the results obtained by the two methods. Deviations are observed in the steep slope sectors.

Table 1 contains a list of the X and Z coordinates of the points of the profiles shown in Fig. 3, obtained by the method described in the present paper.

The analysis of the measurement accuracy which the method ensures shows   that the following error sources exist in the data registration and processing:
- image discretization error (512 x 512 elements);
- error due to "smearing" of the light spot on the bottom;
- error due to violation of the image totality because of "shading" in the   investigation of strongly indented reliefs with comparatively steep slopes;
- error due to the presence of a window between the object and the TV camera;
- error due to coordinate systems transformation.

As seen from the above, in the method's technical realization an attempt   was made to eliminate the influence of these error sources.

For example, the possibility for error due to "smearing" of the light spot is eliminated by the procedure described of averaging the light sectors in each TV line.

The error due to TV scan nonlinearity and the presence of windows between the camera and the object is avoided by means of a precise calibration procedure.

The error due to image "shading" is reduced to a minimum by applying the described approximation procedure in the results processing.

To decrease the absolute error due to image discretization and increase the resolution, a zoom optics was used, but the effect was at the expense of the decrease of observed bottom sector dimensions.

As compared with the method used so far in the BSHC Hydraulic Laboratory for bottom relief measurement, the method experimented on ensures higher measurement accuracy and greater reliability of the results.

4. CONCLUSIONS

The results obtained show that in the conditions of a real hydrotechnical experiment the method described has the following advantages over the method of waterlines photographing:
- the necessity is eliminated of stepwise drainage, during which in the periods of keeping the same level the bottom relief is eroded;
- the time for the experiment is decreased, since the large-volume manual processing of the photographic material is eliminated;
- the measurement accuracy and results reliability are improved;
- conditions are created for measurement automation with all its advantages;
- the method allows direct measurement both of isobates and of cross-sections, and the necessity for extrapolation is eliminated.

A convenience in the method's technical realization is the fact that the measuring device is composed mostly of standard modules (laser, TV camera, videoprocessor, microcomputer), and only the optical head, buffer memory and software need to be developed specially.

As a shortcoming, the comparatively high cost of the equipment can be pointed out.

Discussing the method's future development and perspectives for its application, the central attention should be paid to the possibility for using the method under water and in full-scale investigations.

On the basis of the experience gained it can be asserted that the method is applicable in these conditions as well, without principle changes.

It should be noted that an underwater TV camera is needed for the purpose and, instead of a He Ne laser, e.g. an Argon laser with increased power should be used as source of light.

ACKNOWLEDGEMENTS

The authors express their gratitude to the BSHC management for the all-round help in the method's realization and investigations' organization, as well as to the staff of the BSHC Measuring Equipment and Hydrotechnical Laboratories for the assistance in the software development and experiments conduct in the wave flume.

REFERENCES

1. Kobus H., Editor: "Hydraulic Modelling". German Association for Water Resources and Land Improvement, Bulletin 7, pp. 133-134.
2. Novak P., Cabelka J.: "Models in Hydraulic Engineering". Pitman Publishing Program, 1981.
3. "Fundamental Research on Sediment Transport". Delft Hydraulics Laboratory, Hydro Delft, No 42, March 1976, pp. 1-7.

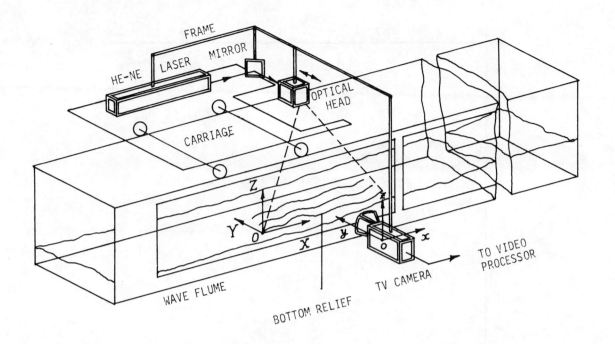

Fig. 1. Experimental set-up

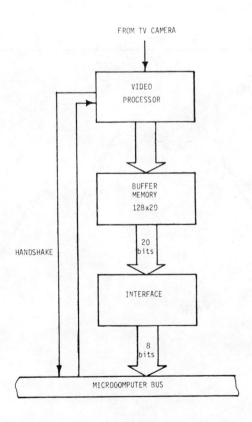

Fig. 2. Block-diagram of the signal processing system

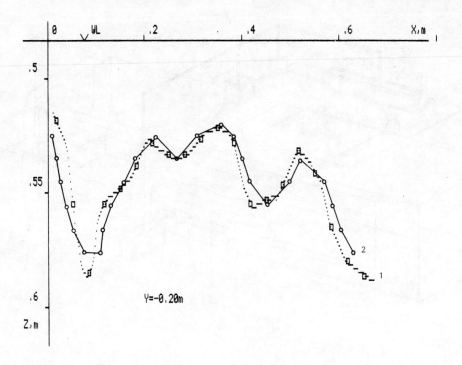

Fig. 3a. Graphical presentation of the measured profile at $Y = - 0.20$ m
     1 - results obtained by the TV method
     2 - results obtained by the stepwise draining technique

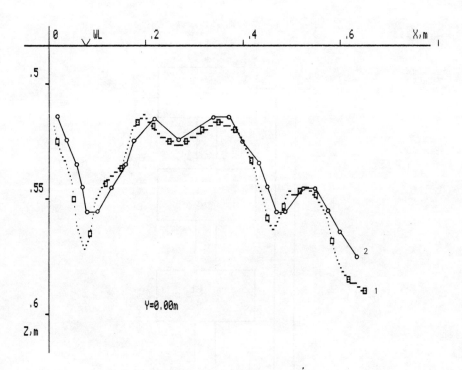

Fig. 3b.

Graphical presentation of the measured profile at the center plane of the flume ($Y = 0.00$ m)
1 - results obtained by the TV method
2 - results obtained by the stepwise draining technique

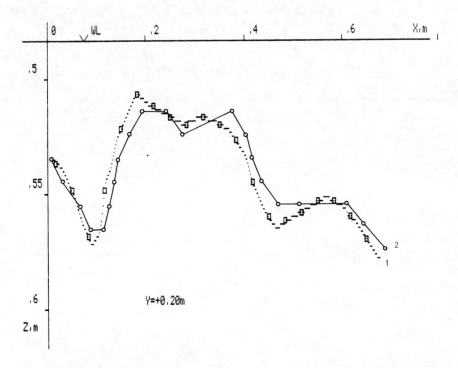

Fig. 3c. Graphical presentation of the measured profile at Y = 0.20 m
1 - results obtained by the TV method
2 - results obtained by the stepwise draining technique

No	X ,m	Z ,m		
		Y=-.20m	Y=.00m	Y=+.20m
1	0.017	0.518	0.525	0.537
2	0.050	0.555	0.550	0.548
3	0.083	0.585	0.565	0.568
4	0.117	0.555	0.543	0.548
5	0.150	0.548	0.537	0.522
6	0.183	0.538	0.517	0.507
7	0.217	0.528	0.518	0.512
8	0.250	0.533	0.525	0.517
9	0.283	0.533	0.525	0.520
10	0.317	0.527	0.520	0.517
11	0.350	0.522	0.517	0.520
12	0.383	0.528	0.520	0.527
13	0.417	0.555	0.533	0.545
14	0.450	0.553	0.558	0.560
15	0.483	0.547	0.553	0.562
16	0.517	0.532	0.547	0.558
17	0.550	0.542	0.548	0.553
18	0.583	0.565	0.568	0.553
19	0.617	0.580	0.585	0.560
20	0.650	0.587	0.590	0.570

Table 1. Object coordinates of the points marked with ∏
in Figures 3a, b, c

INTERNATIONAL CONFERENCE MEASURING TECHNIQUES BHRA,

APRIL 1986, LONDON, ENGLAND

EVALUATION OF MEASURING INSTRUMENTS FOR SUSPENDED SEDIMENT

L.C. van Rijn

Harbours and Coasts Division

Delft Hydraulics Laboratory

A.S. Schaafsma

Instrumentation Department

Delft Hydraulics Lababoratory

Summary

This paper provides information about the performance and accuracy of measuring
instruments for suspended sediment in field conditions. Some of the most popular
instruments for sand transport measurements like the USP-61, the Delft bottle, the
pump-filter sampler and the acoustical sand transport meter (AZTM) are compared based
on field measurements. A similar comparison of commercially available optical
instruments like the Partech Twin-Gap, the Eurcontrol, the Metrawatt GTU and the
Monitek for measuring silt concentrations is presented. New developments are also
reported. Finally, information is given of instruments for the in-situ measurement of
the settling velocity of suspended sediment particles.

Held at Imperial College of Science and Technology, London. Organised and sponsored by BHRA, The
Fluid Engineering Centre. Co-sponsored by the American Society of Civil Engineers and the
International Association for Hydraulic Research.
©BHRA, The Fluid Engineering Centre, Cranfield, Bedford MK43 0AJ, England 1986.

1. INTRODUCTION

Sediment transport measurements in field conditions can be divided into two
categories, being:

- detailed measurements for basic research,
- routine measurements for morphological studies.

Good examples of basic research measurements are those reported by Soulsby et al
(1985), who measured instantaneous sand concentrations in a tidal current by using an
impact-probe. Another example is the study reported by Mulder et al (1985), who
measured instantaneous sand concentrations and transport rates by using an acoustical
sand transport meter (AZTM).

Both categories of measurements require different types of instruments. For basic
research studies sophisticated acoustical or optical instruments enabling the
continuous sampling of the instantaneous variables are of essential importance. High
accuracy and resolution are required. For routine measurements it usually is
sufficient to determine the time-averaged values of the suspended sediment
concentrations. This information can be obtained with sufficient accuracy by using
relatively simple mechanical samplers such as the trap-type, bottle-type and pumping-
type samplers, as will be shown in Par. 2. To select the most appropriate instruments
for a routine field survey, qualitative and quantitative information of the physical
parameters to be measured should be available prior to the actual survey. It is
important to have information of the various transport modes at the sampling site such
as the (relative) value of the wash load and the bed material load, the bed and
suspended load, the flow velocities and sediment concentrations (low or high). Another
important criterion for instrument selection is the purpose and the accuracy required
for the morphological study. This paper presents information of the above mentioned
items. Emphasis is given to measuring errors in relation to the required accuracy.
Simple and sophisticated suspended load samplers for silty and sandy conditions are
described and evaluated. New developments are also discussed.

2. MEASURING ERRORS AND REQUIRED ACCURACY

Inevitably, errors are made in sediment transport measurements due to instrumental
errors and sampling errors. The errors can be systematic as well as random due to the
stochastic nature of the variables to be measured. An example of a systematic
instrumental error is the inefficient trapping of fine sediment particles (< 100 μm)
by the Delft bottle. A good example of a systematic sampling error is the error
introduced by determining the time-averaged suspended sediment transport ($\bar{s}$) as the
product of the time-averaged velocity ($\bar{u}$) and concentration ($\bar{c}$) resulting in $\bar{s}=\bar{u}\,\bar{c}$,

while it should be determined as the time-averaged product of the instantaneous velocity (u) and concentration (c) resulting in $\overline{s} = \overline{uc} = \overline{u}\ \overline{c} + \overline{u'c'}$. The $\overline{u'c'}$- term represents the diffusive transport due to the velocity (u') and concentration (c') fluctuations.

This type of systematic error was recently investigated by Soulsby et al (1985) and Mulder et al (1985).

Soulsby et al measured instantaneous sand concentrations and velocities at 0.13 m and 0.33 m above the bed of a tidal river by using an impact probe. They found that the mean transport $(\overline{u}\ \overline{c})$ at both heights was at least 100 times larger than the diffusive flux $(\overline{u'c'})$. The latter value was generally negative. Their results are shown in Figure 1. They also conclude that the commonly-made assumption that the time-averaged sediment transport is given by the product of the time-averaged velocity and concentration $(\overline{u}\ \overline{c})$ is sufficiently adequate for practical purposes. Similar results were found by Mulder et al (1985), who measured instantaneous sand concentrations and transport rates by using an acoustical sand transport meter (AZTM). The contribution of the diffusive transport $\overline{u'c'}$ was found to be smaller than 5%. The mean value of $\overline{u'c'}$ appeared to be negative.

To get a better understanding of the consequences of the errors of the measured sand transport rates in relation to the errors of the basic hydraulic variables (mean velocity, surface slope, particle size, water depth and width), the total load formula of Engelund-Hansen (1967) is considered. This formula reads as:

$$S = \alpha \frac{u^2 b\ (hi)^{1 \cdot 5}}{\Delta^2\ g^{0 \cdot 5}\ D_{50}} \qquad (1)$$

in which: S = total load transport (m³/s), u = cross-section averaged velocity (m/s), b = width (m), h = water depth (m), i = water surface slope (-), $\Delta = (\rho_s - \rho)/\rho$ = relative density (-), g = acceleration of gravity (m/s²), D_{50} = median particle size of bed material (m), α = coefficient (= 0.05, as proposed by Engelund and Hansen, 1967).

Generally, the coefficient proposed by Engelund-Hansen (α = 0.05) is not the most optimum value for a specific river. Improvement can be obtained by performing special calibration measurements (field survey). Assuming that the basic variables are stochastically independent, the relative error of the α-coefficient can be expressed as:

$$\left(\frac{\sigma_\alpha}{\alpha}\right)^2 = \left(\frac{\sigma_S}{S}\right)^2 + 4 \left(\frac{\sigma_u}{u}\right)^2 + \frac{9}{4} \left(\frac{\sigma_h}{h}\right)^2 + \frac{9}{4} \left(\frac{\sigma_i}{i}\right)^2 + \left(\frac{\sigma_b}{b}\right)^2 + \left(\frac{\sigma_{D50}}{D50}\right)^2 \qquad (2)$$

in which σ = standard deviation of each variable.

Typical relative errors of the velocity, water depth, width and particle size are $\sigma_u/u = 0.1$, $\sigma_h/h = 0.05$, $\sigma_b/b = 0.05$, $\sigma_{D50}/D50 = 0.1$. The relative error of the water surface slope may be rather large, especially in isolated field conditions, say $\sigma_i/i = 0.15$-0.25. Using these values, the relative error of the α-coefficient has been computed. The results are shown in Figure 2.

Accepting an instrumental and sampling (relative) error of 50% in S, the relative error in α is about 65% which is not much larger than the minimum value of about 45%. Thus, it is not very efficient to use a sophisticated measuring instrument with an instrumental error less than say 10%, even when the number of samples is so large that the random error introduced by the stochastic nature of the sediment transport process is small. Usually, however the number of samples to determine the sediment discharge in the entire cross-section is so small, that the stochastic sampling error becomes dominant. This is another reason for applying relative simple measuring instruments. A disadvantage of the simple instruments, especially the bottle-type instruments, is the relative expensive laboratory analysis of the samples. The more sophisticated optical and acoustical instruments donot have this disadvantage and are therefore, attractive although their use may not be necessary for reasons of accuracy.

Not only the instrumental and sampling errors are of importance, but also the required accuracy of morphological predictions. Ideally, one might first wish to determine what level of accuracy would be acceptable in morphological predictions for a specific project and to select a sampling method which would meet this requirement. Thus, the question of required accuracy can not be answered in general, but it should be studied for each project. It largely depends on the nature of the project and the applied prediction methods (mathematical model or physical scale model). Interesting papers on this subject have been presented by De Vries (1982, 1983).

3. SUSPENDED LOAD SAMPLERS FOR SAND PARTICLES (> 50 μm)

3.1 INSTRUMENTS

A wide range of instruments is available from the simple mechanical trap-type and bottle-type samplers to the sophisticated acoustical and impact probes. In this paper only those instruments which are widely used, are considered. These are: the trap, the bottle, the USP-61, the Delft bottle, the pump sampler, the acoustical sampler and the impact sampler.

Trap

The trap sampler consists of a horizontal cylinder with two valves on both ends which can be closed suddenly (by a messenger). Using this sampler, an instantaneous water-sediment sample is collected.

Bottle

The most simple bottle sampler consists of a bottle which is placed vertically in a container. Arrived at the sampling point, the bottle is opened by pulling a rope. The filling time varies from about 20 to 400 seconds depending on the bottle orientation, flow velocity and sampling position (Delft Hydraulics Laboratory, 1980). A cork ball in the bottle can be used to close the bottle after filling.

USP-61

The sampler consists of a streamlined bronze casting ($\simeq$ 50 kg), which encloses a small bottle. The intake nozzle, which can be opened or closed by means of an electrically operated valve, points directly into the approaching flow. The filling time varies between 10 to 50 seconds, depending on the flow velocity (Federal Inter-Agency, Minnesota, USA).

Delft bottle

This sampler ($\simeq$ 20 kg) is based on the flow-through principle, which means that the water entering the intake nozzle leaves the bottle at the backside. The intake nozzle points directly into the approaching flow. As a result of the bottle geometry, there is a strong reduction of the flow velocity inside the bottle resulting in the trapping of sand particles larger than about 100 µm.

Pump-sampler

This sampler consists of an intake nozzle connected to a pump by means of flexible hose. The hose diameter should be as small as possible to reduce the stream drag. Basically, the intake velocity should be equal to the local flow velocity. Measurements in laboratory conditions, however, have shown that a deviating intake velocity does not result in large errors (< 20%), provided that the intake velocity is not smaller than 0.8 times the local flow velocity or not larger than 2 times the local flow velocity (Nelson and Benedict, 1950). Hulsbergen and Bosman (1986) investigated the influence of the intake nozzle orientation to design a pump sampling method for oscillatory flow conditions (coastal zone). Their provisional results are presented in Figure 3.

To avoid the storage of large volumes of water, an in-situ sediment separation method should be used. This can be done by using a filter method (50 µm filter size) or by using a sedimentation method, as shown in Figure 4. Hydro-cyclones have also been used (Ref. 21). The filter method cannot be used in a silty environment because of rapid filter blocking (Crickmore and Aked, 1975; Van Rijn 1979). The sedimentation method also allows the determination of the silt concentration by taking a water sample from the outlet during emptying of the container (Van Rijn, 1980).

Acoustical sampler

An acoustical sand transport meter (AZTM) has been developed at the Delft Hydraulics Laboratory (Jansen 1978, 1981; Schaafsma and Der Kinderen, 1985). This sampling instrument is based on the scattering and attenuation of ultrasound waves by the suspended particles. The scattered energy and the decrease in energy of the continuing beam (attenuation) are related to the particle concentration. For a fixed concentration the sensitivity of the scattered or attenuated signal shows a maximum value when the particles have a size D equal to λ/π (λ = wave length in water). For larger sizes the sensitivity decreases linearly with D, for smaller sizes the sensitivity decreases with D^3. Although this description is somewhat simplified, it emphasizes the main point of the particle size dependence in a semi-quantitative way. Since the spectrum of values of λ/π is relatively wide (1 to 1000 µm) depending on the selected frequency, an optimum choice can be made for each specific field situation. The presently available instrument is based on a frequency equal to 4.5 Mhz to optimize for a grain size of about 100 µm, which is a typical value for the estuaries in the Netherlands. Laboratory calibrations carried out in a flume showed a relative inaccuracy of the concentration of about 10% for particles in the range of 80 to 300 µm. The concentrations were as large as 5000 mg/l. A great advantage of the acoustical samplers is the continuous and simultaneous measurement of velocity, concentration and transport of the suspended sediment particles. A similar instrument as the AZTM has been developed by Lenn and Enever of the Cranfield Institute of Technology, Bedford, England (1985).

Impact sampler

An impact sampler has been developed at the Institute of Oceanographic Sciences (IOS) in Taunton, England (Salkield et al, 1981). The impact sampler is based on the principle of momentum transfer. The high density of the sand particles gives them excess momentum over the surrounding water so that they tend to strike a transducer placed in the stream. The impact transfers some of the momentum of the sand particles to a sensor (piezo-electric ceramic element) which converts it to an electric output signal. The output is a count of frequency of impacts. The detected impact rate depends on the mass, the velocity and the angle of impact of the particles. A separate instrument is necessary to measure the flow velocity, which is a disadvantage of the impact method. Calibration is necessary to relate the impact rate to the sand concentration. Salkield et al (1981) showed results for small concentrations only (< 100 mg/l). Information of the upper measuring limit was not given.
Downing (1981) applied an impact sensor to measure the bed-load transport.

3.2 COMPARATIVE MEASUREMENTS

USP-61, Delft bottle and pump-filter sampler

During May 1979 a field investigation using the USP-61, the Delft bottle and the pump-filter sampler was carried out in the Danube river near Ilok, Yugoslavia (Dijkman and Milisic, 1982). The three instruments were used at the same level above the bed within a few metres from each other. The vertical positioning was done by means of an echo sounder attached to each instrument. Figure 5 shows typical results of the measured sediment transport per size fraction for each instrument at a height of 0.55 m above the bed and a flow velocity of 0.9 m/s.
The sediment transport measured by the USP-61 sampler shows rather large fluctuations, even for the silt particles (< 70 µm). The fluctuations are caused by the relatively short sampling period ($\simeq$ 30 seconds). For individual samples the inaccuracy may be as large as 50%. To obtain a reliable average value, many samples should be collected.

The Delft bottle also shows fluctuations, although the sampling period was rather long (600 to 900 seconds). Probably, the fluctuations are caused by sediment losses through the openings at the backside of the instrument during hoisting of the instrument. Compared with the USP-61 sampler, the results of the Delft bottle are systematically smaller, especially for the 70-150 μm size fraction. For all measurements the average difference between the Delft bottle and USP-61 varies from about 10 to 100%, depending on the grain size fraction and nozzle type. The pump-filter sampler shows relatively small fluctuations (sampling period ≈ 300 s). The filter size was 50 μm. Laboratory analysis, however, showed that almost all particles smaller than 70 μm were lost. For all measurements the results of the pump-filter sampler are about 15% lower than those of the USP-61 sampler.

Pump-filter and pump-sedimentation sampler

Sand concentration measurements were carried out in the New Waterway near Rotterdam (Van Rijn, 1980). The intake nozzles of both samplers were installed next to each other on a carrier body. The sampling period for the pump-sedimentation sampler was about 180 seconds. The sampling period of the pump-filter sampler was only 90 seconds due to rapid filter blocking (silt particles). Figure 6 shows sand concentrations (> 50 μm) at 1.0 m above the bed measured by both samplers. The overall agreement is quite satisfactory.

Pump-filter and acoustical AZTM sampler

Sand concentration measurements using both samplers were carried out in the Eastern Scheldt estuary, the Netherlands. The intake nozzle of the pump-filter sampler was installed close to the probe of the acoustical (AZTM) sampler. Figure 7 shows sand concentrations (> 50 μm) at 1 m above the bed during flood tide. The agreement is quite good. The maximum deviations are about 30% for small concentrations (< 50 mg/l). Figure 8 shows a similar comparison for concentrations in the range of 10 to 300 mg/l. The maximum deviations are about 20%.

3.3 NEW DEVELOPMENTS

Lenn and Enever (1985) showed some new developments of the acoustical sampling methods. They used the ratio of the sound scattering under two different angles (10° and 170°) to obtain a measure of the average grain size. This ratio increases with average grain size for relatively narrow particle size distributions within the range of 60 to 300 μm approximately. They also used the average grain size distribution to eliminate the grain size dependence of the sensitivity of the concentration measurement. However, it is still necessary to make assumptions regarding the shape of the particle size distribution. These assumptions should be checked regularly during field measurements. New developments of acoustical sampling methods in field conditions will also be presented by Crickmore et al during the present conference. Another new development is the acoustical measurement of sand concentrations in

vertical direction based on the echo-sounding principle. Hay and Heffler (1983) have
formulated design considerations for an ultrasonic sediment profiler employing a
multi-frequency (0.7-7 Mhz) range-gated system. A first prototype was built and tested
in the laboratory and in the field (Hay, 1985). Many problems remain to be solved to
obtain a reliable field instrument. Research should be intensified because this
approach is of great value offering a possibility to measure the near-bed
concentrations without disturbing the local hydraulic conditions.

Finally, some new developments regarding acoustical sampling methods at the Delft
Hydraulics Laboratory are presented. The attention is focussed on the multi-frequency
(or broadband) approach. Flammer (1962) already measured the frequency dependence of
ultrasonic attenuation for many different sand size fractions in the sound frequency
range of 2.5 to 25 Mhz. His results were encouraging as regards the possibility of
measuring the concentrations independent of the actual sand size distribution. The
research at the Delft Hydraulics Laboratory is aimed at measuring concentrations of
silt and sand particles in the range of 1 to 1000 μm. For that purpose the frequency
dependence of the ultrasonic attenuation and the scattering, at some selected angles,
will be measured for the relevant sizes in the frequency range of about 1 to 100 Mhz.
The idea is that within a given size interval and depending on the nature of the
particles (sand or silt) the frequency dependence of the ultrasonic properties will
represent a characteristic finger print. A preliminary result is shown in Figure 9
which demonstrates the potential of the multi-frequency approach (Wolthuis and
Schaafsma, 1985).

4. SUSPENDED LOAD SAMPLERS FOR SILT PARTICLES (< 50 μm)

4.1 INSTRUMENTS

The simple mechanical trap, bottle and pump samplers as described in paragraph 3 can
also be used to determine the silt concentrations. However, the mechanical analysis of
the samples is laborious (filtering) and therefore not attractive for economical
reasons. More sophisticated samplers giving a continuous and direct recording of the
silt concertrations are the optical samplers based on the scattering and attenuation
of light waves. The value of λ/π (see par. 3.1) is approximately 0.1 μm for light in
water, which means that for particles larger than this value the concentration
measurement will be linearly dependent on the particle size. Since in natural
conditions many particles sizes do occur, regular calibration measurements are
necessary which is a major disadvantage of the optical samplers. Other problems are
the influence of ambient light, pollution of the sensors and a varying background
turbidity of the fluid. The optimum light path distance in relation to statistical
aspects of the particles in the measuring volume has been studied by Bosman (1981).
Commercially available instruments for field conditions are the Partech Twin-Gap, the
Eurcontrol, the Metrawatt GTU 702, and the Monitek 230 and 350. These instruments have
been used for comparative measurements in a flume.

Partech Twin-Gap

This instrument is based on light-transmission using two light paths with different lengths. The submersible sensor contains a light source with on both sides a light sensitive resistor being part of a wheatstone-bridge. The influence of ambient light is relatively small due to a small aperture angle of the detector.

Eurcontrol

This instrument is based on light-transmission using two light paths with different lengths. The submersible sensor consists of two light sources and two photodiodes. The light sources are used in turn for a period of one second. Two sensor types are available: the TAG 120/25 for the concentration range of 100 to 3000 mg/l and the TAG 30/15 for the range of 100 to 10000 mg/l. The TAG 120/25 sensor, which has been tested, is very sensitive to ambient light. Recently, the manufacturer has developed a new sensor which does not have this deficiency.

Monitek 350

This instrument is based on the light-transmission and the light-scattering principle. The submersible sensor contains two light sources which are used in turn. Two photodiodes are used to detect the attenuated and the scattered light. The sensor is not sensitive to ambient light.

Monitek 230

This instrument is based on the light-transmission and the light-scattering principle. The non-submersible sensor contains a light source that is a simple lamp producing a narrow light beam by means of a system of lenses. Two photodiodes are used to detect the attenuated and the scattered light. This instrument can only be used in combination with a pump sampler to produce a continuous water-sediment flow for the (on-board) optical system.

Metrawatt GTU 702

This instrument is based on the light-transmission and the light-scattering principle using infra-red light. The sensor is non-submersible. The instrument can only be used in combination with a pump sampler to produce a continuous water-sediment flow for the optical system.

4.2 COMPARATIVE MEASUREMENTS

The optical samplers were tested simultaneously in a large recirculating flume of the Delft Hydraulics Laboratory using various types of silt (Der Kinderen, 1982). The silt concentrations were varied in the range of 1 to 1000 mg/l. The silt concentrations were determined from water-sediment samples collected by use of a siphon system (iso-kinetic sampling). The submersible sensors were tested at a height of 0.5 m above the flume bottom. The Monitek 230 and the Metrawatt GTU 702 were used in combination with a pump sampler. The intake nozzle of the pump sampler was located at the same height (0.5 m) above the bed. The output signals of all optical instruments were averaged over 30 seconds (time integration). Figure 10 shows typical results for one type of silt (North-sea silt). Other parameters which have been investigated are: zero-drift, temperature influence, ambient light influence and the influence of sand particles.

The zero-drift of the instruments was tested by performing clear water tests regurlarly during the test period (8 weeks). The influence of water temperature fluctuations was tested by comparing the output signals for a temperature of 10°C and 20°C. The influence of ambient light was tested by darkening the test section of the flume. The influence of sand particles was investigated by measuring the output signal increase after increasing the initial silt concentration with an equal sand concentration.

The Partech-Twin Gap shows a good linearity. The inaccuracy below concentrations of 100 mg/1 is large because of zero-drift fluctuations. The instrument is sensitive to ambient light. The effective measuring range is 100 to 1000 mg/1.

The Eurcontrol TAG 120/25 shows a good linearity and accuracy from 100 to 1000 mg/1. The instrument is very sensitive to ambient light and cannot be used in practice without special equipment to eliminate the ambient light influence. The effective measuring range is 100 to 3000 mg/1.

The Monitek 350 shows linearity from 10 to 400 mg/1. The zero-drift and ambient light influence is minimal. The effective measuring range is from 10 to 1000 mg/1.

The Monitek 230 shows linearity from 100 to 1000 mg/1. This instrument is not sensitive for concentrations smaller than 100 mg/1. The effective measuring range is from 100 to 2000 mg/1.

The Metrawatt GTU 702 shows linearity from 1 to 300 mg/1. The zero-drift and ambient light influence is minimal. The effective measuring range is 10 to 1000 mg/1.

All instruments are insensitive for sand particles. Increasing the initial silt concentration present in the flume with an equal sand concentration resulted in an increase of the output signal of not more than 10%.

4.3 NEW DEVELOPMENTS

Minor improvements of the commercially available optical instruments are possible with respect to the influence of ambient light, sensor pollution and electronic stability of the equipment. However, the basic problem of the strong particle size dependence can not be overcome. Therefore, new developments of the optical samplers are not be expected.

More promising is the application of the acoustical sampling methods for the measurement of the silt concentrations, as reported in par. 3.3.

5. EVALUATION OF SUSPENDED LOAD SAMPLERS

The suspended load samplers for sand and silt particles (described in paragraphs 3 and 4) are evaluated by considering the following characteristics: measuring range, response time, sampling period, minimum cycle period and overall inaccuracy.

The sampling period is the time period during which a sample is collected. For the bottle-type instruments the sampling period is equal to the filling period of the bottle. The sampling period of the Delft bottle is restricted by the volume of the sediment catch which should be small compared with the volume of the sedimentation chamber of the instrument. The sampling period of the pump-filter sampler is restricted by the filter characteristics. A 50 μm-filter size may be blocked rather easily, especially when there is a small background concentration of silt. The sampling period of the optical and acoustical instruments is free.

The minimum cycle period is the minimum time period between two successive measurements in adjacent points in the vertical. The minimum cycle period is of importance to evaluate the total time period needed to complete the measurement of a concentration profile. For tidal flow conditions this latter period should be small in relation to the tidal period.

The overall inaccuracy is an estimate of the overall error in a single measurement due to instrumental and sampling errors. Random sampling errors are introduced by stochastic fluctuations of the measured variables. Eyster and Mahmood (1976) showed that single bottle measurements may result in a sampling error of about 100%. This type of error can only be reduced by collecting many samples or by increasing the sampling period. The accuracy of the optical and acoustical samplers is largely dependent on the accuracy of the calibration curves. For field conditions the scatter usually is relatively large. The evaluation results are summarized in Table A.

	Suspended load samples (point-integrating)	silt (< 50 µm)	sand (> 50 µm)	measured variable	measuring range concentration (mg/1)	submersible sensor	respons time (s)	sampling period (min)	minimum cycle period (min)	overall accuracy (one measurement) silt	sand
Mechanical	Trap	yes	yes	concentration	> 1	-	instantaneous	instantaneous	5	100%	100%
	Bottle	yes	yes	concentration	> 1	-	-	1	5	100%	100%
	USP-61	yes	yes	concentration	> 1	-	-	1	5	100%	100%
	Delft bottle	no	yes (> 100 µm)	concentration, transport	> 10	-	-	5-30	10	-	50%
	Pump-filter	no	yes	concentration	> 10	-	-	5-10	10	-	20%
	Pump-sedimentation	yes	yes	concentration	> 10 (silt) > 50 (sand)	-	-	5-10	15	20%	20%
Optical	Eurcontrol TAG 120/25	yes	no	concentration	100-3000	yes	5	free	5^X	50%	-
	Partech Twin-Gap	yes	no	concentration	100-1000	yes	1	free	5^X	50%	-
	Monitek 350	yes	no	concentration	10-1000	yes	10	free	5^X	50%	-
	Monitek 230	yes	no	concentration	100-2000	no	<1	free	5^X	50%	-
	Metrawatt GTU 702	yes	no	concentration	10-1000	no	10	free	5^X	50%	-
Acoustical	AZTM	no	yes	velocity, concentration, transport	10-5000	yes	0.1	free	5^X	-	20%
Impact	IOS	no	yes	concentration	5-10000	yes	0.2	free	5^X	-	20%

X Assuming a sampling period of 3 minutes

Table A: Summary of measuring instruments

Focussing on the accuracy of the instruments, the pump filter sampler and the acoustical AZTM sampler are the most attractive instruments for sandy conditions. For silty conditions the optical instruments with a submersible sensor are the most attractive instruments. A great advantage of the optical instruments is the possibility of continuous recordings over longer periods using in-situ data storage. For silty and sandy conditions the pump-sedimentation sampler can be used. The traditional bottle-type instruments are still attractive because they are suitable for all type of conditions. It is essential to collect many samples to reduce random sampling errors. A great disadvantage is the handling and analysis of a large series of bottles.

6. IN-SITU DETERMINATION OF SETTLING VELOCITY OF SUSPENDED SEDIMENT

6.1 INSTRUMENTS

In a saline environment it is of essential importance to determine the settling velocity curve of the suspended sediments by using an in-situ analysis method. Applying such a method, the disturbance of the settling flocculated particles is supposed to the minimal. Owen (1976) introduced the bottom-withdrawal tube for field conditions. Basically, this method consists of the sedimentation of particles in a tube with a length of about 1 m and a diameter of about 0.05 m (see Figure 11). The withdrawals are made at the bottom of the tube at preset times.

A suitable schedule for particles in the range of 5 to 100 μm is withdrawals at 3,6,10,20,40,60 and 120 minutes.

The principle of the bottom-withdrawal tube method is that at time T, being the ratio of the settling length (L) of the tube and the settling velocity (W), all particles with a settling velocity smaller than (W) are deposited at the bottom of the tube at a constant rate $(dG/dt)_T$. Consequently, the total weight of particles with a settling velocity smaller than W can be represented by $T(dG/dt)_T$. However, the total weight of all deposited particles is G_T. Hence, the difference represents the weight of the particles with a settling velocity larger than W. Owen (1976) designed an in-situ measuring instrument based on the bottom-withdrawal tube method. For this purpose the tube was equipped with two valves on both ends. The instrument is lowered horizontally with opened valves to the sampling point. After closing the valves, the tube is raised and placed in a vertical position on board of the survey vessel (start of sedimentation process). The bottom-withdrawal tube has two major disadvantages: the complicated removal of the silt particles deposited at the bottom of the tube and the time-consuming procedure to compute the settling velocity curve.

Therefore, a new field instrument was developed at the Delft Hydraulics laboratory. This latter instrument is based on the principle of the Andreasen-Esenwein pipet method (see Figure 12).

The basic principle of the pipet method is that particles having a settling velocity greater than the ratio of the sampling depth (H) and the elapsed time period (T) will have settled below the sampling depth. The weight percentage of the particles with a settling velocity smaller than W = H/T is equal to c/c_o with c = sediment concentration at time T and c_o = initial concentration at time T = 0. Based on the simple principle of the Andreasen-Esenwein Pipet, a new instrument has been designed for making large side-withdrawals (≈ 0.2 l), which is of importance for reasons of accuracy (Van Rijn and Nienhuis, 1985). This new instrument is called the side-withdrawal tube (see Figure 13). Based on pilot experiments, the following dimensions were chosen: tube diameter = 0.12 m, initial settling height (H_o) = 0.2 m, and internal diameter of the withdrawal tube = 0.003 m. For in-situ sampling in field conditions, the tube is equipped with two valves on both ends. The tube is lowered horizontally and raised vertically (see Figure 14).

6.2 COMPARATIVE MEASUREMENTS

Comparative measurements were carried out in a laboratory set-up using a bottom-withdrawal tube (Fig. 11), a side-withdrawal tube with initial settling height H_o = 0.3 m (Fig.13) and a commercially available Sartorius balance-accumulation tube. This latter instrument which is assumed to give the most accurate results, consists of a small tube with a length of about 0.3 m and a diameter of about 0.1 m and is equipped with an under-water balance accurate to 1 mg. For reasons of accuracy (weighing), the initial concentration should not be smaller than about 200 mg/1.

Various types of silts with initial concentrations in the range of 20 to 2000 mg/1 were used. Fig. 15 shows comparative results of the side-withdrawal tube (SW-tube) and the balance accumulation tube (BA-tube). The results of the BA-tube are represented by an "average" curve for all tests (initial concentrations in the range of 200 to 2000 mg/1). As can be observed, the SW-tube shows good agreement with the results of the BA-tube. Even for a small initial concentration of 20 mg/1 the results of the SW-tube are quite good. Figure 16 shows the results of the bottom-withdrawal tube (BW-tube) and those of the balance-accumulation tube (BA-tube). Only for initial concentrations larger than 500 mg/1 the BW-tube shows good agreement. For an initial concentration of 20 mg/1 large deviations can be observed. The poor results of the BW-tube for small concentrations were caused by deficiencies of the bottom-withdrawal system with respect to the sediment particles deposited at the bottom of the tube. Although the flushing velocities generated in the cone and nozzle were rather large during each withdrawal, a small part of the sediment particles was not removed as could be observed visually (perspex tube). In case of small initial concentrations (< 100 mg/1) the amount of unremoved particles is large enough to introduce errors. Field experience with the Owen-tube, as used by the Delft Hydraulics Laboratory, also shows problems with respect to the withdrawal system. The side-withdrawal tube for field (in-situ) conditions (Fig. 14) was used in a laboratory flume to evaluate the accuracy of the instrument. The sediment suspension consisted of water and very fine silicaflour. Two experiments were carried out; one with a sediment concentration of about 150 mg/1 and another with a concentration of 450 mg/1. For comparison a siphon sampler was used to collect water-sediment samples simultaneously with those of the side-withdrawal tube. The nozzle of the siphon sampler was located as close as possible (at the same height above the flume bottom) to the side-withdrawal tube. The siphon samples were transferred to a balance-accumulation tube (Sartorius, 1981) directly after collection of the samples. Figure 17 shows settling velocity curves based on the side-withdrawal tube method and the Sartorius BA-tube. This latter instrument is assumed to give the best results. Good agreement between both methods can be observed for both concentrations (150 and 450 mg/1).

6.3 EVALUATION

Based on comparative measurements in flume conditions, the new side-withdrawal tube shows good results for initial concentrations larger than 20 mg/1. The bottom-withdrawal tube shows good resuls for initial concentrations larger than 500 mg/1, but less good results for concentrations in the range of 20 to 500 mg/1.

7. CONCLUSIONS

The information presented in this paper can be summarized by the following conclusions:

1. It is sufficiently accurate to represent the time-averaged suspended sediment transport by the commonly-made assumption of the product of the time-averaged velocity and concentration ($\bar{u} \ \bar{c}$).

2. For reasons of accuracy it is not necessary to use a sophisticated instrument for routine measurements of suspended sediment transport because of the relatively large measuring errors of the basic flow variables to which the sediment transport is related and because of the limited number of samples which usually is collected resulting in large stochastic sampling errors.

3. Comparative sand transport measurements using various instruments show
   - relatively large concentration fluctuations for the USP-61 because of the short sampling period,
   - transport rates which are systematically too small (10 to 100%) for the Delft bottle because of the inefficient trapping of the particles smaller than about 100 µm,
   - good results for the pump-filter sampler, the pump-sedimentation sampler and the acoustical sampler (AZTM).

4. Comparative silt concentration measurements using various optical instruments show:
   - an effective measuring range of 100 to 1000 mg/l and a rather large zero-drift instability for the Partech Twin-Gap,
   - an effective measuring range of 100 to 3000 mg/l and a rather large influence of ambient light for the Eurcontrol TAG 120/25,
   - an effective measuring range of 10 to 1000 mg/l for the Monitek 350,
   - an effective measuring range of 100 to 2000 mg/l for the Monitek 230,
   - an effective measuring range of 10 to 1000 mg/l for the Metrawatt GTU 702.

5. In-situ determination of the settling velocity of the suspended sediment particles can be best performed by using the side-withdrawal tube method, which shows accurate results for initial concentrations larger than 20 mg/l.

6. For routine transport measurements in sandy conditions the pump samplers are the most attractive. The acoustical AZTM sampler is attractive for economical reasons (no sample analysis) or when continuous recordings of the turbulence characteristics are necessary. For silty conditions the optical instruments with a submersible sensor are the most attractive, especially when long-term recordings are necessary (in-situ data storage). For silty and sandy conditions the pump-sedimentation sampler can be used. The traditional bottle-type instruments are still attractive for all type of conditions, provided that many samples are collected to reduce the stochastic sampling errors.

7. A new development for measuring silt and sand concentrations is the multi-frequency acoustical sampling method.

REFERENCES

1. Bosman, J., 1981: Optical Measurement of Sediment Concentrations. Report R 716 IV, Delft Hydraulics Laboratory, Delft, The Netherlands

2. Crickmore, M.J. and Aked, R.F., 1975: Pump Sampler for Measuring Sand Transport in Tidal Waters, Conf. on Instrumentating Oceanography, Proc. No. 32, Bangor, England.

3. Downing, J.P., 1981: Particle Counter for Sediment Transport Studies, J. of the Hydraulics Division, ASCE, 107, No. HY11, Proc. Paper 16633.

4. Delft Hydraulics Laboratory, 1980: Investigation Vlissingen Bottle (in dutch), Report M 1710, Delft, The Netherlands.

5. Dijkman, J. and Milisic, V., 1982: Investigations on Suspended Sediment Samplers, Delft Hydraulics Laboratory and Jaroslav Cerni Institute, Report S 410, Delft, The Netherlands.

6. Engelund, F. and Hansen, E., 1967: A Monograph on Sediment Transport in Alluvial Streams, Teknisk Forlag, Copenhagen, Denmark.

7. Eyster, C.L. and Mahmood, K., 1976: Variation of Suspended Sediment in Sand Bed Channels, River Sedimentation, Vol. II, Fort Collins, USA.

8. Federal Inter-Agency: Sedimentation Project, Instructions for USP-61, Suspended Sediment Sampler, Minnesota, USA.

9. Flammer, G.H., 1962: Ultrasonic Measurement of Suspended Sediment, Geological Survey Bulletin 1141-A, US Government Printing Office (Washington).

10. Hay, A.E. and Heffler, D., 1983: Design Considerations for an Acoustic Sediment Transport Monitor for the Nearshore Zone, National Research Council Canada, Report No. C2 S2 - 4.

11. Hay, A.E., 1985: Private communication

12. Hulsbergen, C. and Bosman, J., 1986: Tranverse Sediment Suction Sampling, to be published in Coastal Engineering, Amsterdam.

13. Jansen, R.H.J., 1978: The In-Situ Measurement of Sediment Transport by means of Ultrasound Scattering, Delft Hydraulics Laboratory, Publication No. 200, Delft, The Netherlands

REFERENCES (continued)

14. Jansen, R.H.J., 1981: Combined Scattering and Attenuation of Ultrasound, IAHR workshop on Particle Motion and Sediment Transport, Rapperswil, Switzerland.

15. Kinderen, W. der, 1982: Comparison of Optical Silt Concentration Probes (in dutch), Report M 1799-I, Delft, The Netherlands.

16. Lenn, C.P. and Enever, K.J., 1985: The Measurement of the Size and Concentration of Suspended Fine Sands Using Scattered Ultrasound, Proc. 21 st IAHR Congress, Melbourne.

17. Mulder, H.P.J., Kolk, A.C. van der and Kohsiek, L.H.M., 1985: Measurements of Suspended Sand Concentrations and Velocity in the Eastern Scheldt Estuary, The Netherlands, Euromech 192, Munich, West-Germany.

18. Nelson, M.E. and Benedict, P.G., 1950: Measurement and Analysis of Suspended Sediment Loads in Streams, ASCE-proceedings, Vol. 76, USA.

19. Owen, M.W., 1976: Determination of Settling Velocities of Cohesive Muds, Hydraulics Research Station Wallingford, Report No. IT 161, Wallingford, England.

20. Rijn, L.C. van, 1979: Pump-Filter Sampler, Report S 404-I, Delft The Netherlands.

21. Rijn, L.C. van, 1980: Methods for In-Situ Separation of Water and Sediment, Delft Hydraulics Laboratory, Report S 404-II, Delft, The Netherlands.

22. Rijn, L.C. van, and Nienhuis, L.E.A., 1985: In-Situ Determination of Fall Velocity of Suspended Sediment, 21 th IAHR-Congress, Melbourne, Australia.

23. Salkield, A.P., LeGood, G.P. and Soulsby, R.L., 1981: Impact Sensor for Measuring Suspended Sand Concentration, Conf. on Electronics for Ocean Technology, Birmingham, England.

24. Sartorius, 1981: Aufstellungs und Bedienungs-anweisung, Sartorius Werke GMBH, D-3400, Göttingen, West-Germany.

25. Schaafsma, A.S. and der Kinderen, W.J.G.J., 1985: Ultrasonic Instruments for the Continuous Measurement of Suspended Sand Transport, Proc. IAHR Symp. on Measuring Techniques in Hydraulic Research, Delft (Balkema Publ. Rotterdam).

REFERENCES (continued)

26. Soulsby, R.L., Salkield, A.P., Haine, R.A. and Wainwright, B., 1985: Observations of the Turbulent Fluxes of Suspended Sand near the Sea-Bed, Euromech 192, Munich, West-Germany.

27. Vries, M. de, 1982: A Sensitivity Analysis Applied to Morphological Computations, Third Congress APD-IAHR, Bandung, Indonesia.

28. Vries, M. de, 1983: On Morphological Forecasts for Rivers, Second Int. Symp. River Sedimentation, Nanjing China.

29. Wolthuis, A.J.W. and Schaafsma, A.S., 1985: to be published.

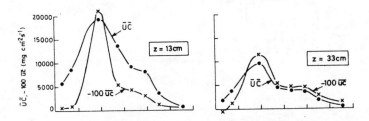

Figure 1   Measured longitudinal sand transport
at a height of 0.13 and 0.33 m above
the bed during flood tide.

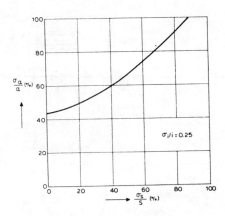

Figure 2   Influence of measuring
errors of the sediment
transport rates.

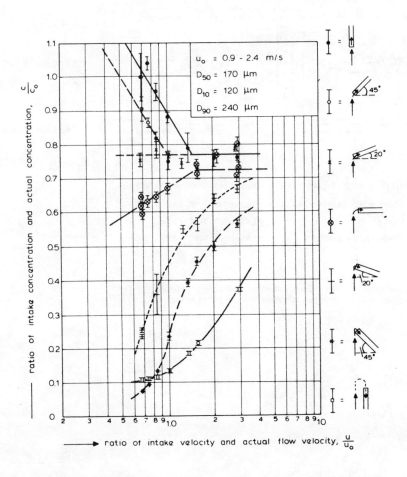

Figure 3   Sampling error as a function of intake
velocity ratio and intake nozzle orientation.

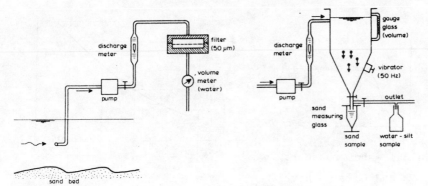

Figure 4
Filter method and
sedimentation method.

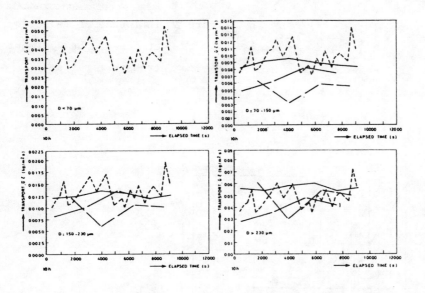

Figure 5
Comparison of USP-61,
Delft bottle and pump-
filter sampler.

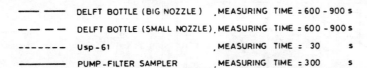

— — — DELFT BOTTLE (BIG NOZZLE) , MEASURING TIME = 600 - 900 s

— — — DELFT BOTTLE (SMALL NOZZLE) , MEASURING TIME = 600 - 900 s

- - - - - - Usp-61   , MEASURING TIME = 30   s

———— PUMP-FILTER SAMPLER   , MEASURING TIME = 300   s

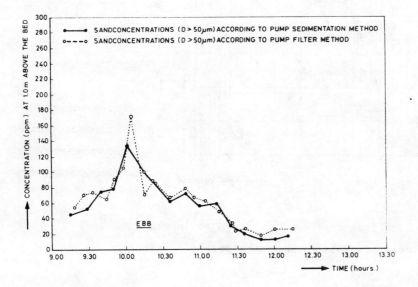

Figure 6
Comparison of pump-filter
sampler and pump-
sedimentation sampler.

420

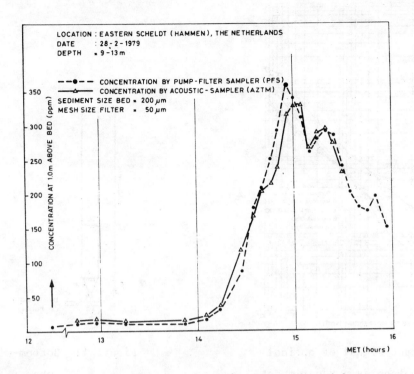

Figure 7   Comparison of pump-filter sampler and
acoustical AZTM sampler in field conditions.

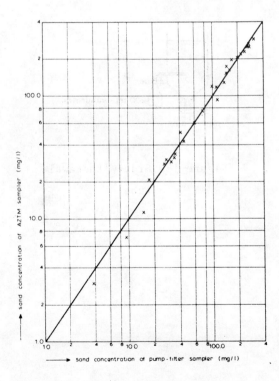

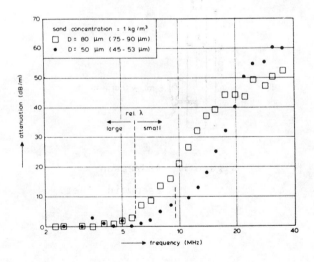

Figure 9   Frequency dependence of
ultrasonic attenuation of sand
suspensions of two different
size fractions.

Figure 8   Comparison of pump-filter
sampler and acoustical AZTM
sampler in field conditions.

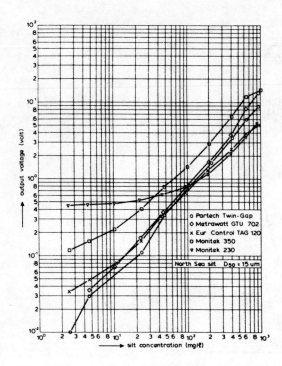

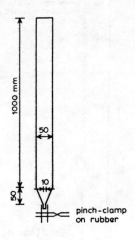

Figure 10   Output signal of optical
samplers as a function of
silt concentration.

Figure 11   Bottom-withdrawal
tube

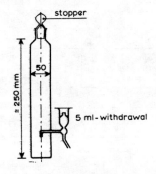

Figure 12   Andreasen-
Esenwein Pipet.

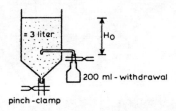

Figure 13   Side-withdrawal
tube

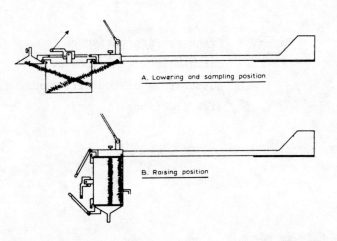

Figure 14
Side-withdrawal tube for field
conditions.

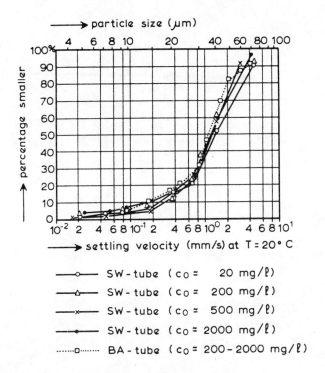

Figure 15 Comparative results of
SW-tube and BA-tube.

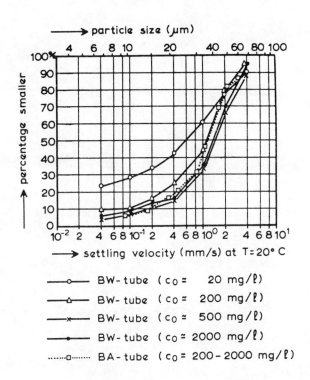

Figure 16 Comparative results of
BW-tube and BA-tube.

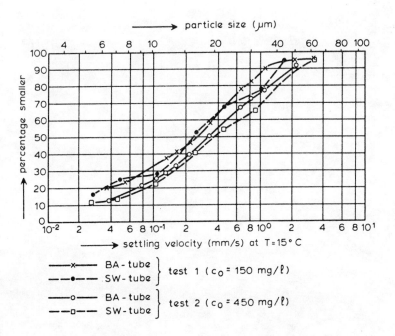

Figure 17 Comparative results of SW-tube and
BA-tube.

423

A FIELD INSTRUMENT FOR MEASURING THE CONCENTRATION AND SIZE OF FINE SAND SUSPENSIONS

M J Crickmore B.A. & I E Shepherd C.Eng, MIERE, MIMC

Hydraulics Research Ltd,
Wallingford Oxon OX10 8BA
England

P M Dore, B.Eng, MSc

D B E Technology
Aldershot, Hampshire
England

Summary

Progress on the development of an instrument suitable for the measurement of fine sand suspensions in rivers, estuaries and coastal waters is described. The principle of extracting information on the concentration and particle size of sand suspensions from measurements of scattered and attentuated acoustic intensity has been demonstrated by experiments on a laboratory rig at the Cranfield Institute of Technology. The present paper reports on the extension of this earlier study into the design and construction of a field prototype and its subsequent testing. The transducer array comprises four ceramic piezoelectric crystals, so configured that one serves as an acoustic source operating at 6 MHz, and the remaining three as receivers for the direct, back scattered and forward scattered ultrasound. The construction of the d.c. powered transducer array, the housing of the associated underwater electronics and cabling are appropriate for operating to an immersed depth of 30m. Results obtained by on-line processing of the attenuated and scattered signals on a BBC microcomputer are presented to show its sensitivity to concentrations up to 5000 mgl^{-1} and its ability to discriminate the particle size of typical suspensions from 60 to 250µ m diameter.

Held at Imperial College of Science and Technology, London. Organised and sponsored by BHRA, The Fluid Engineering Centre. Co-sponsored by the American Society of Civil Engineers and the International Association for Hydraulic Research.

NOMENCLATURE

a_p = pressure attentuation coefficient

D_{50} = median diameter

n = number of 30s records used for the calculation of mean and standard deviation

r = particle radius

λ = acoustic wavelength

σ = backscatter cross-section

1   INTRODUCTION

Our understanding of the sedimentation aspects of many tidal and coastal engineering
questions is frequently hampered by the difficulty of obtaining reliable data on the
concentration and grain size of sediment suspensions. Traditional methods rely on
collecting representative samples of sediment-water mixtures either by sampling bottles and other immersed containment devices or by pumping the mixture to the surface.
The field operations and the subsequent laboratory analysis of the large number of
samples to describe adequately the sediment regime of an estuary are formidable
tasks.

It is possible in cases of fine grained sediment suspensions such as clay and silt to
reduce substantially the number of recovered samples by adopting "in-situ" measurement
of a concentration dependent property of the flow. Optical sensors using light attentuation or light scattering to determine the water's turbidity are commonly used for
clay and silt suspensions provided adequate steps are taken to calibrate the sensor
with some samples of the natural suspension (Ref 1). The required calibration procedures for fine-grained sediment need not be onerous. Although the concentration of
suspensions above muddy beds is flow-dependent, broadly speaking the size grading of
the mud moving in a given reach remains constant. Unlike sand in transport, the size
fractionation of muds over the depth of flow is weak. Thus a single but site-specific
calibration of the sensor over the expected concentration range normally suffices for
a mud study.

Unfortunately the calibration is not so straightforward when the particle size of the
suspension extends into the fine sand range (60 to 200 μm) and above. Mobile sand
suspensions in typical estuarine and coastal environments exhibit not only a strong
dependence of concentration but also of grain size distribution, on both flow strength
and height above the bed. Thus the successful "in-situ" measurement of sand suspensions ideally demands the independent sensing of concentration and particle size.
Normally both characteristics are essential for the making of a full engineering
appraisal of the sedimentary regime. Even when the objective is only to determine
concentration, the chosen sensing principle must be capable of differentiating the
response due to concentration from that due to particle size. Standard turbidity
sensors as used for mud suspensions fail to satisfy this requirement: low concentrations of fine particles can bring about the same optical attentuation and scattering
as very high concentrations of sand. Progress on a field sensor design that can
measure simultaneously the concentration and particle size of fine sand suspensions is
the subject of the present paper.

In 1981 Hydraulics Research Ltd (then Hydraulics Research Station, Department of the
Environment) instigated a desk study by Cranfield Institute of Technology into likely
methods of meeting the above objective. The conclusion of the study favoured a com-
pound sensor using scattered ultrasound operating either in the multi-angular or in
the multi-frequency mode. With the continuing support of Hydraulics Research,
Cranfield Institute of Technology explored the feasibility of the multi- angular
approach by building an adjustable emitter and detector rig on which the acoustic
scattering characteristics from a 8 MHz source in the presence of a sediment suspension could be investigated. Experimental data obtained with the laboratory rig
indicated that a four-transducer array held promise of achieving concentration and
particle size discrimination, with one transducer acting as the acoustic power source
and the other three enabling the detection of backward and forward scattering and of
attenuation within the medium (Ref 2). The results were sufficiently promising to
prompt Hydraulics Research into commissioning the design and construction by D B E
Technology, Aldershot of a prototype field instrument based generally on the findings
of the earlier study. The guiding theoretical principles are outlined only briefly in
the present paper, because the more detailed physical arguments have already been put
in Ref.2.

2   DESIGN OBJECTIVE

The first goal is to produce a battery-powered sensor capable of operating to water depths of 30m but with its power supply and signal processing electronics at the surface, thus permitting operator intervention. The sediment concentration and size range of interest extends from 0 to 5000 mgl^{-1} and 60 to 300 μm particle diameter, respectively. The wish is to determine the former with a repeatability of better than ± 5 percent of reading or 1 mgl^{-1} whichever is the greater, and the mean particle size of the suspension to the nearest 10 μm. Furthermore, the sensor is to be relatively insensitive to sediment finer than 60 μm so that measurement of the sand suspension remains possible in the presence of sediment mixtures that include silt and clay particles, a common feature of many estuaries.

3   ACOUSTIC SCATTERING

A theoretical and detailed experimental investigation by Flammer (Ref 3) demonstrates that the attentuation of a sound wave by waterborne sediment suspensions is a complex function of concentration and particle size of the sediment taken in conjunction with the frequency of the acoustic source. Flammer describes two major loss mechanisms. The first is by viscous dissipation where the particles vibrate in response to but with some lag behind the sound wave and is important when the acoustic wave length (λ) is large compared with particle circumference ie. $\lambda \gg 2\pi r$. The second is the loss of energy by scattering due to a reradiation by the particles of the incident wave; and where again for a given acoustic wavelength the directional pattern and intensity of the reradiated waves are functions of the circumference of the scatterers.

Three domains of scattering can be distinguished. For $2\pi r/\lambda \ll 1$, the Rayleigh region, the scattering pattern is predominantly in the backward directions and its intensity diminishes rapidly with decreasing particle size. For $2\pi r/\lambda \gg 1$, the geometric or shadowing region, scattering is more concentrated in the forward direction. In the transition zone between the two, the so-called diffraction region where $2\pi r/\lambda \cong 1$, the scatter distribution becomes more complex with interference phenomena causing oscillations in the intensity with respect to particle size. The theoretical variation in the backscattered intensity for the three domains can be derived from the non-dimensional curve (Fig 1a) after Urick (Ref 4). Similarly Fig 1b taken from Ref 5 specifically illustrates the theoretical effect of particle size on the attentuation of 6 MHz sound by a suspension of 1000 mgl^{-1}. It is evident from these plots that by selecting an appropriate source frequency such that $2\pi r/\lambda = 1$ is towards the lower end of the sand size band of interest then most of the scattered energy will be from sand. The contribution of finer particles is markedly lower because of their position in the Rayleigh scattering domain, thus enabling in principle the desired selectivity for sand versus silt and clay to be achieved. The present field instrument operates at 6 MHz whereas the earlier laboratory rig utilised a 8 MHz source. The shift to the longer acoustic wavelength was considered necessary to further the instrument's discrimination against silt. At the chosen frequency, $2\pi r/\lambda$ values of 0.75 to 3.8 correspond to the sand sizes of interest. Although advantage is thereby taken of the diffraction domain to minimise the size dependence of backscatter intensity for fine sand (Ref 2), it is not clear whether the oscillations that feature in the domain could lead to undesirable instability in the determination of concentration at these particle sizes.

The changes in the directional distribution of scatter with changing particle size referred to earlier could disclose information about the size of the sediment in suspension. Theoretical polar scatter diagrams given by Lenn (Ref 6) for the distribution around a rigid sphere suggest a strong size dependence of the ratio of forward to backward scatter if measured close to the acoustic beam axis (Fig 1c).

Although of value in guiding our general approach it has to be recognized that the above theory is only an approximate description of scattering from suspended sand particles. First, the theory assumes that particles are sufficiently separated to avoid multiple scattering: namely, that the total scattering intensity is simply proportional to the sum of the individual scattering cross-sections of the particles comprising the population. Secondly, the theory relates to rigid spheres. Sediment particles are not perfect spheres and in the transition zone, $2\pi r/\lambda = 0.5$ to 3.0, Anderson (Ref 7) has shown that their compressibility has a significant effect on scatter intensity.

4   MEASUREMENT CONCEPTS

Jansen (Ref 8) has shown that acoustic backscatter can be used in nature to measure fine sand concentration. However, for optimum performance his gauge needs to be calibrated against numerous samples of the natural sand suspension. It was anticipated (Ref 2) that by adding the detection of forward scatter to that of backscatter, both at small angles to the beam's axis, the present development would advance the technique. As described in the previous section the measurement of forward scatter theoretically provides guidance on particle size which could be exploited to limit the need for excessively detailed site-specific calibration.

Jansen (Ref 9) found it necessary to compensate for the exponential absorption of energy along the transmitted and scattered beam paths. By taking simultaneous readings of attentuation and backscatter and expressing the output as backscattered/attentuated intensity ratios he considerably improved the linearity of output as a function of concentration. At the same time it reduced sensor dependence on the temperature and salinity of the medium and on any variability in the strength of the acoustic source. For the present development a fourth transducer is placed facing the direct acoustic beam to take similar advantage of combining an attentuated signal with the two scattered signals.

Attentuation is determined from the reduction in the intensity of the primary continuous 6 MHz wave incident on the fourth transducer. The strength of the two scattered intensities on the other hand is weak relative to the reception of this direct beam. In order to extract these weak signals in the presence of much direct radiated power advantage is taken of the doppler frequency shift arising from the advection of the waterborne suspension through the beam. It follows that both scattered outputs are based on the intensity of the doppler-shifted spectrum at frequencies ranging from 1 Hz to several kHz.

In brief, it emerges that the requirement is for an instrument radiating a narrow beam of 6 MHz ultrasound and capable of evaluating two ratios:

(a)   backscatter intensity/attenuated intensity, being primarily a function of concentration, but partly particle size-dependent;

and

(b)   forward scatter intensity/backscatter intensity, being a function of the particle size grading within the suspension.

The laboratory detection system developed for testing the basic measuring principles was far from being in a form that could be adapted for field trials. A prototype field instrument was therefore engineered by D B E Technology based closely on the measurement concept established by the earlier study (Ref 2). The outcome is:

(a)   a packaged underwater sensor assembly suitable for lowering on its own 100m power cable from a boat or jetty;

(b)   a surface electronic unit providing the necessary power supply and conditioning for the transducer signals;

(c)   software to be run on a BBC micro-computer with the necessary operating instructions to permit data gathering, processing and performance testing.

The system block diagram and hardware are shown in Fig 2 and photograph 1.

A single transducer is used to transmit continuously at 6 MHz. Three other identical transducers serve as receivers positioned to detect the direct, back scattered, and forward scattered signals. The two scattered signals are combined with the transmit signal in a double balanced mixer outputting the doppler shifted components which are amplified and converted to an equivalent d.c. level. The direct signal from the transmitter is buffered and amplified before being input to an envelope detector to extract signal amplitude. The three d.c. signals are then multiplexed and transferred via the cable to a decoder and finally to a BBC micro-computer for processing. A means is incorporated of driving the receive circuits directly with the transmitter disabled so that the stability of that part of the system can be readily checked.

In the underwater unit as supplied the three d.c. receiver signals were fed to an automatic gain controller (AGC) in parallel with the multiplexer stage. The intention was to control the transmitter power such that the largest of the three receiver outputs, normally the direct or attentuated output, was held constant. In principle this approach could improve the dynamic range of the instrument and at the same time makes the setting of the gains of the three receivers less critical. In practice, however, it was found necessary to disable the automatic gain control loop in order to allow closer inspection of the dependence of the individual signal strengths on sediment concentration and size. In particular it was found that at low concentrations the AGC acting on the high direct signal level limited transmitted power to a degree that scattered signal strengths were negligible. Thus the data presented here are with the AGC disabled and with individual gain settings optimised experimentally.

At the top end of the system the analogue outputs of the three receiver channels are accepted by the BBC micro-computer which collects and stores on disc readings in batches of 100 from each of the channels ie. 30s record. The readings may be displayed on screen or printed either at the time of recording or by retrieving the values from disc. Mean values of the three received signals and of the two ratios, backscatter/ direct and forward scatter/backscatter, are also displayed at the end of each 30s record.

The ceramic piezoelectric crystals of the four transducers are positioned in pairs on a plane which is assumed to be horizontal for standard field applications. The transducer that serves as the direct or attentuated receiver faces the source transducer across a 300mm gap. The backscatter and forward scatter receivers are set alongside the source and direct transducers respectively, on an axis at a 10° displacement from the primary beam axis. All transducers are mounted at the end of short tubular legs projecting beneath the tube holding the underwater electronics.

The intention is to isolate the sensing volume from any flow disturbance by the instrument body when the acoustic path is aligned between 45° to 90° to the flow. The transducers have a measured 3dB beam width of 2.6° at 6MHz. The precise volume of the medium that is sensed in the measuring process is not readily defined. However, in its expected deployment attitude the shallow angle of cut of the direct and scatter paths implies that horizontally the central 50 to 100mm of the path is likely to contribute to the scatter outputs. In the other plane the narrow beam width should allow a vertical resolution of about 10mm.

The aluminium body of the underwater unit is hard anodised and double "0" rings seal the caps at both ends of the electronics housing.

6   CALIBRATION FACILITY

The reproduction of suspensions of known concentration and particle size for calibrating sensor response demands rather special experimental facilities. Flammer (Ref 3) employed a circulatory system in which sediment of known mass and size grading was fully mixed with the circulating water and the acoustic attentuation was measured in a vertical stretch of a continuous loop. The system was completely cleaned between tests whenever the class of sediment was changed. Flammer's measuring crosssection was no more than 76 mm across. Adoption of the same approach in the present case would have required a flow cross-section at least 600mm wide to accommodate the sensor. It was rejected on the grounds that maintenance of a uniform sediment concentration over such a relatively large cross-section implies a high flow rate and there were fears that micro-bubbles would be entrained by the fast flow. Micro-bubbles are particularly strong acoustic scatters and their presence would greatly contaminate the scattered signals.

The alternative approach of comparing sensor output with values obtained by traditional sampling at the same point in open sand-carrying laboratory or natural flows was also not favoured. It is difficult to eliminate micro-bubbles from pumped circulations of typical laboratory channels and it was considered that it would take a long time and be costly to cover a wide range of sand size and concentration by comparative sampling methods in the field.

A more systematic and more easily controlled system of creating suspensions was preferred, whereby acoustic scatter from a cloud of sediment particles was measured as they settled in a confined column through a 1m high, 0.7m diameter chamber of stationary water. The facility used for this purpose is shown in photograph 2, and resembles that employed by Jansen (Ref 8). It had been constructed earlier to assist Lenn in undertaking his investigation (Ref 2). A dry sand feeder, placed above and at the centre of the chamber, consisted of a miniature hopper discharging through an orifice onto a drum rotating on a horizontal axis. The drum rim was perforated by 24 regularly spaced slots so that its rotation interrupted the steady stream of sand from the hopper. Different orifice diameters gave a choice of three rates of hopper feed and several masks altering the mark-space ratio of the drum permitted further variation in the rate that sand was supplied. The sediment fell into a 75mm square cross-section, surface-piercing perspex sleeve extending to about mid-depth. Mechanical stirring was necessary to ensure rapid wetting and mixing of the sediment over the sleeve's cross-section. The acoustic beam was trained on the settling particles as they left the sleeve exit in order to avoid any interruption of the beam by the sides of the sleeve. The concentration of sediment immediately below the exit was readily calculable from the measured rate of feed, the sleeve cross-section, and the fall velocity of the particles. Jansen obtained the last of these parameters by direct measurement of the doppler shift frequency. Less satisfactorily, in the present study it was inferred from an accepted relationship between fall velocity and the size of spheres (Ref 10) and taking the median diameter from sieving as the representative sphere of each tested sand sample.

Several sieved sand fractions within the range 60 to 250 µm were examined in the
rating chamber at nominal concentrations between about 50 and 5000 mgl^{-1}. For a
given particle size and nominal concentration the feed of sand from the injector was
measured immediately before and again at the conclusion of the test. The major
weaknesses of the described procedure were firstly some inconstancy in the feed rates
particularly at the smaller sizes, and secondly the reliance on theoretical rather
than measured fall velocities. Nevertheless the apparatus allowed a systematic
examination of sensor response over a range of concentration and particle size.

7   RESULTS

Operation of the sensor in clean stationary water disclosed zero offsets on both
scattered signals that were attributable to circuit noise within the underwater unit.
These offsets were fortunately reasonably steady throughout the test programme but
prevented full utilisation of the BBC micro-processing software which was developed to
compute automatically the mean ratios referred to at the end of Section 4. However,
for the purposes of the laboratory tests it was deemed sufficiently precise to
calculate the ratios manually after subtracting the offset values from the computed
means of the output values.

Sensor performance for six sand grades and one coarse silt grade is given in Figs 3
and 4. Each plotted point represents the average of up to four individual 30s records
taken when particles of given size and concentration were falling through the beam at
a sensibly steady rate. The family of curves depicting the dependence of forward and
backward scatter on concentration have been fitted by eye. Although there are a
number of rogue points particularly for the 150-180 microns grade, it is clear that
the data are highly sensitive to particle size. It follows that the particle size of
the suspension must be known if a valid estimate of concentration is to be made.

Additional information on size can be obtained from the forward/backscatter ratio as
depicted in Fig 5. The size dependence of this ratio is well in evidence on
photographs 3a and b which illustrate the signal traces for two different sand grades:
the forward scatter exceeding backscatter for the coarser sediment and vice-versa for
the finer sand. The ratio is apparently independent of concentration so that the
results of all tests for a given sand grade were grouped to provide means and standard
deviations. The values plotted on Fig 5 have been adjusted to compensate for the
estimated differences in receiver gain for forward and backscatter and are broadly in
accord with theory (Fig 1c). It would appear that for particle diameters at least
between 100 and 250µm it is feasible to discriminate between sand grades arising from
adjacent standard sieve sizes. The ratio band obtained from a sand made up to
resemble the grading of a typical suspended load is also depicted in Fig 5 and
provides values that are close to the median diameter of that sand. Knowing the size,
the concentration of the suspension can be derived from the appropriate backscatter/
forward scatter curve. It was noticeable under the artificial conditions of the
laboratory tests that the size ratio parameter reduced towards the end of each run as
the more slowly settling finer particles became in the majority.

The correction referred to on Figs 3 and 4 to compensate for acoustic attentuation by
the sediment/water medium was obtained from the measured intensity at the direct
receiver. This correction never exceeded 30 per cent and was less than 10 per cent
for concentrations below 1000 mgl^{-1}.

Although not fully tested in the present series it is normally accepted that the tactic of ratioing measured scatter with the attentuated signal compensates for temperature effects and fluctuations in transmitted power. This being so, the stability of the instruments response to clear water conditions suggests that repeatability is better than 1 per cent of full scale. The non-linearity in scatter response to increasing concentration is advantageous in offering the potential for high reading resolution at concentrations which are most typical for estuarine and coastal applications. For instance, for sieved fractions between 100 and 250 microns diameter 1 per cent of full scale corresponds to better than 50 mgl^{-1} at 1000 mgl^{-1} and is superior to this at lower concentrations.

The apparent size sensitivity of the scatter versus concentration relationship is a fundamental drawback. Size sensitivity is greater for forward scatter than for backscatter, but any uncertainty on which curve to select from the family of curves given in either Fig 3 or Fig 4 leads to widely different estimates of concentration for the same scatter signal. For this reason we believe that it will be essential to support any future field utilisation of the instrument with some suspended-solids sampling by conventional methods. The number of samples that will need to be recovered to provide adequate calibration control is a matter of conjecture but it is expected that a significant reduction can be achieved compared with presentday practice for sediment studies.

The scatter response from silt-sized particles (D_{50} = 6μm) was confirmed to be much less than that from sand. This was particularly so for forward scatter intensity. For instance for equivalent concentrations the forward scatter from silt was only 20 and 6 per cent of that from 105 to 125μm and 150 to 180μm sand grades respectively. Although this represent reasonably effective discrimination against silt, caution will need to be exercised in estuarine environments where silts are frequently present at higher concentrations than sand. In such cases independent checks on the degree of signal contamination by the finer size fraction of the suspended load will be necessary either from sampling or "in-situ" optical sensing of turbidity.

8   CONCLUSIONS

The sensor's performance falls somewhat short of the ambitious target set in Section 2. However, an ultrasonic instrument for field use has been successfully engineered and it's potential for measuring sediment concentration and of discriminating particle size in the fine sand range has been verified. The gain settings of the individual receivers have been adjusted to allow good sensitivity concentrations of suspended sediment at least to 1000 mgl^{-1} throughout the size range 60 to 300μm. Its resolution could be further enhanced by changing the gain settings if interest was confined to sediment sizes that form only part of the above size range.

The results demonstrate that acoustic scatter is more sensitive to particle size than earlier studies suggest and there is little doubt that on field applications the instrument output will need to be supported by traditional sampling at least until a site-specific calibration is established. The technique is highly susceptible to the presence of air bubbles in the medium and is clearly unsuitable for measurements in well-aerated environments such as the breaker zone. However it is anticipated that most estuarine and deeper coastal flows will normally be free of non-sediment scatterers.

The laboratory data presented here do not fully reproduce the geometry of measurements in nature. Some additional attentutation of signal output can be expected in nature because the suspension will be present over the entire acoustic path rather than constrained to the central cross-section used in the calibration chamber. Nevertheless the present data should serve as a valuable framework into which to fit future studies made in the field when spot comparisons will be made between the sensor readings and the results from concurrent pumped recovery of sample suspensions.

9   ACKNOWLEDGEMENTS

This paper describes work carried out under contract No PECD 7/6/56 funded by the Department of the Environment. The DOE nominated officer was Dr R Thorogood. The work was carried out by Mr M J Crickmore and Mr I Shepherd on behalf of Mr M F C Thorn, Head of the Tidal Engineering Department at Hydraulics Research Limited."

10   REFERENCES

1.   Thorn M F C: Monitoring silt movement in suspension in a tidal estuary.   Int. Assoc. Hydraulics Research, Sao Paulo 1975.

2.   Lenn C P and Enever K J: The measurement of the size and concentration of suspended fine sands using scattered ultrasound.   Int. Assoc. Hydraulics Research, Melbourne, 1985.

3.   Flammer G H: Ultrasonic measurement of suspended sediment.   U.S. Geological Survey Bulletin 1141-A, 1962.

4.   Urick R J: Principles of underwater sound.   McGraw Hill, 1975.

5.   Soulsby A L: Sensors for the measurement of sand in suspension.   Institute of Oceanographic Sciences, Taunton. Report No.27, 1977

6.   Lenn C P: The feasibility of ultrasonic methods for measuring fine sand suspension.   Cranfield Fluid Engineering Unit.   Report No. 18423/8, 1982.

7.   Anderson V C: Sound scattering by solid cylinders and spheres.   Acoustical Society America Jour. 22, No.4, 1950.

8.   Jansen R H J: The in-situ measurement of sediment transport by means of ultrasound scattering.   Delft Hydraulics Laboratory Publication No. 203, 1978.

9.   Jansen R H J: Combined scattering and attentuation of ultrasound as a tool for measuring suspended sand transport.   IAHR Workshop on Particle Motion and Sediment Transport, Rapperswil, Switzerland, 1981. ·

10.   Gibbs R J, Matthews M A and Link D A: The relationship between sphere size and settling velocity.   Jour. of Sedimentary Petrology, 41, No.1, 1971.

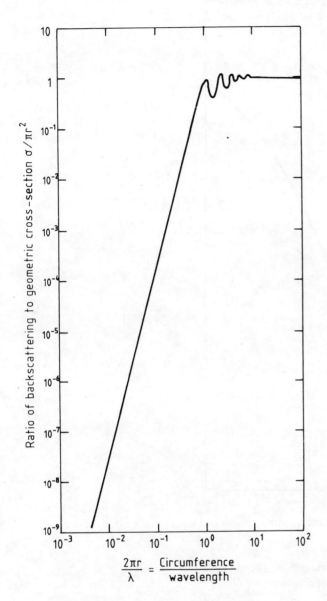

(a) Backscatter from a fixed rigid
sphere of radius r (after Urick, 1975)

(b) Attentuation coefficient for sound of
6MHz in a 1000 mgl⁻¹ suspension of rigid
spheres (after Soulsby, 1977)

1   Theoretical relationships for ultrasound

435

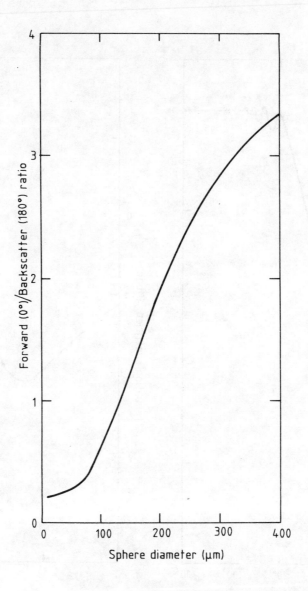

(c) Dependence of forward / backscatter
ratio on rigid sphere diameter 6MHz
(adapted from Lenn, 1982)

1   Theoretical relationships for ultrasound

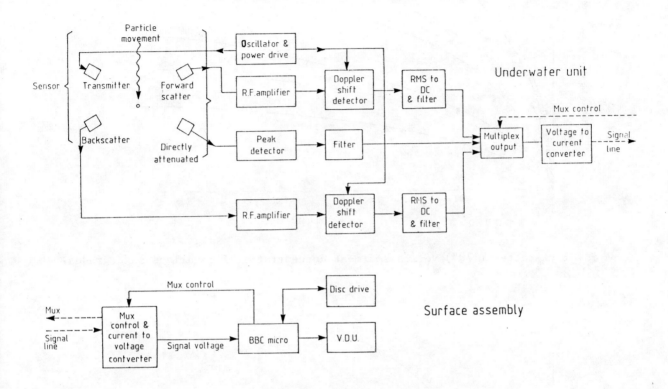

2   Block diagram of system

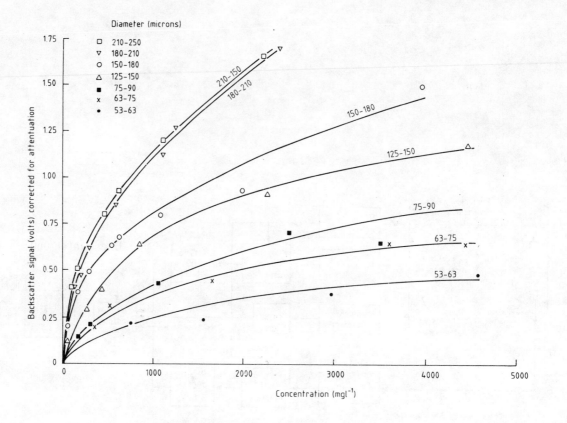

3   Backscatter (170°) versus sediment concentration for various sand grades.

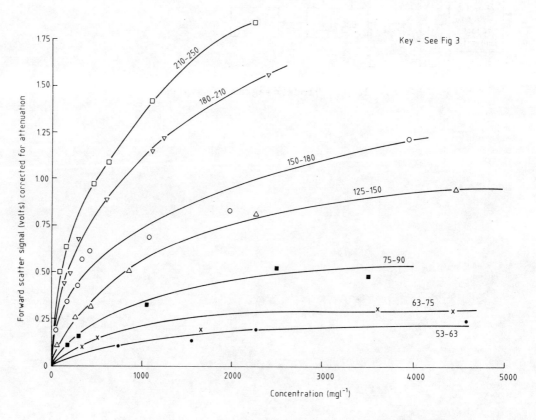

4   Forward scatter (10°) versus sediment concentration for various sand grades.

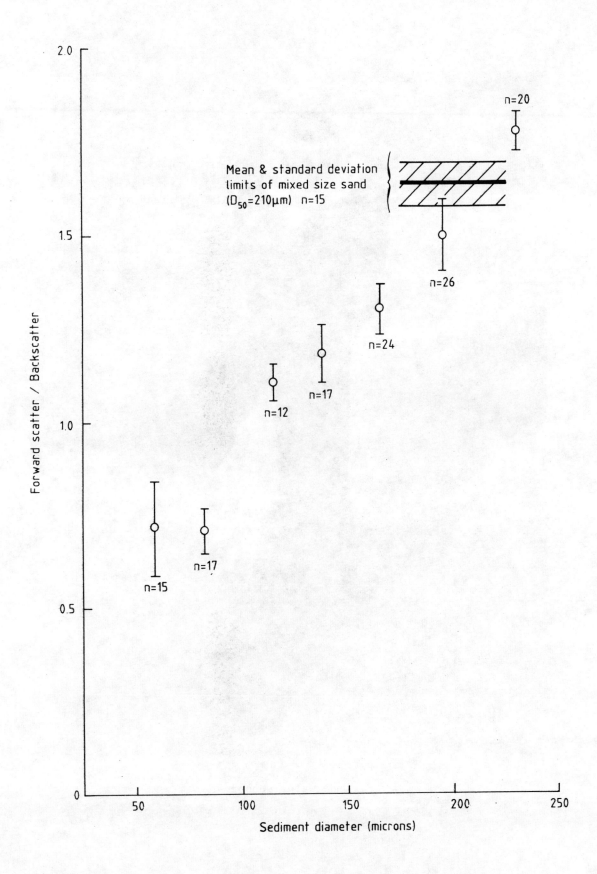

5   Dependence of forward (10°)/backscatter (170°) on particle size

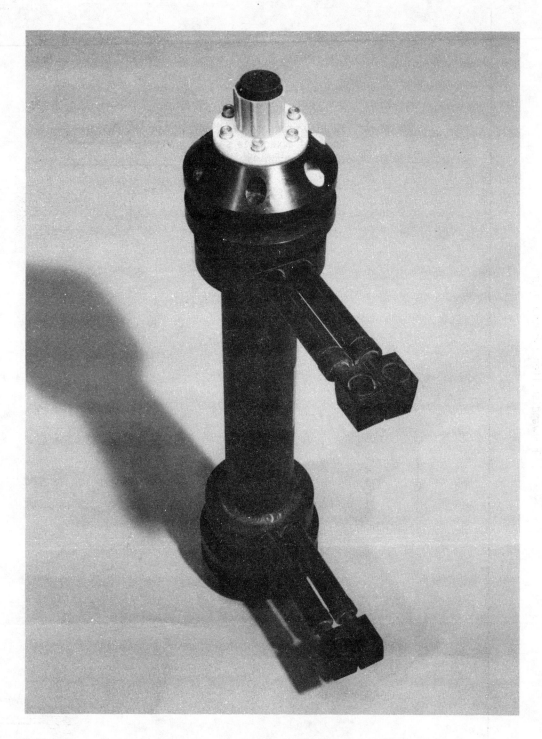

1   Underwater sensor

2   Measuring system with calibration chamber

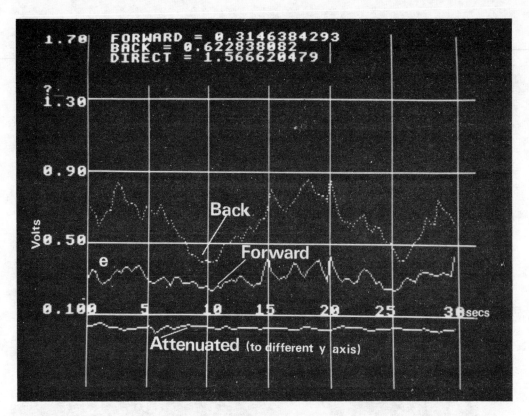

3a   Time trace for 1850 mgl^{-1} of 63 to 75 μm particles

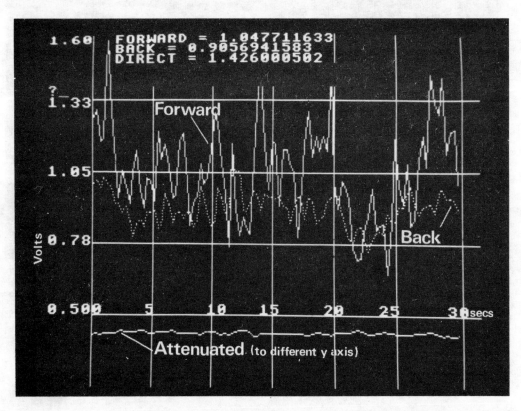

3b   Time trace for 455 mgl^{-1} of 210 to 250 μm particles

THE ACOUSTIC MEASUREMENT OF SUSPENDED SAND IN THE SURF-ZONE

Christopher E. Vincent
School of Environmental Sciences
University of East Anglia
Norwich NR4 7TJ

Daniel Hanes and T. Tamura
Division of Ocean Engineering
Rosenstiel School of Atmospheric and Marine Sciences
University of Miami
FL 33149, USA

Thomas L. Clarke
Atlantic Oceanographic and Meteorological Laboratories
National Oceanic and Atmospheric Administration
Miami FL 33149, USA

Summary

Techniques for measuring the profile of the concentration of suspended sand
close to the sea-bed with a vertical resolution of about 1cm are discussed. A
3MHz Acoustic Concentration Meter (ACM) which uses the intensity of back-
scattered acoustic energy from the suspended sand is described. The ACM has
been used to obtain data on suspended sand in the surf-zone (~2m depth) and in
deeper water on the continental shelf (~10m). In the surf-zone the limitations
of such profilometers are (a) the attenuation of the acoustic beam by high
concentrations of suspended sand in the water column, (b) air bubbles injected
into the water by breaking waves and (c) the variation in the sand size with
height above the sea-bed. In deeper water outside the surf-zone only the third
of the above limitations is important and accuracies of about a factor of two
could be expected in the measurement of the suspended sand concentrations.
Examples are shown of the temporal variations in the profiles of back-scattered
acoustic energy caused by suspended sand which were obtained during the waning
phase of a storm while the ACM was deployed near the surf-zone.

Held at Imperial College of Science and Technology, London. Organised and sponsored by BHRA, The
Fluid Engineering Centre. Co-sponsored by the American Society of Civil Engineers and the
International Association for Hydraulic Research.

1. INTRODUCTION

There are many situations in rivers, estuaries and the oceans where sediment transport is of crucial importance for understanding natural processess or for assessing the influence of structures on the overall system. Accurate field measurements of the material in suspension, particularly of suspended sand, are difficult to make and the empirical and semi-empirical formulae used to calculate rates of sediment transport have often come from tank experiments. The problems encountered with sediment transport in the ocean are essentially the same as those encountered in rivers and estuaries save that the environmental conditions are potentially more destructive to instrumentation (especially in the surf-zone) and oscillatory wave-induced currents, which may dominate in shallower waters, add a further level of complexity to the system.

Several techniques have been employed on the shoreface and in the surf-zone to measure suspended sand under the action of waves and currents. Sediment traps have frequently been used to gain some information on combined bedload and suspended transport but there is considerable discussion concerning exactly what such traps measure and the degree to which they disturb the flow. Pumped-sampling through tubes gives a direct method of sampling the suspended material at particular heights and integrated over a period of time. An impact sensor has been developed (Salkield et al, 1981) which measures sand transport flux, the product of sand concentration and velocity, at one height in the flow but its present configuration restricts its use to unidirectional flow.

Optics and acoustics offer the most versatile methods for measuring suspended sediment. Optical transmissometers are standard instrumentation for marine turbidity measurements but their sensitivity is mainly to the fine sediment in suspension and they give little data on suspended sand. An optical backscattering (OBS) device been successfully used in the near-shore zone to measure suspended sand at a number of discrete heights above the sea-bed (Hanes and Huntley, in press). The OBS is sensitive to sand within the $1cm^3$ volume in front of each sensor and as many sensors as required can be stacked above each other to obtain concentration profiles of the suspended sand. Hanes and Huntley (in press) used 5 sensors spaced approximately logarithmically above the bed.

Acoustical methods allow the measurement of profiles of suspended sediment with spatial and temporal resolutions limited by the wavelength of the sound source and by the ability of the electronic system to process and record the signals. Most echo sounders operate at frequencies of the order of 10kHz (a wavelength of several centimetres), a frequency range chosen to give a reasonably resolution while keeping the absorption of the sound energy by water to reasonable levels. At the high frequencies necessary to give a wavelength of the order of the size of sand grains (a few megahertz) sound attenuation by seawater is very high and the range short (a few 10s of metres). As the majority of sand transport takes place in the lowest metre of the water column the short range is not particularly restrictive. There are however a number of potential problems in interpreting the data obtained by such instruments and this paper discusses these in the light of field experiments carried out over the last few years.

An example of the acoustic backscattered intensities, interpreted as suspended sand, are shown in figure 1. The profiles are at 1s intervals and the water depth was 10m (Vincent et al. 1983).

2. THE 3MHz ACOUSTIC CONCENTRATION METER (ACM)

The acoustic system which was used to obtain the measurements described below was constructed for the NOAA Atlantic Oceanographic and Meteorological Laboratories in Miami, Florida and was first described by Huff and Fiske (1980). The initial tank calibrations were carried out by Young et al. (1982).

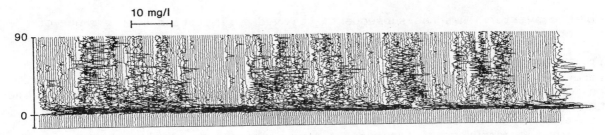

10 mg/l

90

0

Suspended sand profiles one per second

2

0

Time   20 second increments

Figure 1 Profiles every second of suspended sand concentrations (upper) measured at Tiana Beach, Long Island in the 90cm above the bed together with the variation in pressure (lower) at the sea-bed (the units are dynes cm^{-2})

2.1 THE ACM ACOUSTICS

The ACM is basically the transmitter-receiver portion of a sonar system packaged into a small pressure housing, known as the sea-unit. Power input, control signals and receiver signals are communicated through a cable either to a nearby self-contained data logger and battery pack, as was used in the deployment described by Vincent et al. (1982), or directly to a shore station. The transducer operates at 3MHz and is constructed of a specially doped ceramic to produce narrow pencil-like acoustic beam with a spread of $0.8°$. The transmitted pulse has a pulse length of 7μs and a pulse rate of up to 100 pulses per second although for various operational reasons this is usually restricted to 10 pulses per second or less. Power output at the transducer head is about 2 acoustic watts. In the ocean the ACM has usually been deployed at between 1m and 2m above the seabed and the system is set to receive data for up to 4ms following transmitted pulse. The return echo is range-gated to obtain a profile of intensity of back-scattered acoustic energy with a resolution of about 1cm, and the suspended sand concentration is deduced from these back-scattered signals.

2.2 DATA AQUISITION

This has varied with the application but has generally been limited by the speed at which data can be recorded and by the physical limitations of the data storage medium. The first deployment of this system off Tiana Beach, Long Island was a remote deployment with limited data capacity and the ACM was used in burst mode making 256 pulses at 1Hz every 4 hours. The range-gating was set to give 110 values of return echo intensity, at a vertical resolution of 1.2cm.

The surf zone deployments carried out at Stanhope Lane, Prince Edward Island, Canada in October 1984 used a shore-connected system and the data were recorded in real time in analogue form for subsequent analysis. The pulsed analogue signals from the ACM have useful frequency components ranging from DC to 100kHz. The direct recording of such a signal would require special FM instrumentation recorders operating at a speed of 76cm/s and 1100m-tape would last less than 20 minutes. In addition, if the ACM was operating at a 10 pulse per second rate only 4% of the tape would contain useful data! The 4ms of useful data followed by 96ms of 'dead' time offered the opportunity to stretch the signal 20-fold and to reduce the band-width to DC - 5kHz. The signal conditioner (SC) which was built to accomplished this time stretching used a system of double side-band modulation to produce a signal between 5kHz and 15kHz which could be recorded on a standard FM recorder running at 9.5cm/s (Guaraglia and Lauter, 1983). At the time the SC was constructed 8-bit digital technology was all that was commonly available and it was built around an 8-bit

A-to-D converter. This has subsequently proved a limitation to the range of sand concentrations in which this instrument can obtain useful data.

2.3 DEVELOPMENTS OF THE ACM
There are several versions of the ACM under construction but these differ from the NOAA system in the method of data aquisition rather than the acoustics. The ACM was initially a very expensive tool using what were, at the time, the latest electronics. It proved that high-frequency acoustics is a very useful tool for the measurement of suspended sand. Advances in electronics have made rapid digital data aquision cheap and reliable and have stimulated a new generation of acoustic profilometers. Two systems known to the authors are outlined below.

A Canadian version known as the ASP (acoustic sediment profiler) has been built by the Canadian National Research Council and was tested in the field for the first time in October 1984. The ASP uses a fast 10-bit A-to-D converter and microprocessor, both housed in the sea-unit, to average the back-scattered acoustic profile over seven pulses. Although the ASP has a resolution of about 1cm only concentrations in 32 'bins' can be transmitted to shore. The number of levels averaged in each 'bin' is variable and set by software.

A low-cost digital system based around the BBC Microcomputer is under development at the University of East Anglia and the University of St. Andrews based on transmit/receive circuitry developed by the Institute of Oceanographic Sciences at Taunton. A slower ($26\mu s$) 12-bit A-to-D converter is being used and the vertical profile of suspended sand (128 levels at a resolution of 1.1cm) is built up using lagged sampling of 8 rapid pulses at a rate of 100Hz. 5 profiles are averaged and stored on floppy disk once every second. The limitations of this system are the speed at which data can be stored on the disk and the capacity of the disk.

3. CALIBRATION OF THE ACM
The translation of the intensity of the backscattered signal into a suspended sand concentration is probably the most difficult problem which still faces the ACM user. The theoretical attenuation of an acoustic beam by a field of spherical sand grains at a concentration of 1g/l as a function of grain diameter can be seen in figure 2 and is shown for 3 frequencies. Young et al. (1982) undertook the first tank calibrations of the 3MHz system using sand of 3 sizes but found that the strength of the acoustic returns from the suspended sand did not vary as suggested by the theoretical curve for 3Mhz in figure 2. Young et al. (1982) undertook their calibration by stirring known quantities of sieved sand in an oil drum in which the water was kept circulating by a system a vertical propellors, but in all such systems it is difficult to get the sand suspension uniform and to ensure that all the sand was kept in suspension.

The calibration tank used for the ACM prior to the surf zone measurements was a narrow 20cm-diameter tube 1m-long (75cm operational length) with a tapering base leading down to a recirculating pump which pumped the water back to the top of the tank through external tubing (figure 3). Calibrations were performed with both finely sieved sands and with a sand taken from the beach at Stanhope Lane, Prince Edward Island, the site of the surf zone-experiments. The intensities of the backscattered acoustic signals for 5 concentrations of the Stanhope sand as a function of range are shown in figure 4 (Tamura et al., in press). At the higher concentrations the attenuation of the acoustic beam due to the presence of large numbers of scatterers closer to the acoustic head can clearly be seen. At the highest concentration 5.7g/l the acoustic beam has been attenuated to such an extent that little penetrates beyond 50cm at this particular gain setting. Varying the gain of the transducer driving curcuits can extend the range of the ACM when the suspended sand concentration is high. Figure 5 shows the envelopes of the back-scattered acoustic energy obtained in the calibration tank for 3 gain settings at a fixed suspended sand concentration. Although it is possible to "see" the complete profile down the tank using a range of gain settings the interpretation of back-scattered energy in terms of sand concentration is difficult.

4. PROBLEMS IN THE USE OF THE ACM

4.1 BEAM ATTENUATION
The acoustic backscattered intensity at any particular level is not a simple
function of the concentration of scatterers at that level but is also dependant
on the concentration of scatterers at all levels between the transducer and the
level under consideration due to the attenuation of the beam energy. Only at
the lowest concentrations can this effect be ignored: in the surf and nearshore zones the suspended sand concentrations are frequently very high. It can
be seen in figure 4 that measurements of acoustic back-scattered energy by an
instrument gated for only a single range cannot be interpreted in terms of
concentration unless it is certain that the concentration of scatterers is low,
e.g. a low backscattered intensity at a range of 50cm may be due to either a
low sand concentration or to a very high concentration which has attenuated the
signal reaching that distance! The translation of the profile of acoustic
backscatter into a profile of suspended sand concentration involves the
stepwise correction of the intensity of the acoustic beam as, in general, the
sand concentration is rarely uniform in the water column.

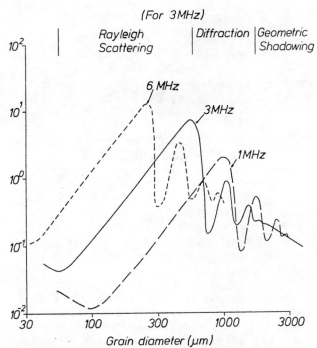

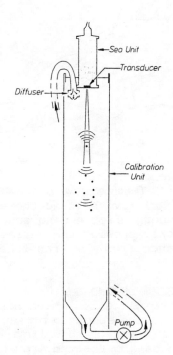

Figure 2 The theoretical attenuation
of an acoustic beam (m^{-1}) by spherical sand grains at a concentration
of 1 g/l.

Figure 3 The tank used for calibration
of the ACM. Rapid recirculation keeps
the sediment concentration approximately uniform.

At high concentrations the acoustic beam can be rapidly attenuated resulting in
a low signal-to-noise ratio close to the seabed. More importantly, any errors
associated with the calculation of suspended sand concentration in the upper
part of the profiles are propagated cumulatively to lower levels resulting in
very large errors in sand concentrations close to the seabed. This is
potentially a severe limitation of the system as the larger sand concentrations
occur close to the seabed.

4.2 AIR BUBBLES IN THE WATER COLUMN
The acoustic impedance of air bubbles injected into the water column by the
action of breaking waves is similar to the acoustic impedance of suspended

sand. Consequently air bubbles can render the ACM useless when significant
numbers of them enter the path of the acoustic beam as has been shown in our
laboratory calibrations. While the ACM was deployed at Stanhope Lane (water
depth ~2m) heavy breaking occurred in the vicinity of the ACM and for a period
of 24hrs it was clear from visual observation that large numbers of bubbles
were present in the water and acoustic signals were received dominantly from
the 30cm closest to the transducer. The ACM records obtained when it was known
that bubble contamination occurred are being analysed with the objective of
establishing criteria which can be used to identify the presence of bubbles
from the ACM records alone. Signal strength distributions and the macro-
structure of the bubble groups are under examination.

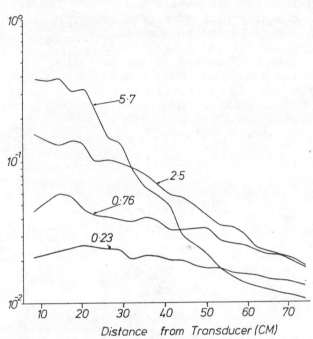

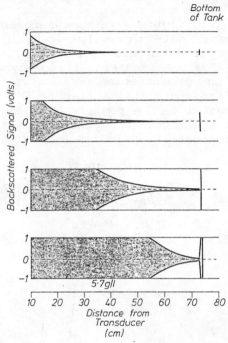

Figure 4   Relative back-scattered
acoustic energy from 4 sand concentrations (g/l) showing the variation
with distance from the transducer.

Figure 5   Envelopes of backscattered
signals from a sand concentration of
5.7g/l showing the effect of altering
the gain of the transducer. Low gain
at the top; high at bottom.

4.3 SAND SIZE DEPENDANCE
A further complication which has thus far been ignored is that of the vertical
variation in the spectrum of sand sizes in suspension. While it is possible to
obtain a well-mixed and uniform suspension in a small recirculating calibration
tank using sand collected from the seabed at the experimental site, it is not
realistic to imagine that this same perfect suspension will occur during
natural sand suspension events! Figure 2 shows how the back-scattered signal
varies with 3 sand diameters and indicates the nature of the problem of dealing
with a real sand suspension rather than an idealised one. Tank calibration
using closely sieved sand have also been performed (Tamura et al., in press)
which confirm the sensitivity of the ACM to sand size. For the Stanhope Lane
measurement below, 95% of the sand was between 0.2 and 0.3mm and calibrations
using sand sizes over this range (0.25ϕ intervals) suggest the sensitivity of
the ACM doubles between 0.2 and 0.3mm.

It may well be possible to obtain data on the vertical variation of the sand
size spectrum by using several transducers operating at different frequencies,
utilising the different backscatter characteristics of sand to sound of
different frequencies (figure 2) but this multifrequency technique has yet to
be tried.

5. SUSPENDED SAND IN THE SURF ZONE

During storms large quantities of sand are moved around on the shoreface and in the surf zone. Beach profiles can change enormously over the course of a single storm with sand being transfered between the shoreface and offshore bars. Usually the sand movements are infered from the results of 'before' and 'after' surveying with little attempt being made to follow the overall movement of sediment during the storm itself. The ACM was deployed in the surf zone at Stanhope Lane, Prince Edward Island, Canada in October 1984 as part of a study designed to examine some of the processes which occurred during storms and to help understand how beaches can be protected from the ravages of storms. It was known that the performance of the ACM would be affected by the injection of air-bubbles into the water column but it was not clear at what stage of the storm this would become a problem or whether it would be possible to distinguish between air bubbles and a high sand concentration.

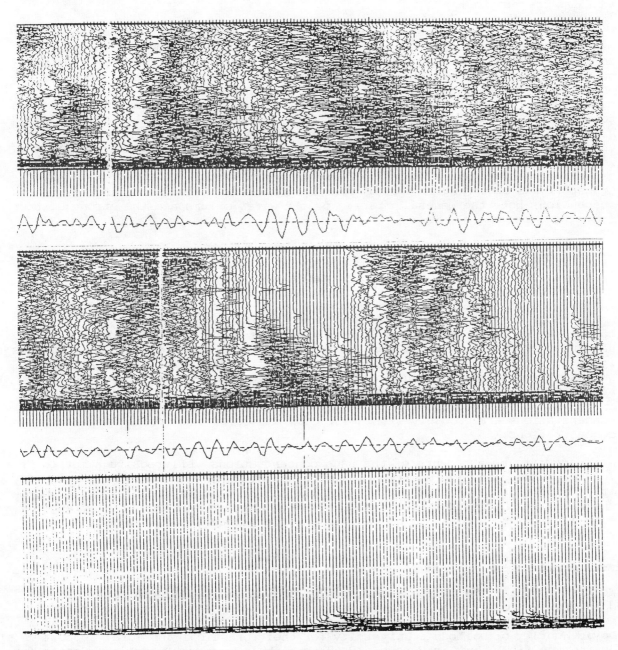

Figure 6 Acoustic backscattered records for the surf-zone. A detailed explanation of the records is given in the text.

The beach at Stanhope Lane faced north into the Gulf of St.Lawence. It is gently shelving sandy beach over a sandstone pebble pavement. The beach is protected by a series of 3 bars at about 100m, 250m and 700m from the shore. The ACM was deployed on bottom mounted 'tripods' on each of the two inner bars for the duration of a single storm. Electromagnetic current meters, a transmissometer and a pressure gauge were also mounted on the tripod. A severe storm occurred while the ACM was deployed on the second bar and for a period of 24hrs the ACM records of suspended sand were contaminated with air bubbles (see 4.2) and, despite adjustments to the transmitter gain, acoustic backscatter from the sea-bed was rare. Three example ACM records are shown in figure 6, obtained during the waning phase of the storm. These records are of the intensities of the back-scattered acoustic energy not of the suspended sand concentrations. Interpretations of these records in terms of the on/off-shore sand transport will be presented elsewhere (Tamura et al., in press; Hanes et al., in preparation).

Figure 6(upper) shows a high concentration period soon after waves stopped breaking over the ACM. Waves were, however, breaking a few metres inshore and some bubbles may have been advected into the ACM beam to produce the high back-scatter region near the ACM transducer. Beneath the profiles is a record of the on-offshore component of the near-bed current measurered by an electromagnetic current meter mounted close to the ACM. The maximum current shown is equivalent to a surface wave of about 2.1m in height. In figure 6 (middle) the waves were still of a significant height ($H_{1/3} \sim 1.5$m) but no breaking was occurring near the ACM and the bubbles can be considered free from major bubble contamination. 24 hours later (figure 6(lower)), the storm had subsided and the wave heights had diminished to around 30cm (no current record shown). At this time sand transport was occurring irregularly as larger waves lifted sand into suspension; note the general clarity of the water column and the strong scatter from the sea-bed.

6 CONCLUSIONS

Despite the limitations exposed by the field experimnets in the surf-zone the technique of acoustic profiling of suspended sand is a valuable one and produces data which are difficult to obtain by any other means. The interpretation of the back-scattered data as suspended sand concentration is not without difficulty: an objective method of identifying potential contamination of the records by air bubbles is a high priority if ACMs are to used near breaking waves. Outside the surf-zone, where there are fewer problems from very high sand concentrations or from bubbles, the data are of high quality and the interpretation of the back-scattered acoustic energy as suspended sand concentrations is straightforward: allowing for the uncertainty in the distribution ofthe sand size with height the calculated sand concentrations are probably within a factor of two of the actual suspended concentrations.

More accurate measurements of the sand concentrations will require a method of measuring the sand size distribution with height. Here a multifrequency-ACM may offer a solution as the sensitivity of the ACM to sand is strongly dependant on the wavelength to sand diameter ratio. However to the best of our knowledge such a technique has yet to be used in the field to make measurements of suspended sand concentrations.

ACKNOWLEDGEMENTS

We wish to thank the many colleaques who have assisted us is obtaining these data. In particular Charles Lauter Jnr. at the Ocean Acoustics Laboratory, NOAA, Miami for his help with the instrumentation, Professors Tony Bowen and David Huntley of Dalhousie University for their practical and moral support during the field work in Canada and the National Research Council of Canada for the opportunity to take part in the Canadian Coastal Sediment Study. We would like to acknowledge the financial support of the Canadian NRC given to us all and that given to CEV by the Royal Society.

REFERENCES

Guaraglia D.O. and Lauter C.A.: "Implementation of a new method to record pulsed analogue signals". Proceedings of the Third Working Symposium of Oceanographic Data Systems, IEEE, 1983,pp. 22-26.

Hanes D. and Huntley D.A.: "Continuous measurement of suspended sand concentration in a wave dominated near-shore environment". Continental Shelf Research, in press.

Huff L. and Fiske D.C.:." Development of two sediment transport systems". Proceedings of the Seventeeth International Conference on Coastal Engineering, American Society of Civil Engineers, New York, 1980.

Salkield A.P., LeGood G.P. and Soulsby R.A.: "An impact sensor for measuring suspended sand concentration". Proceedings of the Conference on Ocean Technology, IERE, London, 1981, pp.37-47.
Tamura T., Hanes D., Clarke T.L. and Vincent C.E.: "High resolution acoustic measurements of suspended sand". American Geophysical Union (abstracts), in press.

Vincent C.E., Young R.A. and Swift D.J.P.: "On the relationship between bedload and suspended sand transport on the Inner Shelf, Long Island, New York". Journal of Geophysical Research, 87, 1982, pp. 4163-4170.

Vincent C.E., Young R.A. and Swift D.J.P.: "Sediment transport on the Long Island shoreface, North American Atlantic shelf; role of waves and currents in shoreface maintenance". Continental Shelf Research, 2, 1983, pp. 163-181.

Young R.A., Merrill J., Proni J.R. and Clarke T.L.: "Acoustic profiling of suspended sediment in the marine boundary layer". Geophyical Research Letters, 9, 1982, pp. 175-178.